GEOTHERMAL SYSTEMS AND ENERGY RESOURCES –
TURKEY AND GREECE

Sustainable Energy Developments

Series Editor

Jochen Bundschuh
University of Southern Queensland (USQ), Toowoomba, Australia
Royal Institute of Technology (KTH), Stockholm, Sweden

ISSN: 2164-0645

Volume 7

Geothermal Systems and Energy Resources

Turkey and Greece

Editors

Alper Baba

Geothermal Energy Research and Application Center, Izmir Institute of Technology, Izmir, Turkey

Jochen Bundschuh

University of Southern Queensland (USQ), Toowoomba, Australia
Royal Institute of Technology (KTH), Stockholm, Sweden

D. Chandrasekaram

Department of Earth Sciences, Indian Institute of Technology, Bombay, India

CRC Press
Taylor & Francis Group
Boca Raton London New York

CRC Press is an imprint of the
Taylor & Francis Group, an **informa** business

A BALKEMA BOOK

Cover photo

Tuzla geothermal power plant in western Turkey (photo: Alper Baba)

CRC Press
Taylor & Francis Group
6000 Broken Sound Parkway NW, Suite 300
Boca Raton, FL 33487-2742

First issued in paperback 2018

CRC Press/Balkema is an imprint of the Taylor & Francis Group, an informa business

© 2014 Taylor & Francis Group, LLC

Typeset by MPS Limited, Chennai, India

No claim to original U.S. Government works

ISBN 13: 978-1-138-00109-1 (hbk)
ISBN 13: 978-1-138-07446-0 (pbk)

Library of Congress Cataloging-in-Publication Data

Applied for

Published by: CRC Press/Balkema
P.O. Box 11320, 2301 EH Leiden, The Netherlands
e-mail: Pub.NL@taylorandfrancis.com
www.crcpress.com – www.taylorandfrancis.com

**Visit the Taylor & Francis Web site at
http://www.taylorandfrancis.com**

**and the CRC Press Web site at
http://www.crcpress.com**

About the book series

Renewable energy sources and sustainable policies, including the promotion of energy efficiency and energy conservation, offer substantial long-term benefits to industrialized, developing and transitional countries. They provide access to clean and domestically available energy and lead to a decreased dependence on fossil fuel imports, and a reduction in greenhouse gas emissions.

Replacing fossil fuels with renewable resources affords a solution to the increased scarcity and price of fossil fuels. Additionally it helps to reduce anthropogenic emission of greenhouse gases and their impacts on climate change. In the energy sector, fossil fuels can be replaced by renewable energy sources. In the chemistry sector, petroleum chemistry can be replaced by sustainable or green chemistry. In agriculture, sustainable methods can be used that enable soils to act as carbon dioxide sinks. In the construction sector, sustainable building practice and green construction can be used, replacing for example steel-enforced concrete by textile-reinforced concrete. Research and development and capital investments in all these sectors will not only contribute to climate protection but will also stimulate economic growth and create millions of new jobs.

This book series will serve as a multi-disciplinary resource. It links the use of renewable energy and renewable raw materials, such as sustainably grown plants, with the needs of human society. The series addresses the rapidly growing worldwide interest in sustainable solutions. These solutions foster development and economic growth while providing a secure supply of energy. They make society less dependent on petroleum by substituting alternative compounds for fossil-fuel-based goods. All these contribute to minimize our impacts on climate change. The series covers all fields of renewable energy sources and materials. It addresses possible applications not only from a technical point of view, but also from economic, financial, social and political viewpoints. Legislative and regulatory aspects, key issues for implementing sustainable measures, are of particular interest.

This book series aims to become a state-of-the-art resource for a broad group of readers including a diversity of stakeholders and professionals. Readers will include members of governmental and non-governmental organizations, international funding agencies, universities, public energy institutions, the renewable industry sector, the green chemistry sector, organic farmers and farming industry, public health and other relevant institutions, and the broader public. It is designed to increase awareness and understanding of renewable energy sources and the use of sustainable materials. It also aims to accelerate their development and deployment worldwide, bringing their use into the mainstream over the next few decades while systematically replacing fossil and nuclear fuels.

The objective of this book series is to focus on practical solutions in the implementation of sustainable energy and climate protection projects. Not moving forward with these efforts could have serious social and economic impacts. This book series will help to consolidate international findings on sustainable solutions. It includes books authored and edited by world-renowned scientists and engineers and by leading authorities in economics and politics. It will provide a valuable reference work to help surmount our existing global challenges.

Jochen Bundschuh
(Series Editor)

Editorial board

Table of contents

Contributors

Alper Baba: Geothermal Energy Research and Application Center, Izmir Institute of Technology, Izmir, Turkey

İlyas Çağlar: Department of Geophysical Engineering, Faculty of Mines, Istanbul Technical University, Istanbul, Turkey

Mehmet Çelik: Department of Geological Engineering, Ankara University, Ankara, Turkey

Walter D'Alessandro: Instituto Nazionale di Geofisica e Vulcanologia—Sezione di Palermo, Palermo, Italy

Ayşen Davraz: Department of Geological Engineering, Suleyman Demirel University, Isparta, Turkey

Mustafa M. Demir: Department of Chemistry, Izmir Institute of Technology, Izmir, Turkey

Irmak Doğan: Department of Chemistry, Izmir Institute of Technology, Izmir, Turkey

Gülden Gökçen Akkurt: Energy Engineering Programme, İzmir Institute of Technology, Izmir, Turkey

Tolga Gönenç: Department of Geophysics Engineering, Engineering Faculty, Dokuz Eylul University, Buca Izmir, Turkey

Nilgün Güleç:Department of Geological Engineering, Middle East Technical University, Ankara, Turkey

Ebru Hancioglu Kuzgunkaya:Geothermal Energy Research and Application Center, Izmir Institute of Technology, Izmir, Turkey

Arif Hepbasli: Department of Energy Systems Engineering, Faculty of Engineering, Yaşar University, Izmir, Turkey

David R. Hilton: Fluids and Volatiles Laboratory, Geosciences Research Division, Scripps Institute of Oceanography, University of California, San Diego, La Jolla, CA, USA

Murat Karadaş: Energy Engineering Programme, İzmir Institute of Technology, Izmir, Turkey

Stylianos Karakatsanis: MSc Geologist, Thessaloniki, Greece

Konstantina Katsanou: Department of Geology, University of Patras, Patras, Greece

Mine Sezgül Kayseri-Özer: Department of Geology Engineering, Dokuz Eylül University, Izmir, Turkey

Konstantinos Kyriakopoulos: Department of Geology and Geo-environment, National and Kapodistrian University of Athens, Athens, Greece

Nicolaos Lambrakis: Department of Geology, University of Patras, Patras, Greece

Halim Mutlu: Department of Geological Engineering, Ankara University, Ankara, Turkey

Erdeniz Özel: Institute of Marine Science and Technology, Dokuz Eylül University, İzmir, Turkey

Tuğbanur Özen: Department of Geological Engineering, Oltu Earth Science Faculty, Atatürk University, Erzurum, Turkey

Maria Papachristou: Department of Geology, Aristotle University of Thessaloniki, Thessaloniki, Greece

Oya Pamukçu: Department of Geophysics, Engineering Faculty, Dokuz Eylul University, Buca Izmir, Turkey

Mahmut Parlaktuna: Petroleum and Natural Gas Engineering Department, Middle East Technical University, Ankara,Turkey

Suzan Pasvanoğlu: Geological Engineering Department, University of Kocaeli, Izmit, Turkey

Bade Pekçetinöz: Institute of Marine Science and Technology, Dokuz Eylül University, Izmir, Turkey

George Siavalas: Department of Geology, University of Patras, Patras, Greece

Petek Sındırgı: Department of Geophysics, Engineering Faculty, Dokuz Eylul University, Buca Izmir, Turkey

Gültekin Tarcan: Department of Geological Engineering, Dokuz Eylül University, Izmir, Turkey

Konstantinos Voudouris: Department of Geology, Aristotle University of Thessaloniki, Thessaloniki, Greece

Preface by Ladislaus Rybach

The geotectonic situation of the Aegean and of Anatolia enables the formation and presence of hydrothermal resources, this mainly due to the back-arc settings related to the subduction the African plate under the Aegean Sea plate in the Aegean and especially to extensional grabens in Anatolia. The area is therefore predestined for geothermal development, i.e. the utilization of the resources for energetic purposes. In Turkey a large number of high-temperature thermal springs demonstrate the large geothermal potential of the country; in fact the country has a long history of thermal spring uses, substantial experience in district heating, and nowadays a rapid increase in installing power generation facilities.

The Volume consists of 15 Chapters, describing various settings and aspects of geothermal exploration and development. Several new results and insights are presented, all relevant for the development and use of geothermal resources in the investigated area. Here the main results of the different Chapters are summarized.

High concentrations of mantle helium in geothermal fluids are associated with the most recently (1692) active Nemrut volcano near Lake Van, eastern Anatolia, present in the Tuzla field/Canakkale, and in the most seismically active western-central segment of the North-Anatolian Fault Zone (NAFZ).

The geothermal fields of Greece are mainly located in back-arc or volcanic arc settings. The most common feature of these fieldsis that there arereductive chemical conditions, favoring the formation of hydrogen sulfide and of reducing forms of nitrogen. Thermal waters from crystalline rocks in back-arc settings are characterized by elevated radon and thoron concentrations.

The geothermal systems of the Aegean Island Arc are characterized by high CO_2 fluxes derived from deep geogenic sources: either directly from the mantle or from the thermo-metamorphism of marine limestones. The majority of deep aquifers, which discharge through springs on Aegean islands, are associated with seawater intrusion and mixing of meteoric waters with fluids of deep origin.

The common characteristic of all thermal and mineral springs in eastern Anatolia is that they are clearly related to young volcanic activity or tectonic block fracturing, whereas their hydrochemical attributes are highly variable – due to different host-rock compositions, flow pathways, and residence times.

Geophysical results identified possible high-enthalpy resources in the Göynük area (located 100 km east of Bursa), especially by pronounced low-resistivity (4–6 Ωm) zones detected by magnetotelluric measurements. Besides fluids, conductive minerals formed by hydrothermal alteration might also contribute to the low resisitivities found. Exploration drilling and testing will be needed to verify the presence of resources.

The geothermal fields of Diyadin and Zilan in easternmost Turkey are represented at the surface by thermal springs (maximum outflow temperatures 64°C and 78°C, respectively). The springs originate from meteoric infiltration on Mount Tendurek, whereas their outflow is located at faults related to the neotectonic regime of the region. Within each thermal water district, a travertine complex created by the thermal waters is covering the upflow zones.

Practical experience with utilization technology is also reported: lessons learned from 16 years of operation with the Balçova district heating system (reservoir pressure decline can only be counteracted by reinjection); exergetic and economic criteria are to be followedfor utilizing shallow geothermal resources by ground-source heat pumps properly.

The deposition of solids in geothermal equipment and components ("scaling") is common in also Turkey; carbonate scaling is most common. Chemical inhibitors are helpful to reduce carbonate scaling; silica scaling is less frequent but also less understood: there is therefore room to acquire more knowledge about the formation mechanism of siliceous scaling and accordingly the methods for its inhibition and/or suppression.

Finally there is remarkable progress now in Turkey in geothermal power generation. Whereas there was no further development after the installation of the Kizildere power plant with 17.8 MWe gross capacity in 1982, the situation completely changed in 2007. New power plants are installed since then on the row; the frequency is remarkable: The following summary (name/year of completion/installed capacity in MWe) is impressive: Dora-1/2006/7.35, Bereket/2007/7.5, Gurmat/2009/47.4, Tuzla/2010/7.5, Dora-2/2010/11.2, Irem/2011/20, Sinem/2012/24, Deniz/2012/24and Dora-3/2013/17. Hopefully this growth will continue in the coming years.

Altogether 30 authors elaborated the various Chapters: 21 from Turkey, 7 from Greece, 1 from Italy, 1 from USA. The Editors took care of harmonizing the contributions, i.e. to include maps in the Chapters, which show the locations treated. The investigations described and the results presented in this Volume shall encourage and initiate further exploration and development in this part of the world that has remarkable potential for geothermal energy utilization.

Zurich, January 2014

Ladislaus Rybach
Emeritus Professor, ETH Zurich, Switzerland
Scientific Advisor, Geowatt AG Zurich

Editors' foreword

People have been using geothermal energy for bathing and for washing clothes since the dawn of civilization in many parts of the world. Turkey and Greece are favored by a large number of thermal springs known since the ancient times. However, it was first in the 20th century that geothermal energy was used on a large scale for direct applications like space heating, industrial processing, domestic water supply, balneology, and for electricity generation. Turkey and Greece belong to one of the most seismically active regions in the world, which has a considerably medium to high enthalpy geothermal energy potential due to its inherent geological and tectonic setting. Turkey and Greece are located within the Alpine-Himalayan seismic belt, whose complex deformation resulted from the continental collision between the African and Eurasian plates. The border of these plates constitutes the seismic belts marked by young volcanics and active faults, while the latter allowing circulation of water as well as heat. The distribution of hot springs in these regions roughly parallels the distribution of the fault systems, young volcanism, and hydrothermally altered areas. In Turkey and Greece, geothermal energy is used in various applications such as electricity generation, greenhouse cultivation, district heating, industrial processes, thermal tourism, and balneology. Most of geothermal sites are low–medium enthalpy fields, which are suitable mostly for direct-use applications such as district heating, greenhouse heating, thermal facilities, and balneology. In addition, many geothermal fields, which are located on the western graben system, have high enthalpy fields. Therefore, many scientists and many companies have been working on geothermal system of Turkey and Greece.

This book seeks to provide knowledge regarding geothermal resources and geothermal energy applications. It is impossible to cover this subject within the space limitations of any normal book. Consequently, a selection of essential topics and case studies has been made. Almost all the developments described, however, are discussed in depth. The fundamental concepts, the technical details, and the economic and environmental issues are described and presented in a simple and logical manner. The book explains in a didactic way the possible applications, depending on local conditions and scales, and it presents new and stimulating ideas for future developments. Additionally, the book discusses the role(s) of possible physicochemical processes in deep hydrothermal systems, the volatile provenance and relative contributions of mantle and crustal components to total volatile inventories and the distribution of mantle volatiles in relation to differing neotectonic provinces, recent seismic activities, and/or the timing and nature of volcanic activity. We believe that this book will provide the reader with a thorough understanding of the geothermal systems of Turkey and Greece and identify the most suitable solutions forspecific tasks and needs. Chapter 1 discusses the chemical and isotopic constraints on the origin of thermal waters in Anatolia, Turkey. In Chapter 2, gas geochemistry of Turkish geothermal fluids, He–CO_2 systematics in relation to active tectonics and volcanism, are explained. Chapters 3 and 4 highlight the geological setting, geothermal conditions, and hydrochemistry of Greece. Chapters 5–9 discuss in depth hydrogeochemical, geophysical, and isotopic properties of different parts of Anatolia (Turkey). Chapter 10 presents district heating system in Turkey. In Chapter 11, geothermal power generation, and in Chapter 12 the scaling problem of the geothermal system in Aegean Region are explained. Chapter 13 highlights exergetic and exergoeconomic aspects of ground-source (geothermal) heat pumps. Chapters 14 and 15 discuss in detail the application of geophysical methods and palaeoenvironmental and palynological of the geothermal system next to the Aeagean Sea.

We hope that this book will help all readers, in the professional and academic sectors, as well as key institutions that deal with geothermal energy. Hopefully, this book will become a reference that is widely used by educational institutions, research institutions, and research and development establishments. The book should also prove to be a useful textbook to senior undergraduate, graduate, and postgraduate students, to engineers involved in geothermal energy, district heating, power generation, and management of geothermal resources, and also to professional hydrologists, hydrogeologists, hydrochemists, environmental scientists, and others trying to address, and understand, geothermal energy.

Alper Baba
Jochen Bundschuh
D Chandrasekaram
(editors)
December 2013

About the editors

Professor Alper Baba, born 1970 in Turkey, holds a degree in geology and a doctorate in the field of hydrogeology from the Dokuz Eylul University, Izmir. He has about 20 years of work experience in hydrogeological and environmental geology problems in different part of the world. Since 2010, he has been a professor at Izmir Institute of Technology as a director of Geothermal Energy Research and Application Center. He teaches and conducts research in the field of groundwater contamination, geothermal energy, and hydrogeology. He has coordinated a variety of national and international R&D projects in cooperation with research institutes and companies, among them NATO funded projects. Dr. Baba has been the recipient of the Turkish Academy of Science Successful Young Scientists Award and the Turkish Geological Engineering Association Gold Medal Award. Dr. Baba is the author of several peer-reviewed scientific publications and contributions to international conferences. Dr. Baba is also the editor of the book *Groundwater and Ecosystems* and *Climate Change and Its Effects on Water Resources, Issues of National and Global Security* (both NATO Science Series, Springer).

Jochen Bundschuh (1960, Germany) finished his PhD on numerical modeling of heat transport in aquifers in Tübingen in 1990. He works in geothermics, subsurface and surface hydrology and integrated water resources management, and connected disciplines. From 1993 to 1999, he served as an expert for the German Agency of Technical Cooperation (GTZ) and as a long-term professor for the DAAD (German Academic Exchange Service) in Argentina. From 2001 to 2008, he worked within the framework of the German governmental cooperation (Integrated Expert Program of CIM; GTZ/BA) as adviser–in-mission to Costa Rica at the Instituto Costarricense de Electricidad (ICE). Here, he assisted the country in evaluation and development of its huge low-enthalpy geothermal resources for power generation. Since 2005, he is an affiliate professor of the Royal Institute of Technology, Stockholm, Sweden. In 2006, he was elected Vice-President of the International Society of Groundwater for Sustainable Development ISGSD. From 2009 to 2011, he was visiting professor at the Department of Earth Sciences at the National Cheng Kung University, Tainan, Taiwan. By the end of 2011, he was appointed as professor in hydrogeology at the University of Southern Queensland, Toowoomba, Australia, where he leads a working group of 26 researchers working on the wide field of water resources and low/middle enthalpy geothermal resources, water and wastewater treatment, and sustainable and renewable energy resources (http://www.ncea.org.au/groundwater). In November 2012, Prof. Bundschuh was appointed president of the newly established Australian chapter of the International Medical Geology Association (IMGA).

Dr. Bundschuh is author of the books *Low-Enthalpy Geothermal Resources for Power Generation* (2008) (Balkema/Taylor & Francis/CRC Press) and *Introduction to the Numerical Modeling of Groundwater and Geothermal Systems: Fundamentals of Mass, Energy and Solute Transport in Poroelastic Rocks*. He is editor of the books *Geothermal Energy Resources for Developing Countries* (2002), *Natural Arsenic in Groundwater* (2005), and the two-volume monograph *Central America: Geology, Resources and Hazards* (2007), *Groundwater for Sustainable Development* (2008), *Natural Arsenic in Groundwater of Latin America* (2008). Dr. Bundschuh is editor of the book series *Multiphysics Modeling*, *Arsenic in the Environment*, and *Sustainable Energy Developments* (all Balkema/CRC Press/Taylor & Francis).

Dornadula Chandrasekharam (Chandra: b1948, India), chair professor in the Department of Earth Sciences, Indian Institute of Technology Bombay (IITB) obtained his MSc in Applied Geology (1972) and PhD (1980) from IITB. He has been working in the fields of geothermal energy resources, volcanology, and groundwater pollution, for the past 30 years. Before joining IITB, he worked as a senior scientist at the Centre for Water Resources Development and Management, and Centre for Earth Science Studies, Kerala, India, for 7 years. He has held several important positions during his academic and research career. He was a Third World Academy of Sciences (TWAS, Trieste, Italy); visiting professor to Sanaa University, Yemen Republic between 1996 and 2001; senior associate of Abdus Salam International Centre for Theoretical Physics, Trieste, Italy, from 2002 to 2007; adjunct professor, China University of Geosciences, Wuhan from 2011 to 2012. Recently he has been appointed as a visiting professor to King Saud University of Saudi Arabia. He received the International Centre for Theoretical Physics (ICTP, Trieste, Italy) Fellowship to conduct research at the Italian National Science Academy (CNR) in 1997. Prof. Chandra extensively conducted research in low-enthalpy geothermal resources in India and is currently the Chairman of M/s GeoSyndicate Power Private Ltd., the only geothermal company in India. He is an elected board member of the International Geothermal Association and has widely represented the country in several international geothermal conferences. He conducted short courses on low-enthalpy geothermal resources in Argentina, Costa Rica, Poland, and China. He has supervised 18 PhD students and published 95 papers in international and 35 papers in national journals of repute and published 5 books in the field of groundwater pollution and geothermal energy resources. His two books on geothermal energy resources—(1) *Geothermal Energy Resources for Developing Countries* by Balkema Pub. (2002) and (2) *Low Enthalpy Geothermal Resources for Power Generation* by Taylor & Francis (2008)—are widely read. Prof. Chandra is currently on the Board of Director of (1) Oil and Natural Gas Corporation, (2) Western Coal Fields Ltd., (3) India Rare Earths Ltd., and (4) Mangalore Refineries and Petrochemicals. He has been appointed as the Chairperson of the Geothermal Energy Resources and Management committee constituted by the Department of Sciences and Technology, Government of India.

Acknowledgements

The editors thank Ladislaus Rybach (Geowatt AG, Zurich, Switzerland), David R. Hilton (Fluids and Volatiles Laboratory, Geosciences Research Division, Scripps Institute of Oceanography, University of California, San Diego, La Jolla, CA, USA), and Ondra Sracek (Department of Geology, Faculty of Science, Palacký University, Czech Republic) for their reviews. The editors are grateful to Ebru Koç and Dr. Adil Caner Sener for the revision of an earlier version of the manuscripts and Dr. Ebru Kuzgunkaya for assistance. They have to also thank the institutes and companies that provided funds for printing the book in full color. The editors thank the University of Southern Queensland (Toowoomba, Australia), the Director IITB and the Dean IRCC IITB, Tuzla Geothermal Energy (Izmir, Turkey) and Deren Chemicals (Istanbul, Turkey) for providing partial funds for printing color figures in this book. The editors and authors thank also the technical people of Taylor & Francis Group, for their cooperation and the excellent typesetting of the manuscript.

CHAPTER 1

Chemical and isotopic constraints on the origin of thermal waters in Anatolia, Turkey: fluid–mineral equilibria approach

Halim Mutlu, Nilgün Güleç & David R. Hilton

1.1 INTRODUCTION

Turkey, located on a tectonically and magmatically active part of the Alpine-Himalayan belt, is characterized by widespread geothermal activity, evident by numerous geothermal areas scattered throughout the country (Mutlu and Güleç, 1998). Geothermal exploration studies in Turkey were started in 1962 by the General Directorate of Mineral Research and Exploration (MTA). By the end of 2010, a total of 498 wells with a total depth of 242,500 m had been drilled with 3881 MW$_t$ (megawatt thermal) heat energy extracted from these wells. The present geothermal power generation capacity in Turkey is about 162 MW$_e$ (megawatt electrical: electric power) while that of direct use installations is around 795 MW$_t$ (Serpen et al., 2010). MTA's projection for 2015 is to increase the installed capacity to about 600 MW$_e$.

In this study, chemical and isotopic compositions of thermal waters from selected geothermal sites in Anatolia, Turkey, are presented using data from previously published works (Mutlu, 2007; Mutlu and Kılıç, 2009; Mutlu and Sarıız, 2000; Mutlu et al., 2008; 2011; 2012). An overall assessment is made on fluid–mineral equilibria with regard to various geothermal minerals and derived reservoir temperatures from a number of tectonically different regions.

1.2 GEOLOGICAL SETTING

The closure of the Neotethyan Ocean corresponds to the beginning of the Neotectonic period in Turkey. The convergence between the African and Eurasian plates in the late Cretaceous and the Arabian Plate in the Miocene gave rise to emplacement of huge ophiolite nappes along the Izmir–Ankara–Erzincan suture in northern Anatolia (Şengör and Yılmaz, 1981). The termination of subduction was followed by a compressional–extensional tectonic regime in the Miocene. The ongoing convergence along the Bitlis Suture Zone (BSZ) has resulted in uplift of the Turkish-Iranian Plateau (Şengör and Kidd, 1979). Two major strike-slip faults, the North Anatolian Fault Zone (NAFZ) and the East Anatolian Fault Zone (EAFZ), were formed in response to the north-ward movement of the Arabian plate (Fig. 1.1). As a result of the change of tectonic regime from E–W trending strike-slip activity to NE–SW trending strike-slip faulting, coupled with noteworthy development of normal faults, north-western Anatolia has experienced extensive crustal extension and lithospheric thinning, which resulted in the formation of a number of graben systems in most parts of western Anatolia (Şengör, 1987; Taymaz et al., 1991) (Fig. 1.1).

1.3 WATER CHEMISTRY

The results of chemical analysis of geothermal waters from five regions—NAFZ, Balıkesir, Eskisehir, eastern Anatolia and western Anatolia (see Fig. 1.1)—are presented in Table 1.1. Temperatures of the thermal waters lie between 22.8°C and 98.5°C. The samples' pH varies

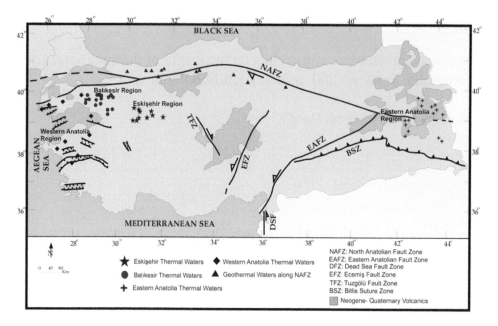

Figure 1.1.   Sampling sites in relation to major tectonic and volcanic features in Anatolia.

Table 1.1.   Chemical analyses of Anatolian thermal waters (mg/L).

Sample	T (°C)	pH	HCO_3	Cl	SO_4	Na	K	Ca	Mg	SiO_2	TDS
NAFZ											
Efteni	43.2	6.5	1709	200	3	339	13	139	154	141	2557
Yalova	61.2	7.6	45	94	820	262	5	160	8	45	1393
Bolu	42.0	6.2	756	11	447	41	16	338	39	37	1648
Mudurnu	39.8	6.2	731	8	27	24	7	159	44	40	1000
Seben	71.4	6.6	1186	61	91	439	32	47	14	34	1871
Hamamözü	42.5	7.4	238	36	24	46	6	38	18	27	405
Gözlek	39.1	7.8	249	15	31	77	5	20	8	20	404
Resadiye	41.2	6.4	1817	804	118	800	52	304	89	45	3982
Kursunlu	57.4	7.0	5942	775	67	2462	261	67	24	44	9598
Balıkesir											
G-8	57.8	7.4	387	249	469	495	29	70	3	57	1758
G-16	77.5	7.0	347	256	453	488	29	49	3	75	1699
EKS	42.7	7.3	192	9	17	12	2	51	13	31	327
MK-2	34.4	6.9	439	85	44	89	11	140	11	33	852
PMK-1	64.9	7.6	217	219	357	376	20	47	4	109	1350
PMK-2	55.5	7.3	278	132	291	318	14	33	3	91	1160
BHS-1	94.6	7.3	1052	207	382	706	74	8	12	119	2559
BHS-2	82.9	7.0	1110	207	350	678	71	30	12	120	2578
SHS-1	98.5	7.0	577	82	97	322	21	22	3	114	1237
SHS-2	97.3	7.4	566	82	94	322	21	66	3	113	1266
SHS-3	95.1	7.0	581	83	93	318	21	22	3	114	1234
SHS-4	87	6.7	567	82	93	314	21	23	4	111	1215
EDR-1	57.6	7.8	49	60	507	273	5	50	1	43	987
EDR-2	44.7	7.6	108	52	391	199	4	70	3	31	858

(*Continued*)

Table 1.1.   Continued.

Sample	T (°C)	pH	HCO$_3$	Cl	SO$_4$	Na	K	Ca	Mg	SiO$_2$	TDS
GDR	56.6	8.4	50	61	493	306	6	22	0	56	994
BLY-1	59.0	8.2	79	79	351	264	4	17	0	61	855
BLY-2	58.1	8.1	75	79	357	270	4	16	0	61	862
SLK-1	31.8	7.1	340	11	120	63	8	83	30	25	679
SLK-2	72.8	6.4	595	57	190	380	13	22	1	106	1365
Eastern Anatolia											
AYR	37.0	7.0	1619	360	150	606	169	200	30	63	2474
BUG	34.3	7.6	805	17	23	84	24	99	71	133	863
CAM-1	22.8	6.8	2034	608	274	894	172	136	121	23	3734
CAM-2	34.0	7.2	2114	561	255	901	158	122	112	31	3534
CAY	53.5	7.6	5508	853	537	2624	171	181	69	99	6970
CKR	34.5	6.5	1545	47	1	299	52	127	94	123	1238
DVT	64.2	6.6	917	115	102	116	52	130	84	45	2312
DYD	53.7	7.0	989	117	156	155	59	183	69	45	1449
GRM	46.8	9.2	205	142	113	210	4	20	2	31	701
HAD	65.0	6.5	1229	2469	422	1876	188	325	64	3	3924
KOK	16.3	6.3	1214	27	852	137	26	388	186	0	2389
NHL	51.6	6.9	1060	23	0	323	33	69	15	34	1079
PAT	18.1	6.2	397	193	34	182	20	92	12	45	911
TAS-1	71.3	6.5	817	18	2	21	10	205	12	24	5375
TAS-2	50.1	7.1	1175	643	463	734	70	224	75	57	3028
TAT	14.2	10.4	899	32	4	116	13	104	62	0	809
TUT	24.6	6.4	1809	558	68	772	46	146	110	0	3038
YUR	25.4	6.7	1128	14	40	40	4	179	124	0	1615
Eskişehir											
Inönü	27.7	7.7	255	7	3	7	1	59	18	13	369
Sakarılıca	56.0	6.3	1514	55	60	288	12	217	46	56	2255
Eskişehir	46.0	7.4	280	9	13	10	1	54	22	18	413
Kızılinler	38.9	6.9	597	25	25	119	4	70	22	28	894
Asagı Ilıca	27.4	7.1	477	10	33	16	3	111	35	12	705
Çifteler	23.6	7.4	347	12	50	15	3	101	21	19	580
Yalınlı	30.7	6.5	611	14	44	33	8	98	66	25	905
Yarıkçı	40.6	6.5	1635	67	34	161	8	153	184	53	2297
Sivrihisar	35.1	7.2	402	41	16	35	2	98	16	17	631
Çaltı	37.7	6.5	1165	31	83	352	17	76	30	32	1791
Western Anatolia											
Ömer-Gecek	79.5	7.2	1135	1723	504	1460	120	146	17	102	5217
Gazlıgöl	64.0	7.1	2934	157	16	1088	88	40	16	61	4417
Hisaralan	94.0	8.3	1129	213	398	650	68	30	16	118	2630
Manyas	49.8	6.6	497	278	84	260	30	141	11	39	1339
Gönen	60.4	7.2	383	250	468	479	29	67	3	58	1737
Çan	47.3	7.5	226	220	1398	475	24	252	3	33	2631
Kestanbol	69.0	6.0	317	9700	170	5100	630	800	54	86	16857
Tuzla	90.0	7.5	128	46250	240	21500	2000	3848	201	98	74265
Balçova	84.6	7.7	622	176	143	395	24	27	3	124	1514
Seferihisar	66.0	6.8	689	4941	375	2750	254	235	100	93	9437
Germencik	41.0	8.6	1376	1819	133	1750	105	5	1	286	5475
Kızıldere	n.d.	8.5	1880	117	773	1300	138	2	1	120	4331

(NAFZ from Süer *et al.*, 2008; Balıkesir region from Mutlu, 2007; eastern Anatolia from Mutlu *et al.*, 2011; Eskişehir from Mutlu and Sarız, 2000; western Anatolia from Mutlu *et al.*, 2008).

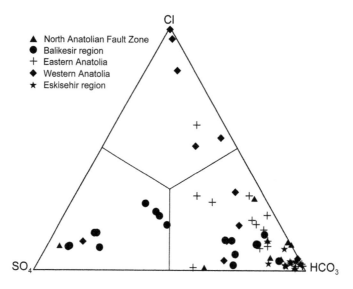

Figure 1.2.   Relative Cl, SO_4, and HCO_3 contents of the Anatolian thermal waters.

over a wide range, from 6.0 to 10.4. The TDS (total dissolved solids) content ranges from 327 to 74,265 mg/L. The results of the chemical analysis are assessed based on the Cl–SO_4–HCO_3 diagram (Fig. 1.2). Generally, samples from the regions studied plot close to the HCO_3 apex. Balıkesir samples show both HCO_3- and SO_4-rich characters. Sulfate in these waters is most probably derived from Neogene evaporate deposits. Most of the western Anatolian waters are of Na–Cl type that is characteristic of coastal areas (e.g. Tuzla, Balçova, and Seferihisar) and their compositions suggest possible seawater intrusion into the reservoir (Mutlu and Güleç, 1998). The HCO_3 nature of most samples is attributed to the interaction between waters and carbonate-type reservoir rocks. This is also supported by the fact that Paleozoic marbles, Mesozoic recrystallized limestones, and Tertiary carbonate deposits comprise the reservoir rocks in most parts of Turkey (Mutlu and Güleç, 1998).

1.4   GEOTHERMOMETRY APPLICATIONS

In this section, reservoir temperatures of Anatolian thermal waters are estimated using the Na–K–Mg diagram and other chemical geothermometers. The results obtained from these various approaches are discussed.

1.4.1   *Chemical geothermometers*

In order to estimate host reservoir temperatures of the Anatolian thermal waters, various silica and cation geothermometers were used. In Figure 1.3, silica contents of waters vs. their discharge temperatures are plotted. With the exception of some samples from the eastern Anatolian region and the NAFZ, these two parameters are strongly and positively correlated. Also displayed in this figure is the range of reservoir temperatures calculated from quartz and chalcedony geothermometers. In general, reservoir temperatures estimated by the quartz geothermometer (Arnórsson, 1985) are about 20–30°C higher than those by the chalcedony geothermometer (Fournier and Potter, 1982). Reservoir temperatures computed by the silica geothermometers attain a maximum temperature of around 130°C. Since chalcedony, rather than quartz, controls the silica solubility at temperatures below 180°C (Fournier, 1991), it seems that the chalcedony geothermometer better reflects the reservoir temperatures of the waters. The average temperatures computed by the chalcedony geothermometer (Fournier, 1977) are 105°C for the western Anatolian region, 91°C

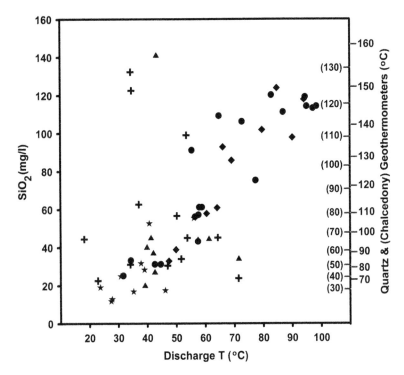

Figure 1.3. Silica–temperature relationships for Anatolian thermal waters. Values in parenthesis mark the scale of the chalcedony geothermometer (symbols as in Fig. 1.2).

for the Balıkesir region, 73°C for the eastern Anatolian region, 64°C for the NAFZ, and 40°C for the Eskisehir region (Fig. 1.3) which is consistent with the results of Mutlu and Güleç (1998) and Güleç *et al.* (2002).

The reservoir temperatures estimated from the cation geothermometers for each sample are usually higher than those obtained from silica geothermometers. The use of ratios, rather than absolute abundances of the ions, makes the Na–K geothermometer less sensitive to secondary processes (e.g., mixing and boiling). Since Na/K ratios of waters are more sensitive to changes in temperature compared to K/Mg ratios (Giggenbach, 1988), the temperature range obtained from the Na–K geothermometer is considerably greater than that from the K–Mg geothermometer (Fig. 1.4a,b). The Na–K geothermometer of Giggenbach (1988) gives anomalously high temperatures (>260°C), whereas the same author's K–Mg geothermometer yields a maximum temperature of around 180°C. Therefore, results of the K–Mg geothermometer are considered more consistent with temperature estimates from the silica (particularly chalcedony) geother- mometer. In recent studies carried out for geothermal waters in the Nevsehir region, central Turkey (Pasvanoğlu and Chandrasekharam, 2011) and in the Salihli region, western Turkey (Özen *et al.*, 2012), Na–K geothermometer is suggested to yield higher reservoir temperatures than the K–Mg geothermometer.

1.4.2 *Na–K–Mg diagram*

A further evaluation of the cation geothermometers is made using the Na–K–Mg diagram (Fig. 1.5) proposed by Giggenbach (1988). The model is based on the following geothermometers assuming that activities of minerals are close to unity:

$$t_{kn} = 1390/[1.75 - \log(C_K/C_{Na})] - 273.15$$

$$t_{km} = 4410/[14.00 - \log(C_K^2/C_{Na})] - 273.15$$

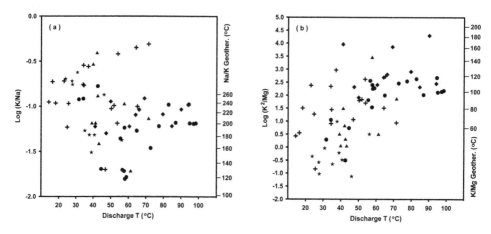

Figure 1.4.   Cation–temperature relationships for Anatolian thermal waters: (a) Na–K geothermometer and
(b) K–Mg geothermometer (symbols as in Fig. 1.2).

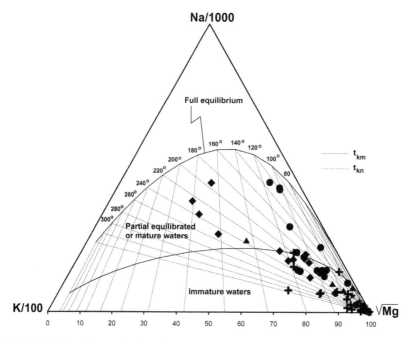

Figure 1.5.   Na–K–Mg diagram for Anatolian thermal waters (from Giggenbach, 1988) (symbols as in
Fig. 1.2).

As shown in Figure 1.5, none of the Anatolian waters attains water-rock equilibrium and most
of the samples plot in the field of immature waters (shallow or mixed waters). Some waters from
western Anatolia and Balıkesir region plot in the mature waters field. Such behavior might indicate
that waters gained their salinity through simple rock dissolution processes and that water-rock
equilibrium is not achieved due to mixing of the deep thermal water with cold water en route to the
surface. For the samples plotting far into the immature waters field, the application of both K–Na
and K–Mg, and indeed any type of cation geothermometers, is doubtful and the interpretation
of the temperature predictions of such waters should be made with caution (Giggenbach, 1988).

However, for samples plotting in the partial equilibrium field, temperature estimates can be performed with greater confidence.

1.5   STABLE ISOTOPES

1.5.1 $\delta^{18}O$–δD compositions

Stable isotope compositions of Anatolian thermal waters are presented in Table 1.2. The $\delta^{18}O$ values of samples lie between −13.5 and −4‰ (VSMOW), while δD values vary in a wide range from −95.4 to −23.0‰ (VSMOW). On the $\delta^{18}O$–δD diagram (Fig. 1.6), all the samples are consistent with a meteoric origin. Most of Balıkesir waters plot close to the Mediterranean meteoric water line (MMWL; $\delta D = 8\delta^{18}O + 22$) (Gat and Carmi, 1970), while waters from NAFZ, eastern and western Anatolia, and Eskişehir region generally plot in the vicinity of the global meteoric water line (GMWL; $\delta D = 8\delta^{18}O + 10$). Significantly enriched ^{18}O–D compositions of some western Anatolian waters are attributed to recharging at low elevations (reference). On the other hand, most NAFZ and eastern Anatolia waters with notably depleted $\delta^{18}O$–δD values indicate recharging from high elevations (Fig. 1.6).

1.5.2 $\delta^{34}S$–$\delta^{13}C$ compositions

In order to investigate the origin of sulfur and carbon in the Balıkesir and eastern Anatolian thermal waters, their $\delta^{34}S$ and $\delta^{13}C$ contents were evaluated. Analyses were conducted on sulfate for $\delta^{34}S$ and dissolved inorganic carbon (DIC) for $\delta^{13}C$ (Table 1.2). The $\delta^{34}S$ values (VCDT) of Anatolian thermal waters range from −5.5 (sample EKS-2) to 45.7‰ (sample GRM). Apart from these two extreme samples, the $\delta^{34}S$ ratios vary between 7.7 and 29.5‰.

The $\delta^{34}S$ ratio of SO_4 in marine sediments in the geological past varied from 10 to 35‰ depending on the time of deposition (Clark and Fritz, 1997; Krouse, 1980). Negative $\delta^{34}S$ values are related to diagenetic environments, which are represented by reduced sulfur compounds. The $\delta^{34}S$ ratio of sulfate from oxidation of magmatic sulfur falls in the range −5 to +5‰ (Clark and Fritz, 1997). Sulfur isotope values of Balıkesir and eastern Anatolian thermal waters are consistent with marine sulfate deposits (Fig. 1.7a). However, since there is no sulfate deposit of marine origin in the studied regions, marine deposits cannot be the source of sulfur in these samples (Mutlu, 2007; Mutlu *et al.*, 2012). Indeed, nonmarine Neogene evaporate deposits (e.g., borate and gypsum) occur widely in northwestern Anatolia including most parts of the Balıkesir region (Helvacı *et al.*, 1993; Palmer and Helvacı, 1997). As regards to eastern Anatolia, continental clastic and carbonate rocks, together with evaporites, occur in varying extents (Okay *et al.*, 2010). Structural substitution of sulfate (CAS: carbonate-associated sulfate) into marine carbonates exerts a major control on sulfur to having a marine source (Kampschulte and Strauss, 2004). In this respect, upper Cretaceous flysch deposits exposed along the Tethyan suture zone might contribute to the marine sulfur endmember. The mixing of sulfate from dissolution of marine carbonates and terrestrial evaporite units is an alternative process behind the observed sulfate isotope compositions of the Balıkesir and eastern Anatolian thermal waters (Mutlu *et al.*, 2012).

Carbon dioxide derived from the decay of organic material and from the dissolution of carbonates is the major source of carbon contributing to DIC in the waters. The $\delta^{13}C$ (DIC) values of the Balıkesir and eastern Anatolian thermal waters vary from −17.7 to 5.6‰ (VPDB). In the HCO_3–$\delta^{13}C$ diagram (Fig. 1.7b), waters from the Balıkesir and eastern Anatolian regions represent two distinct groups. The $\delta^{13}C$ (DIC) values of the Balıkesir thermal waters vary over a wide range, from −14.7 to +0.7‰; however, those of eastern Anatolian waters, with the exception of sample GRM, vary over a narrow range, from 1.8 to 5.6‰. It is likely that carbon isotope ratios of the eastern Anatolian waters closely resemble those of marine limestones which are represented by $\delta^{13}C$ values of about −3 to +3‰ (Clark and Fritz, 1997). Limestones of Mesozoic age, which comprise the basement in most parts of the area around Lake Van, are the source of carbon in these waters (Mutlu *et al.*, 2012).

Table 1.2.   Stable isotope compositions of Anatolian thermal waters (‰).

Sample	$\delta^{18}O$ (VSMOW)	δD (VSMOW)	$\delta^{13}C$ (VPDB)	$\delta^{34}S$ (VCDT)	Sample	$\delta^{18}O$ (VSMOW)	δD (VSMOW)
Balıkesir					Western Anatolia		
G-7	−12.12	−76.8	0	11.5	Ömer-Gecek	−13.1	−86.2
G-8	−12.78	−77.1	0	12.3	Gazlıgöl	−13.1	−89.9
G-16	−12.50	−77.4	0.7	12.5	Hisaralan	−11.61	−68.5
EKS-1	−11.94	−60.9	−1.7	n.a.	Manyas	−11.0	−60.9
EKS-2	−11.21	−55.9	−17.7	−5.5	Gönen	−12.0	−76.9
MK-1	−9.91	−61.4	−3.6	16.1	Çan	−10.5	−65.4
MK-2	−10.77	−58.1	−8.4	7.7	Kestanbol	−8.0	−42.1
PMK-1	−10.67	−63.8	−8.9	7.7	Tuzla	−4.0	−23.0
PMK-2	−9.92	−58.5	−11.2	12.1	Balçova	−7.0	−41.0
BHS-1	−9.94	−71.1	−2.0	25.2	Seferihisar	−6.1	−36.0
BHS-2	−9.86	−71.8	−1.2	24.2	Germencik	−4.7	−45.0
SHS-1	−11.61	−68.5	−4.7	15.5	Kızıldere	−7.7	−61.7
SHS-2	−11.78	−69.0	−4.8	16.8	NAFZ		
SHS-3	−11.53	−69.3	−4.7	17	Efteni	−11.3	−82.9
SHS-4	−11.73	−68.9	−3.4	16.2	Yalova	−11.3	−75.8
EDR-1	−10.05	−55.2	−10.5	11.1	Bolu	−11.5	−82.8
EDR-2	−10.03	−53.4	−14.7	10.2	Mudurnu	−12.0	−84.7
GDR-1	−9.61	−52.1	−7.2	10.3	Seben	−12.1	−89.0
BLY-1	−12.50	−76.8	−6.6	13.6	Hamamözü	−12.0	−87.4
BLY-2	−12.91	−75.9	−6.7	13.4	Gözlek	−12.8	−94.2
SLK-1	−11.05	−63.2	−1.4	18.6	Reşadiye	−12.7	−92.9
SLK-2	−12.36	−72.4	−1.8	18.8	Kurşunlu	−8.5	−88.3
Eastern Anatolia					Eskisehir		
AYR	−11.4	−89.7	5.2	19.6	Inönü	−11.8	−83.6
BUG	−13.5	−91.0	2.9	12.2	Sakarılıca	−12.6	−95.4
CAM-1	−9.8	−77.4	5.2	24.6	Eskisehir	−10.9	−80.9
CAM-2	−10.7	−81.5	3.6	24.6	Kızılinler	−11.9	−84.6
CAY	−3.4	−72.2	1.8	19.8	Asagıllıca	−11.4	−83.9
CKR	−10.0	−68.8	2.6	n.a.	Çifteler	−10.8	−81.0
DVT	−11.3	−90.9	4.6	20.0	Yalınlı	−10.1	−80.1
DYD	−11.5	−92.1	4.8	13.7	Yarıkçı	−12.2	−88.8
GRM	−10.9	−72.4	−17.5	45.7	Sivrihisar	−11.2	−81.0
HAD	−10.1	−79.0	4.0	18.2	Çaltı	−12.3	−91.8
KOK	−10.7	−68.9	4.0	29.5			
NHL	−9.6	−64.5	1.9	n.a.			
PAT	−11.4	−82.7	3.1	15.6			
TAS-1	−11.4	−81.3	4.4	18.1			
TAS-2	−11.3	−82.3	5.6	18.1			
TAT	−9.8	−65.8	2.3	n.a			
TUT	−12.1	−86.5	4.6	15.4			
YUR	−12.7	−89.4	3.1	20.3			
BUG	−13.5	−91.0	2.9	12.2			

(Balıkesir region from Mutlu, 2007; NAFZ from Süer et al., 2008; eastern Anatolia from Mutlu et al., 2012; Eskişehir from Mutlu and Sarıız, 2000; western Anatolia from Mutlu et al., 2008).

Carbon isotope composition of Balıkesir thermal waters shows multiple sources of carbon. Carbon in high-temperature waters might originate from dissolution of marine carbonates, an interpretation supported by carbon isotope compositions of marine carbonate rocks in the Balıkesir region (Mutlu, 2007). The $\delta^{13}C$ values of low-temperature waters, however, are consistent with an organic source.

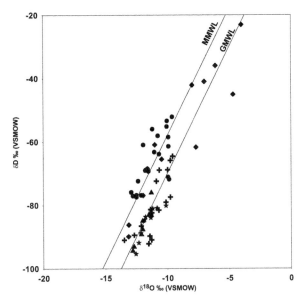

Figure 1.6. $\delta^{18}O-\delta D$ diagram for Anatolian thermal waters. MMWL: Mediterranean meteoric water line ($\delta D = 8\delta^{18}O + 22$; Gat and Carmi, 1970); GMWL: global meteoric water line ($\delta D = 8\delta^{18}O + 10$; Craig, 1961) (symbols as in Fig. 1.2).

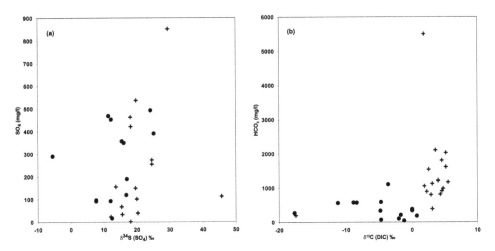

Figure 1.7. (a) $SO_4-\delta^{34}S$ diagram and (b) $HCO_3-\delta^{13}C$ diagram for the Balıkesir and eastern Anatolia thermal waters (symbols as in Fig. 1.2).

1.6 MINERAL EQUILIBRIUM CALCULATIONS

Calcite scaling is a serious problem during the utilization of CO_2-rich thermal waters as the CO_2 expelling facilitates waters become oversaturated with calcite. High silica content of waters may also cause supersaturation to several silica minerals, particularly chalcedony. That is why saturation states of calcite and chalcedony minerals in the studied water samples were examined.

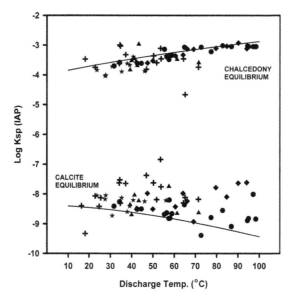

Figure 1.8.   Saturation states of chalcedony and calcite in waters. Chalcedony and calcite equilibrium lines are from Arnórsson *et al.* (1982) (symbols as in Fig. 1.2).

Saturation calculations were performed with the PHREEQC computer code of Parkhurst and Appelo (1999). In Figure 1.8, measured water temperatures are plotted against their ionic activity products (IAP). Most of the samples are found to be in equilibrium or oversaturated with respect to both chalcedony and calcite. Chalcedony saturation of waters is almost similar to or slightly greater than the chalcedony equilibrium line. In contrast, probably due to dissolution/precipitation of carbonate minerals, calcite saturation of samples is quite variable and might attain as much as two orders of magnitude greater than the calcite equilibrium line (Fig. 1.8). This is also revealed by widespread travertine deposits around most of the sampling sites of the studied thermal waters.

ACKNOWLEDGMENTS

This study was financially supported by the Eskisehir Osmangazi University under grant nos. 1999/12, 2005/15016, and 2009/15017 and by the Scientific and Technical Research Council of Turkey (TUBITAK) under grant no. 100Y097.

REFERENCES

Arnórsson, S.: The use of mixing models and chemical geothermometers for estimating underground temperatures in geothermal systems. *J. Volcanol. Geoth. Res.* 23 (1985), pp. 209–335.
Arnórsson, S., Gunnlaugsson, E. & Svavarsson, H.: The chemistry of geothermal waters in Iceland—I.
Calculation of aqueous speciation from 0° to 370°C. *Geochim. Cosmochim. Acta* 46 (1982), pp. 1513–1532.
    Clark, I.D. & Fritz, P.: *Environmental isotopes in hydrogeology.* Lewis Publishers, CRC Press, New York, 1997.
Craig, H.: Isotopic variations in meteoric waters. *Science* 133 (1961), pp. 1702–1703.
Fournier, R.O.: Chemical geothermometers and mixing models for geothermal systems. *Geothermics* 5 (1977), pp. 41–50.
Fournier, R.O.: Water geothermometers applied to geothermal energy. In: F. D'amore (ed): *Application of geochemistry in geothermal reservoir development.* UNITAR, USA, 1991.
Fournier, R.O. & Potter, R.W. II: A revised and expanded silica quartz geothermometer. *Geotherm. Resourc. Counc. Bull.* 11:10 (1982), pp. 3–12.

Gat, J.R. & Carmi, I.: Evolution in the isotopic composition of atmospheric waters in the Mediterranean Sea area. *J. Geophys. Res.* 75 (1970), pp. 3039–3048.

Giggenbach, W.F.: Geothermal solute equilibria. Derivation of Na–K–Ca–Mg geoindicators. *Geochim. Cosmochim. Acta* 52 (1988), pp. 2749–2765.

Güleç, N., Hilton, D.R. & Mutlu, H.: Helium isotope variations in Turkey: relationship to tectonics, volcanism and recent seismic activities.*Chem. Geol.* 187 (2002), pp. 129–142.

Helvacı, C., Stamatakis, M., Zagouroglou, C. & Kanaris, J.: Borate minerals and related authigenic silicates in northeastern Meditarranean Late Miocene continental basins. *Explor. Mining Geol.* 2 (1993), pp. 171–178.

Kampschulte, A. & Strauss, H.: The sulphur isotopic evolution of Phanerozoic seawater based on the analysis of structurally substituted sulfate in carbonates. *Chem. Geol.* 204 (2004), pp. 255–286.

Krouse, H.R.: Sulfur isotopes in our environment. In: P. Fritz & J.CH. Fontes (eds): *Handbook of environmental isotope geochemistry* I, *The terrestrial environment*, A. Elsevier, Amsterdam, The Netherlands, 1980.

Mutlu, H.: Constraints on the origin of the Balıkesir thermal waters (Turkey) from stable isotope (d18O, dD, d13C, d34S) and major-trace element compositions. *Turk. J. Earth Sci.* 16 (2007), pp. 13–32.

Mutlu, H. & Güleç, N.: Hydrogeochemical outline of thermal waters and geothermometry applications in Anatolia, Turkey. *J. Volcanol. Geoth. Res.* 85 (1998), pp. 495–515.

Mutlu, H. & Kılıç, A.: Geothermometry applications for the Balıkesir thermal waters, Turkey. *Environ. Geol.* 56 (2009), pp. 913–920.

Mutlu, H. & Sarıız, K.: Geochemical and isotopic characteristics of Eskisehir thermal waters, Turkey. In: O.Ö. Dora, I. Özgenç & H. Sözbilir (eds): *Proceedings of IESCA 2000 (International Earth Science Colloquium on the Aegean Region)*, 25–29 September 2000, Dokuz Eylül University, Izmir, Turkey, 2000, pp. 189–196.

Mutlu, H., Güleç, N. & Hilton, D.R.: Helium-carbon relationships in geothermal fluids of western Anatolia, Turkey. *Chem. Geol.* 247 (2008), pp. 305–321.

Mutlu, H., Güleç, N., Hilton, D.R. & Aydın, H.: Helium and stable isotope systematics of eastern Anatolian geothermal fluids, Turkey. Eskisehir Osmangazi University, Research Project Report, Project No. 2009-15017, Eskisehir, Turkey, 2011 (unpublished; in Turkish).

Mutlu, H., Güleç, N., Hilton, D.R., Aydın, H. & Halldórsson, S.A.: Spatial variations in gas and stable isotope compositions of thermal fluids around Lake Van: implications for crust-mantle dynamics in eastern Turkey. *Chem. Geol.* 300–301 (2012), pp. 165–176.

Okay, A.I., Zattin, M. & Cavazza, W.: Apatite fission-track data for the Miocene Arabia–Eurasia collision. *Geol. Soc. Am.* 38 (2010), pp. 35–38.

Özen, T., Bülbül, A. & Tarcan, G.: Reservoir and hydrogeochemical characterizations of geothermal fields in Salihli, Turkey. *J. Asian Earth Sci.* 60 (2012), pp. 1–17.

Palmer, M.R. & Helvacı, C.: The boron isotope geochemistry of the Neogene borate deposits of western Turkey. *Geochim. Cosmochim. Acta* 61 (1997), pp. 3161–3169.

Parkhurst, D.L. & Appelo, C.A.J.: User's guide to PHREEQC (Version 2) — A computer program for speciation, batch-reaction, 1 D transport, and inverse geochemical calculations. US Geological Survey, Water Resources Investigations Report, USA, 1999.

Paşvanoğlu, S. & Chandrasekharam, D.: Hydrogeochemical and isotopic study of thermal and mineralized waters from the Nevsehir (Kozakli) area, central Turkey. *J. Volcanol. Geoth. Res.* 202 (2011), pp. 241–250.

Sengör, A.M.C. & Kidd, W.S.F.: The post-collisional tectonics of the Turkish-Iranian Plateau and a comparison with Tibet. *Tectonophysics* 55 (1979), pp. 361–376.

Şengör, A.M.C.: Cross-faults and differential stretching of hanging walls in regions of low-angle normal faulting: examples from western Turkey. In: M.P. Coward, J.F. Dewey & P.L. Hancock (eds): Continental extensional tectonics. *Geol. Soc. London, Spec. Publ.* 28, 1987, pp. 575–589.

Şengör, A.M.C. & Yılmaz, Y.: Tethyan evolution of Turkey — a plate tectonic approach. *Tectonophysics* 75:3–4 (1981), pp. 181–241.

Şerpen, U., Aksoy, N. & Öngür, T.: 2010 Present status of geothermal energy in Turkey. *Proceedings, Thirty- Fifth Workshop on Geothermal Reservoir Engineering*, 1–3 February 2010, Stanford University, Stanford, CA, 2010.

Süer, S., Güleç, N., Mutlu, H., Hilton, D.R., Çifter, C. & Sayın, M.: Geochemical monitoring of geothermal waters (2002–2004) along the North Anatolian Fault Zone, Turkey: spatial and temporal variations and relationship to seismic activity. *Pure Appl. Geophys.* 165 (2008), pp. 17–43.

Taymaz, T., Jackson, J. & McKenzie, D.: Active tectonics of the north and central Aegean Sea. *Geophys. J. Int.* 106 (1991), pp. 433–490.

CHAPTER 2

Gas geochemistry of Turkish geothermal fluids: He–CO$_2$ systematics in relation to active tectonics and volcanism

Nilgün Güleç, Halim Mutlu & David R. Hilton

2.1 INTRODUCTION

Gas geochemistry has been widely used over the last decades to investigate the origin of geothermal fluids and thus aid the understanding of reservoir processes, as well as in monitoring studies in tectonically and volcanically active regions. In this respect, the coupled systematics of CO$_2$ and He, including both their relative abundances and isotopic compositions (^{3}He/^{4}He, δ^{13}C), is a well-proven approach for defining volatile provenance in various tectonic settings (e.g., Hilton, 1996; Italiano et al., 2001; Kennedy and Truesdell, 1996; Marty and Jambon, 1987; O'Nions and Oxburgh, 1988; Sano and Marty, 1995; Van Soest et al., 1998).

The present chapter is centered upon utilizing the combined He–CO$_2$ characteristics of geothermal fluids in Turkey in a geodynamic context, taking advantage of the published database of previous studies (e.g., Güleç et al., 2002; de Leeuw et al., 2010; Mutlu et al., 2008; 2012). The major objectives are to assess (i) the role(s) of possible physicochemical processes in deep hydrothermal systems, (ii) the volatile provenance and relative contributions of mantle and crustal components to total volatile inventories, and (iii) the distribution of mantle volatiles in relation to differing neotectonic provinces, recent seismic activities, and/or the timing and nature of volcanic activity.

2.2 FRAMEWORK OF TECTONIC, VOLCANIC, AND GEOTHERMAL ACTIVITIES

Turkey comprises an integral segment of the Alpine–Mediterranean orogenic belt. Recent tectonic activity in Turkey has been governed by convergence between the Arabian and Anatolian plates, and is manifested along by four major structures: Bitlis Suture Zone (BSZ), North Anatolian Fault Zone (NAFZ), East Anatolian Fault Zone (EAFZ), and Western Anatolian Graben System (WAGS) (Fig. 2.1). Bounded by these structures, four major neotectonic provinces are recognized in Turkey: (i) eastern Anatolian contractional province (to the north of BSZ along which convergence between the Arabian and Anatolian plates is still continuing), (ii) western Anatolian extensional province (where tectonic activity is concentrated in a number of approximately E–W trending grabens), (iii) central Anatolian "ova" province (characterized by large sediment filled basins, "ovas", and transfer of compression in the east to extension in the west), and (iv) the north Turkish province to the north of NAFZ (characterized by limited E–W shortening) (Şengör et al., 1985).

Closely associated with neotectonic evolution is Neogene–Quaternary volcanism, the products of which cover extensive areas in Turkey. These products comprise lava flows, domes, and pyroclastics. There are several volcanoes and stratovolcanoes to be found in central and eastern Anatolian provinces (Fig. 2.1). In the eastern Anatolian volcanic province, the most recent eruptions occurred at Nemrut volcano in about AD 1400 (Tchalenko, 1977), while the latest eruptions in central and western Anatolian regions are reported to have occurred 20 and 25 ka ago, respectively (Bigazzi et al., 1993; Ercan et al., 1985).

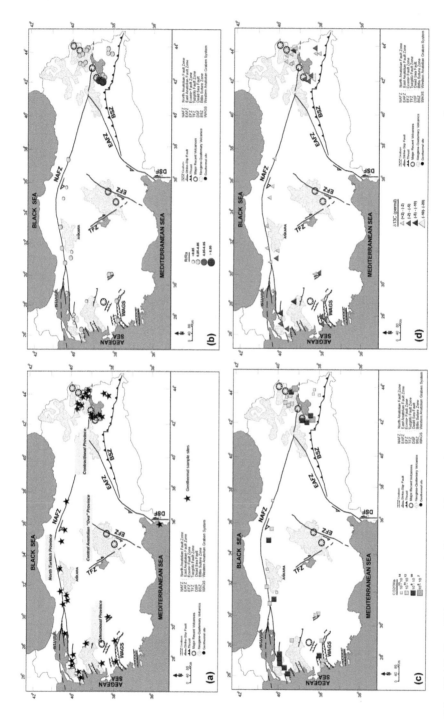

Figure 2.1. Distribution of (a) neotectonic features, Neogene–Quaternary volcanics and geothermal sample sites, (b) R/Ra values, (c) $CO_2/^3He$ ratios, and (d) $\delta^{13}C$ values.

Geothermal activity accompanying tectonic and volcanic activity in Turkey manifests itself in the form of hot springs and fumaroles emerging along major structural zones (Fig. 2.1a). The majority of thermal waters are Na–Ca–HCO$_3$ type, although Na–Cl type waters characterize the coastal sites in western Anatolia reflecting seawater intrusion into the aquifers (Mutlu and Güleç, 1998). The reservoir temperatures estimated *via* the use of various geothermometers and/or measured as bottom-hole temperatures indicate that the western Anatolian province has the highest geothermal potential in Turkey (Mutlu and Güleç, 1998).

2.3 He–CO$_2$ SYSTEMATICS

Our previously published data on He and CO$_2$ systematics (isotope compositions and relative abundances) of gases associated with geothermal fluids from western, eastern, and northern Anatolian provinces are presented in Table 2.1. As can be seen from Table 2.1, He-isotope compositions reported as R/Ra values (measured ^{3}He/^{4}He ratio (R) divided by the ^{3}He/^{4}He in air, $Ra = 1.4 \times 10^{-6}$) ranges from 0.27 (Çan, western Anatolia) to 7.76 (Nemrut caldera, eastern Anatolia). C-isotope compositions cover the range between −14.59‰ (Germav, eastern Anatolia) and +1.52‰ (Yalova, NAFZ). CO$_2$/^{3}He ratios are highly variable and span a range from 2.4×10^5 (Germav, eastern Anatolia) to 26×10^{13} (Kurşunlu, NAFZ). The spatial distribution of He–CO$_2$ characteristics of Turkish geothermal fluids are shown in Figure 2.1b–d.

Regarding the He-isotope compositions, the recorded R/Ra ratios point to the presence of mantle-derived helium throughout Turkey. Assuming simple binary mixing between a MORB-type mantle-He (R/Ra: 8, Farley and Neroda, 1998) and crustal-He (R/Ra: 0.02, Mamyrin and Tolstikhin, 1984), mantle-derived helium is found to range from 3 to 97% in Turkish geothermal fluids.

C-isotope compositions encompass values characteristic of mantle-derived carbon ($\delta^{13}C = -6.5$‰, Sano and Marty, 1995) and crustal-derived carbon: the latter displaying a wide range of end-member compositions from $\delta^{13}C = +2$‰ (recent marine carbonates: Veizer *et al.*, 1999), through $\delta^{13}C = 0$‰ (limestone: Sano and Marty, 1995), to $\delta^{13}C = -30$‰ (organic carbon: Sano and Marty, 1995).

CO$_2$/^{3}He ratios vary over eight orders of magnitude from 10^5 to 10^{13}. The majority of samples span the range between those considered to be typical for mantle-derived (2×10^9; Marty and Jambon, 1987) and crustal-derived (10^{11}–10^{12}, O'Nions and Oxburgh, 1988) volatiles, although the presence of values as high as 10^{13}, and as low as 10^7 to 10^5 probably reflect the effects of physicochemical processes (hydrothermal degassing, preferential CO$_2$ loss, and mixing between volatiles from different gas provenance) in deep hydrothermal systems. Therefore, in the following sections, the discussion is first focused on the impacts of such hydrothermal processes, in order to eliminate their effects on the assessment of relative contributions of mantle and crustal components to the total volatile inventory and the spatial distribution of mantle volatiles.

2.4 DISCUSSION

In this section, He–CO$_2$ characteristics of Turkish geothermal fluids are investigated in terms of the effects of various possible physicochemical processes in deep hydrothermal systems, relative roles of mantle and crustal contributions to the volatile inventory, and the spatial distribution of mantle volatiles with respect to differing neotectonic provinces and volcanic activities.

2.4.1 *Physicochemical processes in hydrothermal systems*

In order to evaluate the effects of deep hydrothermal processes, the He–CO$_2$ data are plotted on a ternary diagram (Fig. 2.2). Also shown in the diagram are the binary mixing trajectories between mantle-derived (R/Ra: 8, CO$_2$/^{3}He: 2×10^9) and various crustal volatile end-members,

Table 2.1.   He–CO_2 systematics of Turkish geothermal fluids (northern Anatolia (NAFZ) from de Leeuw *et al.*, 2010; western Anatolia from Mutlu *et al.*, 2008; eastern Anatolia from Mutlu *et al.*, 2012) (s: spring/bubbling pool, w: well).

Sample	Latitude (°N)	Longitude (°E)	Air-corrected R/Ra	$CO_2/^3He(\times 10^9)$	$\delta^{13}C\,(CO_2)$ (‰V-PDB)
NAFA[1]					
Efteni (s)	40°45′	31°01′	0.76	8700	−1.55
Yalova (s)	40°35′	29°16′	0.28	32	+1.52
Bolu (w)	40°38′	31°30′	1.16	1300	−0.18
Mudurnu (w)	40°24′	30°59′	2.19	55	−2.46
Seben (s)	40°25′	31°32′	0.58	11000	−0.34
Hamamözü (w)	40°47′	35°01′	0.69	35	−0.54
Gözlek (w)	40°24′	35°48′	0.30	120	−0.04
Reşadiye (s)	40°23′	37°20′	0.97	77000	−1.92
Kurşunlu (w)	40°50′	37°20′	1.16	260000	+0.92
Eastern Anatolia					
Ayrancı-Çaldıran (AYR) (s)	39°07′	43°51′	1.05	0.089	−9.07
Buğulu-Çaldıran (BUG) (s)	39°02′	43°59′	1.03	89	1.39
Çamlık-Başkale (CAM) (s)	37°56′	44°05′	1.00	16622	−1.43
Çaybağı-Saray (CAY) (s)	38°30′	44°10′	0.85	37742	−1.53
Çukur-Güroymak (CKR) (s)	38°39′	42°01′	6.36	84	−3.97
Diyadin (DVT) (s)	39°28′	43°39′	1.91	3660	+1.30
Diyadin (DYD) (s)	39°29′	43°39′	0.93	28779	−0.19
Germav-Hizan (GRM) (s)	38°17′	42°11′	2.40	0.00024	−14.59
Kokarsu-Bitlis (KOK) (s)	38°23′	42°15′	3.63	83	−2.88
Nemrut Caldera (NHL) (s)	38°38′	42°14′	7.76	20	−2.08
Patnos-Ağrı (PAT) (w)	39°14′	42°50′	2.42	14708	−0.83
Taşkapı-Erciş (TAS) (s)	39°16′	43°25′	3.68	889	−0.25
Tutak-Van (TUT) (s)	39°33′	42°46′	1.83	114	−0.93
Yurtbaşı-Gürpınar (YUR) (s)	38°13′	43°47′	0.86	10292	−2.97
Western Anatolia					
Tuzla (s)	39°34′	26°10′	1.44	45.5	+0.35
Kestanbol-1 (w)	39°44′	26°11′	0.80	1.66	−3.59
Kestanbol-2 (s)	39°44′	26°11′	0.79	15.7	−3.37
Çan (w)	40°01′	27°02′	0.27	1.72	−4.10
Gönen (w)	40°06′	27°39′	0.31	33.6	−1.52
Manyas (w)	40°03′	27°54′	0.41	34.3	−6.61
Hisaralan (s)	39°16′	28°19′	0.66	220	−8.04
Seferihisar (s)	38°04′	26°54′	0.39	892	−3.79
Balçova (w)	38°23′	27°01′	0.56	95.1	−3.65
Germencik (s)	37°55′	27°37′	1.06	3865	−1.89
Germencik (s)	37°55′	27°37′	1.03	1934	−2.35
Kızıldere (w)	37°57′	28°50′	1.67	305	−0.62
Gazlıgöl (w)	38°56′	30°29′	0.30	5450	−2.58
Ömer (w)	38°50′	30°24′	1.06	23540	−0.66

[1]The values for samples associated with NAFZ represent the mean of measurements taken in the period 2000–2004 during the course of a monitoring program.

along with the general trends expected from addition and/or loss of a particular volatile phase. As can be seen from the diagram, the majority of Turkish samples represent products of variable amounts of mixing between mantle-derived and crustal volatiles. It is also worth noting that some samples plot at or close to the CO_2 apex (with exceptionally high $CO_2/^3He$ ratios), while others plot along the base of the diagram (with extremely low $CO_2/^3He$ values). Hydrothermal degassing

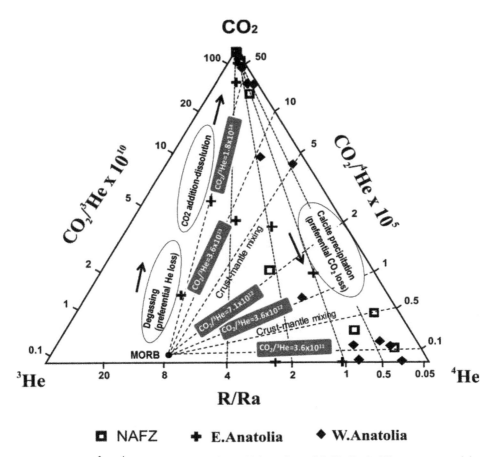

Figure 2.2. CO_2–^{3}He–^{4}He ternary diagram for Turkish geothermal fluids. Dashed lines represent mixing between mantle (MORB: $R/Ra = 8$, $CO_2/^3He = 2 \times 10^9$) and various crustal end-members ($R/Ra = 0.02$ and $CO_2/^3He = 10^{11}$–10^{14} as labeled). Dotted lines indicate effects of various physiochemical processes, which act to modify $CO_2/^3He$ ratios.

seems to be a plausible mechanism to explain the high $CO_2/^3He$ ratios as He (being less soluble than CO_2 in aqueous phase) is preferentially lost into the first formed gas phase, leading to an increase in the $CO_2/^3He$ ratio of the residual fluid. A more speculative alternative possibility is the addition/dissolution of pure CO_2 into geothermal fluids (e.g., directly from carbonate lithologies without accompanying He) which would result in an increase in the $CO_2/^3He$ ratios. The group of samples plotting along the base of the diagram, on the other hand, appears to be consistent with preferential CO_2 loss, possibly associated with calcite precipitation, which is a widespread phenomenon in most geothermal fields as evidenced by travertine deposition around hot springs and/or calcite scaling in drilling wells. Included in the first group of samples (at or close to CO_2 apex) are CAM, CAY, DYD, DVT, PAT, TAS, and YUR from eastern Anatolia, Germencik, Gazlıgöl, and Ömer samples from western Anatolia, and Seben, Reşadiye, and Kurşunlu samples from NAFZ.

   The second group of samples (at the base of the diagram) are AYR and GRM from eastern Anatolia, and Kestanbol (well) and Çan from western Anatolia. It should be noted here that not only $CO_2/^3He$ ratios but also $\delta^{13}C$ values can be significantly affected by degassing, CO_2 addition/dissolution, and preferential CO_2 loss and/or calcite precipitation processes. To eliminate the potential masking effects of these hydrothermal processes, the aforementioned samples are not utilized in the following discussion concerned with the provenance of CO_2 gas.

2.4.2   *Volatile provenance: relative contributions of mantle and crustal components*

To determine the relative contributions to the volatile inventory from possible end-members, two approaches are used in this section: one is concerned with the assessment of mantle-He percentage based on the assumption of a simple binary mixing between mantle and crustal components, the other is the assessment of provenance of CO_2 based on three-component mixing model of Sano and Marty (1995) (Fig. 2.2).

In the first approach, for the assessment of the mantle-He percentage, the crustal-He is assigned $R/Ra = 0.02$ (Mamyrin and Tolstikhin, 1984), while the mantle-He component is assigned an R/Ra value of 8 (Farley and Neroda, 1998) based on the isotope composition of volcanics which suggests involvement of both subduction-modified and asthenospheric mantle components beneath western and eastern Anatolia (Aldanmaz *et al.*, 2006; Güleç, 1991; Keskin, 2007; Özdemir, 2011). It should be noted here that although $CO_2/^3He$ ratios and $\delta^{13}C$ values are significantly affected by fractionation associated with deep hydrothermal processes (e.g., degassing, calcite precipitation), $^3He/^4He$ ratios are not, and hence the mixing calculations are performed here for all the samples. Table 2.2 lists the calculated mantle-He percentages and shows that the mantle-He component ranges from 3% up to 97% of the total He content in a single sample.

The second approach, the three-component mixing model, could be performed only for samples not significantly affected by deep hydrothermal processes as explained in Section 2.4.1. In this model, a mantle component (M) and two crustal components, limestone (L) and sedimentary organic carbon (S), are used as end-members. The compositions assigned to these end-members are $\delta^{13}C = -6.5‰$ for M, 0‰ for L, $-30‰$ for S, and $CO_2/^3He = 2 \times 10^9$ for M, 10^{13} for L and S (Sano and Marty, 1995). A graphical representation of the mixing model is given in Figure 2.3. Along with data points, also given in the diagram are the trajectories for binary mixing between (i) mantle (M) and limestone (L), (ii) mantle (M) and sedimentary organic carbon (S), and (iii) limestone (L) and sedimentary organic carbon (S) components. In the diagram, the distinction between mantle and crustal components is essentially defined by $CO_2/^3He$ ratios, whereas the distinction between the two crustal sources (limestone and sedimentary organic carbon) is based on $\delta^{13}C$ values. Although three samples plot beyond the boundaries of the mixing curves in Figure 2.3 (pointing to a minimal contribution from sedimentary organic source), the majority of the samples require a contribution from all three end-members to explain their $CO_2/^3He$–$\delta^{13}C$ (CO_2) characteristics. The percentages of the contribution of these three end-members are given in Table 2.2. The table shows that the principal contributor to the carbon inventory is limestone comprising >72% of the total CO_2 budget. This conclusion is in agreement with the fact that carbonate levels are widespread constituents of the basement metamorphics and comprise the reservoir lithologies in most geothermal fields in Turkey (Mutlu and Güleç, 1998; Mutlu *et al.*, 2012).

2.4.3   *Spatial distribution of mantle volatiles: relation to tectonic and volcanic activities*

The spatial distribution of mantle-derived volatiles in Turkey is shown in Figure 2.4 in relation to various neotectonic provinces and Neogene–Quaternary volcanism. Because of the effects of fractionation on $CO_2/^3He$ ratios and $\delta^{13}C$ values, which can mask the original source signature, the data pertinent to CO_2 in Figure 2.4 are not adequate to make an overall correlation between the distribution of mantle-He and mantle-C. Nevertheless, the association of the highest mantle-He and the highest mantle-C values near Nemrut volcano (eastern Anatolia) is the most striking feature of the figure. This is of no great surprise given the fact that Nemrut is the most recent volcanic center in Turkey with the latest eruption at AD 1441 (Tchalenko, 1977). The recent geophysical data from Eastern Turkey Seismic Experiment (ETSE) point to asthenospheric upwelling beneath

Table 2.2.   Proportions of mantle *vs.* crustal volatile contributions to each sample (w: well, s: spring).

Sample	Air-corrected R/Ra	Mantle-derived He (%)	Mantle	CO₂ (%) Limestone	Sediments
NAFZ					
Efteni (s)	0.76	8.9	0.00	94.50	5.50
Yalova (s)	0.28	2.9	6.15	93.85	0.00
Bolu (w)	1.16	14.0	0.45	98.50	1.05
Mudurnu (w)	2.19	26.9	3.65	88.50	7.85
Seben (s)	0.58	6.7			
Hamamözü (w)	0.69	8.1	5.85	93.05	1.10
Gözlek (w)	0.30	3.1	1.83	98.17	0.00
Reşadiye (s)	0.97	11.6			
Kurşunlu (w)	1.16	14.0			
Eastern Anatolia					
Ayrancı-Çaldıran (AYR) (s)	1.05	12.6			
Buğulu-Çaldıran (BUG) (s)	1.03	12.3	2.23	97.77	0.00
Çamlık-Başkale (CAM) (s)	1.00	11.9			
Çaybağı-Saray (CAY) (s)	0.85	10.1			
Çukur-Güroymak (CKR) (s)	6.36	79.4	2.36	84.92	12.72
Diyadin (DVT) (s)	1.91	23.4			
Diyadin (DYD) (s)	0.93	11.1			
Germav-Hizan (GRM) (s)	2.40	29.6			
Kokarsu-Bitlis (KOK) (s)	3.63	45.0	2.40	88.51	9.09
Nemrut Caldera (NHL) (s)	7.76	97.0	9.98	85.25	4.76
Patnos-Ağrı (PAT) (w)	2.42	29.8			
Taşkapı-Erciş (TAS) (s)	3.68	45.7			
Tutak-Van (TUT) (s)	1.83	22.4	1.73	95.54	2.72
Yurtbaşı-Gürpınar (YUR) (s)	0.86	10.2			
Western Anatolia					
Tuzla (s)	1.44	17.5	4.37	94.58	26.60
Kestanbol-1 (w)	0.80	9.4			
Kestanbol-2 (s)	0.79	9.3			
Çan (w)	0.27	2.8			
Gönen (w)	0.31	3.3			
Manyas (w)	0.41	4.5			
Hisaralan (s)	0.66	7.7	0.89	72.50	26.60
Seferihisar (s)	0.39	4.3	0.22	87.20	12.58
Balçova (w)	0.56	6.4			
Germencik (s)	1.06	12.7	0.03	93.68	6.29
Germencik (s)	1.03	12.3			
Kızıldere (w)	1.67	20.4	0.64	97.43	1.93
Gazlıgöl (w)	0.30	3.1			
Ömer (w)	1.06	12.7			

eastern Anatolia where crustal thickness is at its minimum (38 km) near Nemrut volcano (Al-Lazki *et al.*, 2003; Sandvol *et al.*, 2003; Şengör *et al.*, 2003; 2008). Geochemical studies also suggest interplay of asthenospheric and lithospheric mantle-derived melts in the generation of Nemrut volcano (Özdemir *et al.*, 2006). Furthermore, the N–S trending fissure zones associated with the current compressional–extensional regime in the region (Koçyiğit *et al.*, 2001) are more dense around Nemrut and seem to have provided channelways for mantle-volatiles to reach the surface (Mutlu *et al.*, 2012).

The occurrence of a relatively high mantle-C input (4.4%) in the Tuzla field in western Anatolia is also a striking feature of Figure 2.4, despite the relatively moderate mantle-He (17.5%)

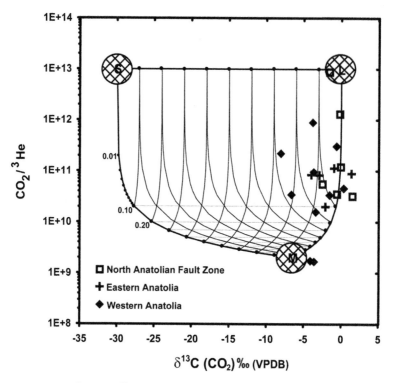

Figure 2.3. Plot of $CO_2/^3He$ *vs.* $\delta^{13}C$ (CO_2) for Turkish geothermal fluids. The end-member compositions for sedimentary organic carbon (S), mantle carbon (M), and limestones (L) are $\delta^{13}C$ $(CO_2) = -30‰$, $-6.5‰$, and $0‰$; and $CO_2/^3He = 1 \times 10^{13}$, 2×10^9, and 1×10^{13}, respectively (Sano and Marty, 1995).

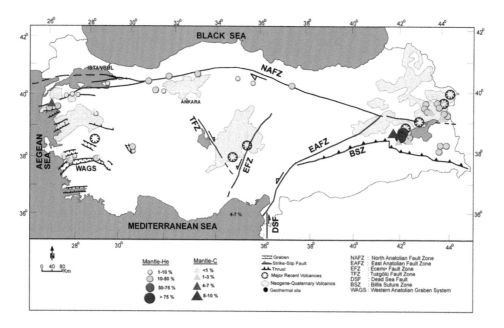

Figure 2.4. Spatial distribution of mantle-He and mantle-C contributions in Turkey.

contribution at this site. It is worth noting here that the Tuzla geothermal field is spatially asso-ciated with the Ezine volcanics that comprise the products of the late, alkali basaltic phase of Neogene–Quaternary volcanism in western Anatolia, and is associated with adiabatic melting (asthenospheric upwelling) related to the presently active extension in this province (Aldanmaz, 2002; Aldanmaz *et al.*, 2000; Güleç and Hilton, 2006). Additionally, Tuzla is located at the western extension of the NAFZ and, hence, is under the influence of not only the extensional graben but also strike-slip tectonics. It appears then that mantle-volatiles, brought to crustal levels by young magmatic activity, have been scavenged and transferred to the surface by deep water circulation facilitated by deep fault planes (Mutlu *et al.*, 2008). Although the mantle-C contribution is rather low, the high mantle-He component (12–20%) recorded in areas like Kızıldere and Germencik further suggests that mantle-derived melts might have been emplaced at crustal levels without any surface volcanic equivalents (Güleç and Hilton, 2006).

Along the NAFZ, the highest mantle-He contributions are recorded at the western-central segment (Mudurnu, Seben, Kurşunlu) which has been the seismically, most active segment in recent years. In fact, a 3-year monitoring program conducted along NAFZ between 2001 and 2004 revealed a possible link between the variability of $CO_2/^3He$ ratios and seismicity at some specific locations (Yalova and Efteni) (de Leeuw *et al.*, 2010). This, in turn, suggests that changes in the mantle-volatile input are related to changes in permeability along this segment of the fault zone.

In summary, the distribution of mantle-derived volatiles points to a close relation to neotectonic activities, such as recent seismicity and the timing and nature of volcanism in Turkey. The transfer of volatiles to crustal levels seems to have been governed by young magmatic activity, while their transfer to the surface is aided by fault zones influenced by current tectonic activity.

2.5 CONCLUSIONS

The combined He–CO_2 systematics of geothermal fluids from the eastern Anatolian contrac-tional and western Anatolian extensional provinces, as well as those from the NAFZ, have been investigated in terms of the possible effects of physicochemical processes on deep hydrother-mal systems, relative roles of mantle- and crustal-components in the total volatile inventory, and the relation to differing neotectonic provinces and the age and nature of Neogene–Quaternary volcanism. He-isotope compositions display a range from 0.28 to 2.19 *Ra* for NAFZ, 0.85 to 7.76 *Ra* for eastern Anatolian, and 0.27 to 1.67 *Ra* for western Anatolian fluids, reflecting a range of 3–97% for the mantle-derived He component. ^{13}C values are between −1.92 and +1.52‰ for NAFZ, −14.59 and +1.30‰ for eastern, and −8.04 and +0.35‰ for western Ana-tolian fluids, with a range of $CO_2/^3He$ ratios of 32×10^9–26×10^{13}, 2.4×10^5–3.7×10^{13}, and 1.66×10^9–2.35×10^{13}, respectively. Hydrothermal degassing (preferential He-loss) and/or CO_2 addition from/to geothermal fluids, and preferential CO_2 loss associated with calcite precip-itation seem to have significantly affected geothermal samples resulting in the fractionation of CO_2/He elemental ratio. Quantitative assessment of the volatile inventory suggests lime-stone as the major carbon contributor (comprising >72% of the total budget) in agreement with the composition of basement metamorphics (containing marble) and reservoir lithologies (marble, limestone) in most fields. The highest mantle-He contribution (97%), accompa-nied by the highest mantle-C (9.98%), is recorded in eastern Anatolia close to the most recently active Nemrut volcano. In western Anatolia, relatively high mantle-He and mantle-C contributions in the Tuzla field suggest a genetic link with the spatially associated young, alkaline volcanics (Mutlu *et al.*, 2008). Along the NAFZ, the highest mantle contributions are recorded at the most seismically active western-central segment (Güleç *et al.*, 2002). These observations collectively suggest that mantle volatiles are added to crust by young magmatic activity and that fault zones of these tectonic provinces provide pathways for volatiles to reach the surface.

ACKNOWLEDGMENTS

Grants from the Scientific and Technical Research Council of Turkey (TUBITAK-100Y097), Middle East Technical University (ODTU-AFP 99-03-09-02), Eskişehir Osmangazi University (1999/12, 2005/15016 and 2009/15017), and NSF (EAR-9724189, 0229508) are gratefully acknowledged.

REFERENCES

Aldanmaz, E.: Mantle source characteristics of alkali basalts and basanites in an extensional intracontinental plate setting, western Anatolia, Turkey: implications for multi-stage melting. *Int. Geol. Rev.* 44 (2002), pp. 440–457.
Aldanmaz, E., Pearce, J.A., Thirlwall, M.F. & Mitchell, J.G.: Petrogenetic evolution of late Cenozoic volcanism in western Anatolia, Turkey. *J. Volcan. Geoth. Res.* 102 (2000), pp. 67–95.
Aldanmaz, E., Köprübaşı, N., Gürer, Ö.F., Kaymakçı, N. & Gourgaud, A.: Geochemical constraints on the Cenozoic, OIB-type alkaline volcanic rocks of NW Turkey: implications for mantle sources and melting processes. *Lithos* 86 (2006), pp. 50–76.
Al-Lazki, A., Seber, D., Sandvol, E., Turkelli, N., Mohamad, R. & Barazangi, M.: Tomographic Pn velocity and anisotropy structure beneath the Anatolian plateau (eastern Turkey) and the surrounding regions. *Geophys. Res. Lett.* 30:24 (2003). doi: 10.1029/2003GL017391.
Bigazzi, G., Yeğingil, Z., Ercan, T., Oddone, M. & Özdoğan, M.: New data for the chronology of central and northern Anatolia by fission track dating of obsidians. *Bull. Volcanol.* 55 (1993), pp. 588–595.
de Leeuw, G.A.M., Hilton, D.R., Güleç, N. & Mutlu, H.: Regional and temporal variations in $CO_2/^3He$, $^3He/^4He$ and $\delta^{13}C$ along the North Anatolian Fault Zone, Turkey. *Appl. Geochem.* 25 (2010), pp. 524–539.
Ercan, T., Satır, M., Kreuzer, H., Türkecan, A., Günay, E., Çevikbaş, A., Ateş, M. & Can, B.: Batı Anadolu Senozoyik volkanitlerine ait yeni kimyasal, izotopik ve radyometrik verilerin yorumu. *TJK Bülteni* 28 (1985), pp. 121–136.
Farley, K.A. & Neroda, E.: Noble gases in the Earth's mantle. *Annu. Rev. Earth Pl. Sci.* 26 (1998), pp. 189–218.
Güleç, N.: Crust-mantle interaction in western Turkey: implications from Sr and Nd isotope geochemistry of Tertiary and Quaternary volcanics. *Geol. Mag.* 125 (1991), pp. 417–435.
Güleç, N. & Hilton, D.R.: Helium and heat distribution in western Anatolia, Turkey: relationship to active extension and volcanism. *Geol. Soc. Am. Spec. Paper*, 409, 2006, pp. 305–319.
Güleç, N., Hilton, D.R. & Mutlu, H.: Helium isotope variations in Turkey: relationship to tectonics, volcanism and recent seismic activities. *Chem. Geol.* 187 (2002), pp. 129–142.
Hilton, D.R.: The helium and carbon isotope systematics of a continental geothermal system: results from monitoring studies at Long Valley Caldera (California, U.S.A.). *Chem. Geol.* 127 (1996), pp. 269–295.
Italiano, F., Martinelli, G. & Nuccio, P.M.: Anomalies of mantle-derived helium during the 1997–1998 seismic swarm of Umbria-Marche. Italy. *Geophys. Res. Lett.* 28 (2001), pp. 839–842.
Kennedy, B.M. & Truesdell, A.H.: The Northwest Geysers high-temperature reservoir: evidence for active magmatic degassing and implications for the origin of The Geysers geothermal field. *Geothermics* 25 (1996), pp. 365–387.
Keskin, M.: Eastern Anatolia: a hot spot in a collision zone without a mantle plume. In: G.R. Foulhrt & D. Jurdy (eds): Plates, plumes, and planetary processes. *Geol. Soc. Am. Spec. Paper* 430, 2007, pp. 693–722.
Koçyiğit, A., Yılmaz, A., Adamina, S. & Kuloshvili, S.: Neotectonics of East Anatolian Plateau (Turkey) and Lesser Caucasus: implication for transition from thrusting to strike-slip faulting. *Geodin. Acta* 14 (2001), pp. 177–195.
Mamyrin, B.A. & Tolstikhin, I.N.: Helium isotopes in nature. *Developments in Geochemistry*, vol. 3. Elsevier, Amsterdam, 1984.
Marty, B. & Jambon, A.: $C/^3He$ in volatile fluxes from the solid Earth: implication for carbon geodynamics. *Earth Planet. Sci. Lett.* 83 (1987), pp. 16–26.
Mutlu, H. & Güleç, N.: Hydrochemical outline of thermal waters and geothermometry applications in Anatolia, Turkey. *J. Volcanol. Geoth. Res.* 85 (1998), pp. 495–515.
Mutlu, H., Güleç, N. & Hilton, D.R.: Helium-carbon relationships in geothermal fluids of western Anatolia, Turkey. *Chem. Geol.* 247 (2008), pp. 305–321.

Mutlu, H., Güleç, N., Hilton, D.R., Aydın, H. & Halldórsson, S.A.: Spatial variations in gas and stable isotope compositions of thermal fluids around Lake Van: implications for crust-mantle dynamics in eastern Turkey. *Chem. Geol.* 300–301 (2012), pp. 165–176.

O'Nions, R.K. & Oxburgh, E.R.: Helium volatile fluxes and the development of continental crust. *Earth Planet. Sci. Lett.* 90 (1988), pp. 331–347.

Özdemir, Y.: *Volcanostratigraphy and petrogenesis of Süphan stratovolcano.* PhD Thesis, Middle East Technical University, Ankara, Turkey, 2011.

Özdemir, Y., Karaoğlu, Ö., Tolluoğlu, A.Ü. & Güleç, N.: Volcanostratigraphy and petrogenesis of the Nemrut stratovolcano (East Anatolian High Plateau): the most recent post-collisional volcanism in Turkey. *Chem. Geol.* 226 (2006), pp. 189–211.

Sandvol, E., Türkelli, N. & Barazangi, M.: The eastern Turkey seismic experiment: the study of a young continent-continent collision. *Geophys. Res. Lett.* 30 (2003), doi: 10.1029/2003GL018912.

Sano, Y. & Marty, B.: Origin of carbon in fumarolic gas from island arcs. *Chem. Geol.* 119 (1995), pp. 265–274.

Şengör, A.M.C., Görür, N. & Şaroğlu, F.: Strike-slip faulting and related basin formation in zones of tectonic escape: Turkey as a case study. *The Society of Economic Paleontologists and Mineralogists*, Special Publication No. 37, 1985, pp. 227–264.

Şengör, A.M.C., Özeren, S., Zor, E. & Genç, T.: East Anatolian high plateau as a mantle-supported, N–S shortened domal structure. *Geophys. Res. Lett.* 30 (2003), 8045, doi: 10.1029/2003GL017858.

Şengör, A.M.C., Özeren, M.S., Keskin, M., Sakınç, M., Özbakır, A.D. & Kayan, İ.: Eastern Turkish high plateau as a small Turkic-type orogen: implications for post-collisional crust-forming processes in Turkic-type orogens. *Earth-Sci. Rev.* 90 (2008), pp. 1–48.

Tchalenko, J.S.: A reconnaissance of seismicity and tectonics at the northern border of the Arabian plate (Lake Van region). *Rev. Geogr. Phys. Geol. Dyn.* 19 (1977), pp. 189–208.

Van Soest, M.C., Hilton, D.R. & Kreulen, R.: Tracing crustal and slab contributions to arc magmatism in the Lesser Antilles island arc using helium and carbon relationships in geothermal fluids. *Geochim. Cosmochim. Acta* 62 (1998), pp. 3323–3335.

Veizer, J., Ala, D., Azmy, K., Bruckschen, P., Buhl, D., Bruhn, F., Carden, G.A.F., Diener, A., Ebneth, S., Godderis, Y., Jasper, T., Korte, C., Pawellek, F., Podlaha, O.G. & Strauss, H.: $^{87}Sr/^{86}Sr$, $\delta^{13}C$, and $\delta^{18}O$ evolution of Phanerozoic seawater. *Chem. Geol.* 161 (1999), pp. 59–88.

CHAPTER 3

Geothermal fields and thermal waters of Greece: an overview

Nicolaos Lambrakis, Konstantina Katsanou & George Siavalas

3.1 PROLOGUE–HISTORICAL BACKGROUND

Greece is favored by a large number of thermal springs known since antiquity. It was believed that the springs possessed supernatural and healing properties and thus were also called healing springs. The Temple of Artemis in Lesvos island (NE Aegean Sea) is built over one of these springs. Hippocrates (460–375 BC), who is considered to be the founder of medical science and the "father" of hydrotherapy, extensively studied natural water and distinguished it into (i) stagnant, including water from mires and lakes, (ii) rainwater, and (iii) water rising through rocks, called mineral (metallic) water. He described the latter type as warm water containing sulfur, iron, copper, silver, gold, and other metals. The Roman physician Galen cured his patients by offering them off-season fruits, presumably cultivated in a primitive greenhouse heated by thermal waters.

Strabo (1st century BC) in his work "Iliaka" (book 8, chapter 3), Pausanias (2nd century BC) in his work "Messiniaka" (book 5, chapters 5 and 6), as well as many other Roman travelers and authors referred to the famous Kaiafas springs. During those times, the area surrounding the springs were considered the sanctuary of Anigrides Nymphs, and people suffering from dermal diseases would sacrifice animals, bath in the mud of the springs and then in the water of river Anigros, to heal themselves.

According to Strabo, the myth attributes the stench of the Kaiafas springs to the Centaurs, who used the spring's water to clean their wounds obtained after Hydra's venomous bites or according to Pausanias, inflicted by the poisoned arrows of Hercules. During Roman and Byzantine times, many physicians and healers, namely Erofilos, Erasistratos, Asklipiades, Agathenos, Galenos, Horevasios, Pavlos the Aiginitis, and others used to recommend hydrotherapy and healing bathing to their patients. Moreover, one of Agathenos' scholars recorded the healing properties of healing springs in 1st century AD and stated that it is impossible to accurately determine and describe the mechanism that develops these properties for each spring without long-term monitoring and experiments. In Greece there are 750 recorded healing springs: the thermal springs at Thermopylae have been flowing at a rate of 30 kg/s for at least 2500 years (Fytikas et al., 2005). However, the number of those officially recognized by the state is significantly lower reaching 76, with 15 of them considered of great touristic interest.

In the present review, the term "thermal" is used to describe spring water displaying a temperature higher than the mean annual air temperature of the respective site up to the boiling point. Additionally, the term mineral spring is used for springs with total dissolved salt content exceeding 1000 mg/L or when gas content is higher than the usual levels of natural water.

3.2 INTRODUCTION

The systematic exploration for the discovery and exploitation of geothermal fields in Greece commenced in 1970 and was carried out by the Institute of Geological and Mineral Exploration (IGME). The first exploration project was funded in 1971 and the areas of interest were the islands of Melos, Nisyros and Lesvos, and the continental sites of Methana, Sousaki, Kamena Vourla,

Thermopyles, Ypati, and Aidipsos (Fytikas and Andritsos, 2004). A number of very important reports from the results of this exploration project were published by IGME during the 1980s including a brief description of the geological background and approximately 1300 chemical analyses from the recorded thermal and mineral springs (Gioni-Stavropoulou, 1983; Karydakis, 1983; Karydakis and Kavouridis, 1983; 1989; Kavouridis *et al.*, 1982; Kolios, 1986; Orfanos, 1985; Sfetsos, 1988; Traganos, 1991). During the same period, the first law for geothermal energy exploitation was passed (Law no. 1575/84) by the Greek state. Today this law has been replaced by Law no. 3175/2003, with private initiatives for geothermal energy are subject to Law no. 2601/1998. Moreover, the natural resources with healing properties and healing-oriented tourism are regulated by Law no. 3498/2006.

At present, geothermal energy exploitation in Greece is connected to the use of thermal and mineral water. It is estimated that 1% of the energy that derives from the exploitation of the geothermal fields is used for domestic heating, 7% in fish farms, 21% is used in healing facilities, 30% is consumed in greenhouses, and 41% in geothermal heat pumps (Hatziyannis, 2011). The total installed thermal capacity in Greece is estimated to exceed 130 MW$_t$ (Arvanitis, 2011).

The extensive formation of geothermal fields and thermal waters in Greece is due to the active tectonics and recent volcanism, which commenced in the beginning of Tertiary as a result of collision between the Eurasian and African plates in the region. High-enthalpy geothermal fields (fluid temperature: >150°C) developed along the Aegean volcanic arc, whereas low-(fluid temperature: <90°C) and medium-enthalpy fields (fluid temperature: 90–150°C) are related to the fault systems of central Aegean and the Tertiary sedimentary basins of Macedonia Prefecture.

A large number of studies in the Greek and international literature deal with the geological setting of the geothermal fields in the country, as well as with the origin and chemical composition of the thermal water. The present study attempts to combine the widely accepted views on the geological background and formation conditions of the Greek geothermal fields with previously unpublished data. This newly released data concern mostly the minor and trace element concentrations in water samples from geothermal fields, along with radon and thoron concentrations.

3.3 THE PALEOGEOGRAPHICAL SETTING OF GREECE

The geological formations in the Greek territory are classically divided into the so-called isopic zones, which reflect the paleogeographical setting of Greece during the Mesozoic (Fig. 3.1). These zones are further subdivided into the internal zones (eastern Greece) and the external zones (western Greece) with pre-Cenozoic and post-Eocene tectonism, respectively. The external zones are considered a passive margin at the eastern part of the Apulian microplate, whereas at the western part the Pindos Ocean developed separating the Apulian from the Pelagonian microplate starting from the Middle Triassic until the early Eocene (Robertson, 1991). In the modern era, Greece is on the collision margin between the Eurasian and African plates, which is expressed with the formation of the Hellenic trench, an arc-shaped deep basin extending from west of Zakynthos island to the south of Crete in the Mediterranean Sea, formed due to the subduction of the African plate. This trench is crosscut by the Kefallonia transform fault at the western part and by the SE Aegean strike-slip faults at the southern part causing differential subduction rate along the trench (Doutsos and Kokalas, 2001). Toward the center of the Greek peninsula, the modern Aegean volcanic arc is developed following the direction of the Hellenic trench, whereas at the north the geotectonic regime is dominated by the North Anatolian Fault (NAF), a strike-slip fault extending from the east of Turkey, where together with the East Anatolian Fault (EAF) separates the Arabian from the Eurasian plate, to the Sporades islands and to Sperchios basin in central Greece.

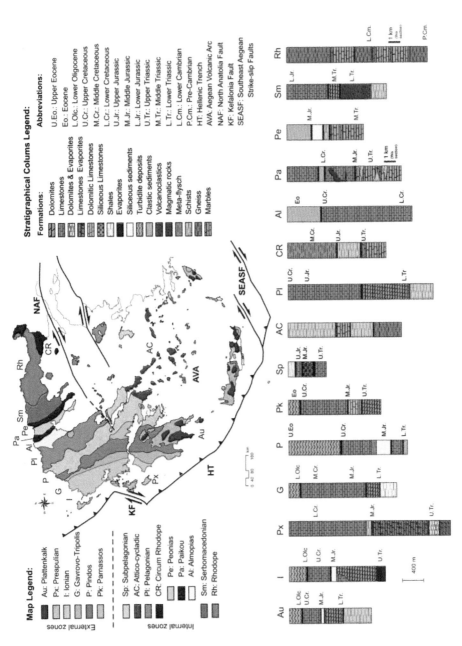

Figure 3.1. Major geotectonic features of the Greek peninsula, geographic distribution, and lithostratigraphic columns of the "isopic" zones of Greece (modified from Mountrakis, 1985).

3.4   VOLCANISM OF GREECE

The Aegean region has been subject to orogenic volcanism since the Tertiary. Fytikas *et al.* (1984) distinguish two major belts of volcanism; the "north" and "south" ones displaying distinct features. The two belts differ on both their formation age and their relative position in the region. Their major differences are the distribution of their petrographic units (e.g., shoshonitic rocks), the continually enhanced potassic character of the volcanism of the northern belt and its shift toward southern directions. These resulted in the very rare occurrence of mafic rocks combined with an increase in K_2O content in all the members of this belt. In the northern belt, the volcanic forehead developed along the margin of the Rhodope massif (Fig. 3.1), where the crust is relatively thick, around 40 km (Makris, 1977) in contrast to the southern belt, where the volcanism was built up on a relatively thinned crust, 25–30 km (Makris, 1977). Evidence for the migration of volcanism in the northern belt is also provided by the observation that in its northernmost part, the Prefecture of Thrace, only volcanic rocks of Oligocene age outcrop, whereas at the southern parts the age of the volcanic rocks is Miocene. In general, from Eocene to Oligocene, the southern margins of the Eurasian plate, represented by the Rhodope massif, were collided with the Apulian microplate, which was between the Eurasian and the African plates, whereas subduction processes triggered the volcanism in the northern belt between Oligocene and Lower Miocene.

Since the Middle Miocene tectonic plate movements, such as the progression of the Arabian plate into the Eurasian, which caused the divergence of Anatolian and Iranian blocks, resulted in the development of diverse, often contrasting features in the Aegean microplate. The southern margin is a typical converging area, whereas the northern and western limits are more like an extension of north Anatolian strike-slip fault. The eastern margin is characterized by a series of E–W trending grabens. Thus, the more recent and relatively weaker volcanism between the Upper Miocene to the Quaternary is limited between these margins. The volcanic rocks are mainly of alkaline character, and usually associated with non-convergent margins and also pointing to tensional tectonics favoring magma formation and rise.

The volcanism of the southern belt commenced at the end of Lower Pleistocene, along an arc ranging from Sousaki-Loutraki (continental Greece) to the islands of Aigina, Melos, Santorini, Nisyros, and Kos (Fig. 3.2a) and is related to the modern volcanic arc of the Aegean Sea, parallel to the subduction zone of the African plate.

3.5   THE DISTRIBUTION OF HEAT FLOW AND CAUSE OF GEOTHERMAL ANOMALIES IN GREECE

In Table 3.1, the potential and the confirmed geothermal fields of Greece are shown, as indicated in the Greek Law no. 3175/2003.

Regions with increased heat flow in Greece are mainly divided into three categories (Fig. 3.2a) that are further discussed in the following paragraphs.

3.5.1   *Back-arc regions*

The back-arc regions include the Tertiary sedimentary basins of Strimonas as well as Nestos and Evros deltas in northern Greece. These regions are low to medium enthalpy back-arc geothermal fields with extensional tectonics, which presumably also results in crust thinning. This type of tectonics trigger the formation of faults, which favor the quick rise of thermal fluids justifying the increased heat flow observed at these regions. An additional cause for the increase of heat flow is the existence of felsic plutonism as it is the case of Chrysochorafa, Eraklias, and Aggistro in the Prefecture of Serres and of N. Erasmios Maggana in the Prefecture of Xanthi.

The regions of north Euboea and Sperchios river drainage basins also belong to the back-arc geothermal fields with higher heat flow rates than the previous group. This is mostly due to the

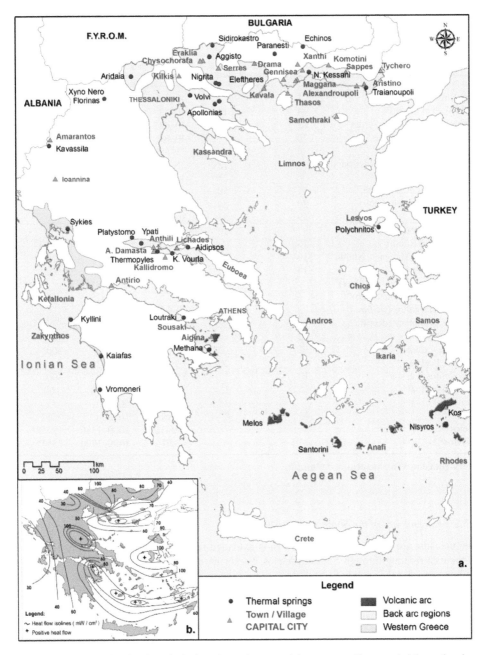

Figure 3.2. (a) Map showing the principal geothermal areas and the corresponding sampled thermal springs of Greece and (b) preliminary heat flow map of Greece (Fytikas and Kavouridis, 1985).

effect of the North Anatolian Fault (NAF, Fig. 3.2) in northern Turkey, which is extended through the NE Aegean Sea (between Samothraki and Lemnos) and ends up in the Sperchios basin (central Greece).

East Aegean Sea islands, Samothraki, Lemnos, Lesvos, Chios, and Ikaria also belong to the back-arc regions and their increased heat flow is due to active tectonics. The activity of large active faults is also responsible for the increased heat flow of Lagadas and Volvi fields (Fig. 3.2).

Table 3.1. The thermal fields of Greece (Law no. 3175/2003).

Geothermal fields	Area (km^2) Confirmed	Potential	T (water) (°C)	Depth (m)	Discharge (m^3/h)
Aggistro, Serres*	1.5		40–48	100–300	80
Sidirokastro, Serres*	4	11	40–75	30–500	200
Lithotopos Hrakleias, Serres*		~45	40–62	300–450	–
Therma Nigritas, Serres*	10		40–64	70–500	~1000
Lagada*	6		33–40	>210	300
Anthemoudas basin, Thessaloniki*		13	25–40	>100	~15
Elaiohorion, Chalkidiki*		25	42	~250	–
Sanis-Afytou, Kassandra*	5	50	35–45	500	100
Aristinou, Alexandroupoli	20	50	30–90	150–450	200
Sappes*	9	51	30–40	50–380	100
Limnis Mitrikou*	7	170	30–40	350–500	–
Nea Kessani, Xanthi	25		40–83	160–500	>300
Neo Erasmiou-Maggana, Xanthi*	16	24	27–68	350–500	250
Eratino-Kavala*	14	93	67–70	650	300
Akropotamos, Kavala (shallow)*	6.9		45–90	100–185	415
Akropotamos, Kavala (deep)*				240–515	
Sousaki, Korinthos (shallow)*	3		60–75	50–200	600
Sousaki, Korinthos (deep)*			≥75	600–900	
Sykies, Arta*	10		32–51	>320	100
Argenou, Lesvos*	1	3	90	≥150	300
Stypsis, Lesvos		20	~90	150–220	–
Polychnitos, Lesvos*	10		65–95	50–200	300
Neniton, Chios*	5	13	78–82	300–500	~60
Santorini*		25	30–65	50–250	–
Melos*	63	87	60–99	50–200	750
Melos**	50		280–320	1000–1380	139***
Nisyros**	3.5		>350	1400–1900	75***

*Low-enthalpy, **high-enthalpy, ***tons/h.

3.5.2 *Volcanic arc of the South Aegean Sea*

The volcanic arc of the South Aegean Sea is extended from Methana in the west to Kos island to the east running through Sousaki, Melos, Santorini, and Nisyros volcanoes (Fig. 3.2a). The main reason for the increased heat flow along the arc is the presence of magma at relatively low depths. They are high-enthalpy regions, which are more or less consistent with the theory of the normative geothermal system (Goguel, 1953). This system includes (i) a heat source at a depth between 3 and 10 km along with fractured metamorphic rocks allowing the circulation of fluids to the surface, (ii) the existence of a reservoir with high primary and secondary hydraulic conductivity between 0 and 3 km depth, and (iii) an impermeable cap rock. The cap rock might be primarily absent and can be formed later through the hydrothermal alteration of parent rocks and clay minerals formation. An example is the case of Melos island.

3.5.3 *Western Greece*

The heat flow in western Greece is substantially lower (Fig. 3.2b) with the geothermal fields of Amarantos, Kavassila, Sykies Artas, Antirio, Kyllini, Kaiafas, and Vromoneri formed under a normal geothermal grade of 30°C/km.

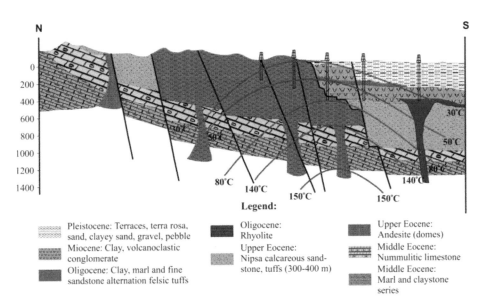

Figure 3.3. Geological cross section of the Aristinos geothermal field (modified from Arvanitis *et al.*, 2008).

3.6 GEOLOGICAL SETTING OF THE MAJOR GEOTHERMAL FIELDS

3.6.1 *Back-arc geothermal fields of Greece*

The back-arc geothermal fields of the country include the fields of the Tertiary sedimentary basins of northern Greece, the fields of north Euboea, the Sperchios basin, and the East Aegean Sea islands. The Aristinos geothermal field (Fig. 3.3) is considered representative for the eastern-most geothermal fields of Greece, those in Alexandroupolis basin; however, the most typical geothermal model for the back-arc fields is shown in Figure 3.4 and is that of Strimonas basin.

The Aristinos geothermal field is a medium enthalpy geothermal field composed of claystone and nummulitic limestone of Middle Eocene age (Fig. 3.3). Andesitic domes and tuffites of Upper Eocene overlay these rocks. Rhyolite along with alternating claystone, marl, and sandstone were deposited during the Oligocene, whereas the formations of Miocene age include volcanogenic claystone and conglomerate. Terra rossa and river terraces of Pleistocene age are the youngest sediments in the region. The thermal water is found in wells at depth between 200 and 300 m in the andesite and rhyolite rocks of the area. Tectonics, mainly characterized by NE–SW trending normal faults, play an essential role in the circulation of the geothermal fluids (Innocenti *et al.*, 1984). The geothermal gradient in the area is calculated at 16°C/100 m (Arvanitis *et al.*, 2008). The adjacent geothermal field of Fylakta is more or less governed by the same geological setting (Tranos, 1995); however, the thermal fluids are located in rhyolite rocks of the region (Dotsika *et al.*, 1997).

The fields of Nea Kessani and Gennisea are developed in the west of the previously mentioned fields, in the Xanthi-Komotini Tertiary basin. This basin consists of Palaeogene and Neogene clastic sediments with a total thickness of 1700 m overlaying a semi-metamorphic basement of sericitized schist with quartzite and marble breccia. The Palaeogene and Neogene sediments consist of alternating impermeable claystone and marl with permeable conglomerate and sandstone. The thermal water reservoir is hosted in Palaeogene clastic formations, where a dense fracture system is developed as a result of extreme tectonic stress. The geothermal grade in the area is estimated to roughly exceed the normal values (Kolios, 1985).

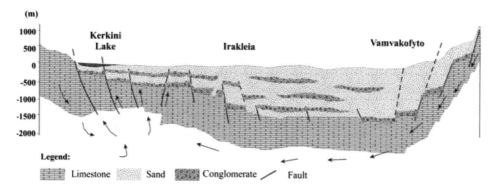

Figure 3.4.   Geological cross section of the Strimonas basin (modified from Fytikas and Andritsos, 2004).

The Nestos river basin, located to the west of the previous geothermal fields, hosts two geothermal fields, namely the Maggana and Nea Eratini fields. The basin is composed of a similar
sequence of Palaeogene and Neogene formations with a total thickness of 2000 m, overlying a
gneiss basement. The geothermal grade in this basin was calculated at 11°C/100 m (Kolios, 1986).

The Strimonas river basin (Fig. 3.4) includes the Therma Nigritas, Sidirokastro, and Aggistro
geothermal fields (Fig. 3.2a). The basin is developed in the form of a NW–SE trending tectonic
graben on the margins of the Serbo-Macedonian and Rodhope massifs. The sedimentary fill of
the basin is approximately 4000 m thick and includes Neogene lacustrine, lagoonal, and terrestrial
sediments comprising sandstone, limestone, conglomerate, and marl alternations, overlying the
crystalline basement rocks (Gramann and Kockel, 1969). Granite intrusion during the Miocene-
Pliocene boundary resulted in the subdivision into smaller subbasins. A major thermal fluid
reservoir is constituted by the Neogene base conglomerate, to which the thermal fluids rise from the
crystalline basement rocks through NW–SE trending normal faults and fracture zones developed
in the crystalline rocks (Mountrakis and Kilias, 1992). The geothermal grade is calculated between
25 and 35°C/100 m (Kavouridis et al., 1982).

At the margins of the basin, several thermal springs like the Aggistro spring were developed
through outcrops of the crystalline basement. The Paranesti and Thermes Xanthis springs have a
similar function mechanism.

In the south of the Strimonas river basin, the Mygdonia basin is located. This basin includes
the geothermal fields of Lagadas and Apollonia. This basin is a NW–SE trending tectonic graben
developed on the contact between the Axios geotectonic zone and the Serbo-Macedonian massif
(Kockel et al., 1971). The graben is filled with Neogene sediments consisting of conglomerate
at the base and sandstone, conglomerate, marl and claystone alternations in the upper horizons.
The basement of the basin consists of metamorphic rocks, mainly schist and gneiss. The thermal
fluid reservoir is hosted in the conglomerate at the base of the sedimentary filling, whereas the
supply of thermal fluids from greater depths is accomplished through active E–W trending normal
vertical faults (Traganos, 1991).

The Sperchios tectonic graben (Fig. 3.5) is considered the extension of the Anatolia strike-
slip fault, which in turn is thought to create favorable conditions for geothermal anomalies. The
geological background of Sperchios basin is composed of the formations of Sub-Pelagonian and
Parnassos-Ghiona isopic zones and formed through the activity of WNW–ESE trending faults
(Georgalas and Papakis, 1966; Marinos et al., 1973), which contribute into both the infiltration of
meteoric water and the uprise of thermal water (Fig. 3.5). The graben hosts the geothermal fields
of Ypati, Thermopyles, Kallidromo, and Kamena Vourla. The Ypati reservoir, which outflows
through Sperchios fluviatile deposits forming travertine pipes, is considered to be hosted in the
Cretaceous limestone of the Sub-Pelagonian geotectonic zone. The same formation is considered
to be the reservoir for the Kallidromo and Thermopyles springs, which are discharged along a

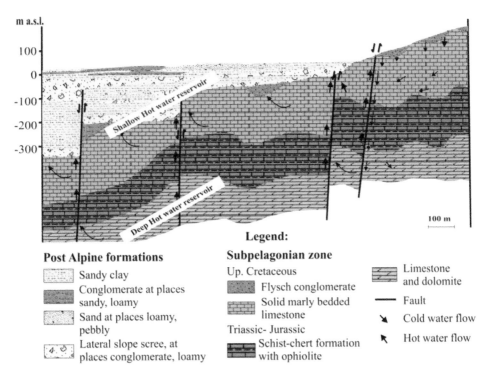

Figure 3.5.   Geological cross section of Damasta in Sperchios basin (modified from Metaxas, 2010).

faulted zone on the contact between alluvial deposits and the Cretaceous limestone. The Kamena Vourla spring is developed under similar conditions on the contact between alluvial deposits and underlying Triassic–Jurassic carbonate formations. The Aidipsos geothermal field (Euboea island) is located opposite of Kamena Vourla field across the sea and belongs to the same group of geothermal fields. The thermal water emerges at the contact between Neogene sediments and phyllite. Phyllite is the basement in the region, whereas Cretaceous limestone of the Sub-Pelagonian zone is believed to be the thermal water reservoir. The active WNW–ESE trending faults of the area play an important role in the circulation of thermal fluids.

Lesvos island with a heat flow approximately 10 times higher than the normal, along with many thermal springs developed through numerous extensional active faults, comprises an important medium-enthalpy geothermal field. The major fault directions are NW–SE, NE–SW, and E–W. The thermal water is hosted in volcanic formations and due to the increased geothermal grade typical for back-arc regions rather than the existence of magmatic chambers at shallow depth or any recent activity (Fytikas *et al.*, 1989). The geological basement of the island is composed of crystalline volcanic rocks, phyllite with marble lenses, and two series of volcanic pyroclastic rocks. Significant secondary hydraulic conductivity is developed in the older series of volcanic rocks, which includes pyroclastic sediments and andesitic to basaltic lava flows, hosting thermal water aquifers. The younger volcanic series consists of shoshonitic lava flows, which were deposited during the Neogene.

3.6.2   *Volcanic arc of South Aegean Sea*

The geological background of the geothermal field of Methana, which is located on the western margin of the volcanic arc (Fig. 3.2), consists of Triassic limestone underlying volcanic rocks and reaching a depth of 1000 m b.s.l. (below sea level) (Volti, 1999). The volcanic rocks consist

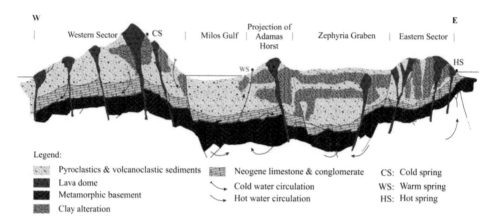

Figure 3.6.   Sketch of the geothermal model of Melos island (modified from Vrouzi, 1985).

of andesitic and dacitic lava domes (Pe-Piper and Piper, 2002). The region is tectonically active with seismic activity being distributed along faults of W–E and SW–NE direction (Makris *et al.*, 2004). The thermal springs develop along the planes of these faults.

To the north of the Methana field is the geothermal field of Sousaki-Loutraki. The geological background of this field includes the formations of the Sub-Pelagonian geotectonic zone of Middle Triassic to Lower Cretaceous age, as well as ultramafic rocks mainly consisting of peridotite, serpentinite, and gabbro. Lagoonal to marine sediments of Plio–Pleistocene age overly the Mesozoic formations. These sediments consist of alternating marl, clayey marl, sandstone, and conglomerate layers. Volcanic rocks, mainly dacites, have intruded in the lowermost horizons of these sediments. Fluvial conglomerate of Pleistocene age, as well as Holocene fluviatile deposits, constitute the most recent sediments in the region. The Loutraki thermal spring is located on the contact between modern talus and limestone. E–W trending faults formed during the Triassic and recently reactivated (Brehm, 1985) are the dominant tectonic feature.

Melos geothermal field (Fig. 3.6), which is located close to the center of the volcanic arc, is characterized by an anticline with its axis trending E–W. Volcanic rocks, lava flows, and pyroclastic deposits, mainly tuffs, overlay an intermediate series of Neogene conglomerate and limestone (Fytikas, 1977; Vrouzi, 1985). Metamorphic rocks consisting of schist form the basement and outcrop at the SE part of the island. The thermal fluid reservoir is developed in these fractured metamorphic rocks and is located at a depth between 1000 and 3000 m with temperatures ranging from 280 to 320°C. The major part of the thermal fluids originates from seawater, which enters the reservoir from the south, where the southern leg of the anticline is developed. The uprise of thermal fluids is accomplished through NNW–SSE trending marginal faults at the northern leg of the anticline (Papanicolaou *et al.*, 1990). In Figure 3.6, the geological conditions, which dominate in Melos, are present and are considered representative of the geological setting of the whole area.

The geological background of Santorini island, located to the east of Melos, is composed of a semi-metamorphic, greenschist-phase system with the intercalation of greywacke, conglomerate, and crystalline limestone of post-Cretaceous age (Tataris, 1964). This complex is strongly fractured by NW–SE and N–S trending major faults. Crystalline limestone of Upper Triassic age is overthrusted on the rocks of this complex. Lava flows, pyroclastic deposits, volcanic glass, scoria, and volcanic ash deposits originated from the recent volcanic activity are the youngest rocks of the area. This region lacks the typical features of a high-enthalpy geothermal system, as except for the heat source all the other prerequisites (permeable rocks, impermeable cap rock) for the formation of a reservoir are not met. A small-scale circulation and storage of thermal fluid takes place within the metamorphic rocks and the highly deformed greywacke, conglomerate, and crystalline limestone. The geothermal gradient is calculated at 16°C/100 m (Kavouridis *et al.*, 1982).

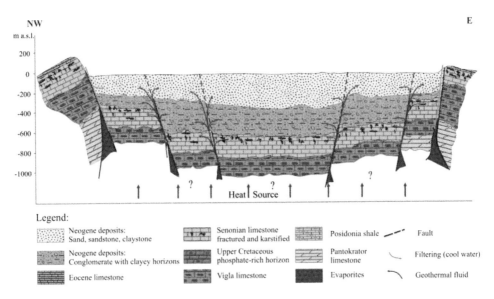

Figure 3.7.   Geological cross section of the Sykies geothermal field (modified from Hatziyannis, 2011).

Nisyros island, located at the eastern part of the volcanic arc, perhaps constitutes the most interesting volcanic system in Greece. According to Papanicolaou *et al.* (1991), a sequence of stratified volcanogenic formations is developed around a volcanic cone, which collapsed to form a caldera and through younger volcanic post-calderan activity, to form massive rhyodacitic domes. Four fault systems affected the morphology of the island. The geothermal field is located in the central part of the island, inside the volcano's caldera. The reservoir is developed at a depth between 1400 and 1900 m and the fluid temperature is thought to exceed 350°C.

3.6.3   *Low–enthalpy geothermal fields of western Greece*

The geothermal fields of western Greece belong to the low-enthalpy group. The Sykies Artas geothermal field developed in alluvial sediments and covering an area of $10\,km^2$ is a typical geothermal field for this group (Fig. 3.7).

Apart from the alluvial deposits, Neogene sediments comprising alternations of sand, clay, and sandstone also participate. Carbonate rocks of the Ionian zone are underlain to these sediments, whose thickness increases from 280 in the west to 480 m in the east due to the differential tectonic setting of the basin. The carbonate rocks constitute the thermal water reservoir. Tectonics, mainly characterized by NNW–SSE trending normal faults and more recent E–W perpendicular faults, play an important role in the infiltration of meteoric water at greater depths, as well as in the thermal water circulation and its uprise (Vriniotis and Papadopoulou, 2004). The geothermal grade is calculated between 5 and 11°C/100 m (Hatziyannis, 2011).

Other low-enthalpy geothermal fields of western Greece, such as the Kaiafas and Kyllini fields, are developed under more or less similar geological conditions.

3.7   CHEMICAL COMPOSITION OF THERMAL AND MINERAL WATERS

3.7.1   *Materials and methods*

Sampling took place on September 2007. A representative sample was selected for each geothermal field. It was collected at the spring's outlet and was kept in two polyethylene bottles. The first bottle of 0.5 L volume was filtered at the site through 0.45 μm pore size Millipore filters. It was then acidified to $\sim$2 pH with 65% ultrapure HNO_3 and used for major cation analysis and trace element determination. The second non-acidified aliquot (1 L) was retained to determine non-metal ions. Temperature (T), electrical conductivity (EC), Redox potential (Eh), pH, and sulfide concentration (S^{2-}) were measured in untreated samples *in situ* using an ion/EC meter (Consort® C533) with combined electrodes. Since dissolved oxygen could not be measured, Eh was used as the main tool to identify aerobic and anaerobic conditions. The major cations Ca^{2+}, Mg^{2+}, Na^+, and K^+ were determined by atomic absorption spectroscopy (GBC Avanta). Trace element analysis was conducted using ICP-MS Perkin Elmer®, ELAN 6100. For major cation and trace element determinations, accuracy was controlled using appropriate laboratory standards.

Alkalinity was measured by volumetric titration using bromocresol green-methyl red indicator by Hach® titration kits. Similarly, chloride (Cl^-) was measured using the $AgNO_3$ method by Hach® titration kits, whereas sulfate (SO_4^{2-}), nitrate (NO_3^-), and ammonium (NH_4^+) were determined using a spectrophotometer (Hach®, DR/4000). All analyses were conducted at the Laboratory of Hydrogeology, University of Patras immediately after collection. The charge balance error for all chemical analyses was within acceptance range ($\pm$5%).

Chemical analyses results of samples from the Greek geothermal fields are given in Tables 3.2a and b. Data shown for the volcanic arc islands Santorini, Melos, Nisyros, and Kos were taken from Minnisale *et al.* (1997) and Bencini, *et al.* (1981).

3.7.2   *Hydrochemistry of geothermal fields*

3.7.2.1   *Hydrochemistry of back-arc geothermal fields*
The temperature of thermal waters varies between 32 and 63°C, while most of the samples display high salt content.

The majority of back-arc geothermal fields and also of other geothermal fields display very low redox potential indicating the prevalence of reducing conditions (Tables 3.2a and b). This is also supported from the presence of reduced sulfur and nitrogen species, along with the absence of high nitrate concentrations, representative of freshwater influence in the chemistry of thermal fluids.

Anomalously high radon concentrations in groundwater have long been associated with faults (Katsanou *et al.*, 2010; King, 1986; 1990; Richon *et al.*, 2007) and with certain rocks including granites, certain volcanic, and sedimentary rocks with high phosphate content, black shales, phosphorites, some sedimentary rocks with high phosphate content, and metamorphic rocks derived from these rocks (Katsanou *et al.*, 2012). As already mentioned, back-arc regions were subjected to severe tectonic stress and deformation, which resulted in the formation of large faults and large Tertiary grabens. In these regions, radon concentrations range from 0.48 to 84 kBq/m^3, whereas radon concentrations in the adjacent seawater are 200–430 Bq/m^3 (Tables 3.2a and b). High concentrations ($>$15 kBq/m^3) were measured in water samples from Traianoupolis, Paranesti, Echinos, Aggistro, Aridaia, Kamena Vourla, and Polychnitos geothermal fields. At the latter, which is located in Lesvos, radon concentration reaches up to 300 kBq/m^3 (Lambrakis and Stamatis, 2008). It is worth mentioning that in all fields except for Kamena Vourla, the thermal fluid circulation takes place in crystalline rocks. Garagounis (1976) suggests that high radon concentrations at Kamena Vourla are related to the adjacent old volcano of Lichades islands.

The ternary Piper plot and various scatterplots were used for the evaluation of chemical analysis results (Fig. 3.8). Additionally, in scatterplots (Fig. 3.8b–e), the simple mixing lines were drawn between the precipitation end-member and the seawater end-member, the so-called seawater dilution line.

According to the Piper diagram, the waters from back-arc geothermal fields comprise two distinct hydrochemical types, sodium-bicarbonate (Na–HCO$_3$) water and sodium-chloride (Na–Cl) water.

Table 3.2a. Chemical analyses of thermal waters. EC is given in mS/cm and Eh in mV.

Back-arc regions-southern Greece

	Traianoupolis	Gennisea Nea Kessani	Echinos	Paranesti thermo	Eleftheres	Nigrita therma	Sidiro kastro	Aggistro	Lagada	Apollonia	Aridaia	Florina Xyno	Platystomo	Ypati	Thermopyles	Kamena Vourla
$T°C$	50.5	60.5	48.0	53.6	37.4	36.3	42.3	39.6	34.4	54.8	35.8	14.2	31.8	32.0	39.4	41.3
EC	15.59	8.53	1.526	2.28	4.66	1.728	1.863	0.537	0.805	1.586	1.173	1.326	0.648	15.03	14.96	28.60
Eh	−255	−3	231	−5	−230	71	215	198	224	282	250	230	−358	−308	−262	23
pH	6.63	6.90	6.55	6.28	5.98	6.15	6.44	7.29	7.09	7.63	6.54	6.35	8.91	6.41	6.25	6.28
CO_2*	92	180	180	272	640	500	284	34	48	28	330	368	0	440	524	370
S^{2-}****	0.221	0.001	0	0	0.201	0	0	0	0	0.299	0	0	4.442	9.81	2.38	0
$\overline{Rh}$****	38.00	1.90	16.00	19.00	3.10	2.70	59.00	26.00	7.00	24.00	38.00	6.20	0.93	0.48	3.50	84.00
Error***	1.4	6.4	2.1	2	4.8	5.3	1.5	1.7	3.8	2.3	1.4	6.7	9	12	4.9	1.1
$\overline{Th}$****	4.80	0.44	1.50	1.50	0.58	0.14	1.50	0.68	0.36	1.10	7.50	0.88	0.18	0.17	0.78	5.60
Error***	5.6	17	10	10	11	30	10	11	14	11	4.8	12	29	35	1.5	6.1
HCO_3*	176.9	1430.0	573.4	1049.2	1098.0	1896.7	878.4	152.5	274.5	290.4	664.9	835.7	31.8	2346.1	829.6	666.1
Cl*	4360.0	1480.0	7.2	4.0	775.0	162.0	6.6	3.5	19.2	18.0	3.0	5.3	124.0	4080.0	4040.0	9060.0
SO_4*	542.0	282.0	186.0	142.0	126.5	95.0	226.0	106.5	168.0	389.5	24.3	5.8	20.2	45.0	481.0	1107.5
NO_3*	1.0	4.0	0.0	1.0	2.0	0.0	1.0	2.0	3.0	1.0	0.0	0.0	0.0	12.0	5.0	4.0
Na*	2599.0	1472.0	90.9	402.5	625.6	579.8	250.7	63.3	123.9	303.6	36.8	5.5	105.2	2028.6	2403.5	4592.6
K*	113.1	156.0	15.2	33.9	49.9	78.0	37.8	3.1	11.7	7.8	0.4	1.2	1.6	222.3	120.9	380.0
Mg*	13.2	18.2	18.6	3.8	13.8	117.6	24.6	7.4	2.4	7.2	21.1	30.6	1.2	174.0	190.8	256.4
Ca*	536.0	131.2	159.6	84.4	276.0	120.0	150.0	35.6	42.0	10.0	172.0	222.8	4.8	940.0	562.0	662.0
NO_2*	0.009	0.015	0.013	0.018	0.019	0.019	0.017	0.012	0.050	0.015	0.010	0.028	0.012	0.019	0.015	0.009
NH_4*	3.05	0.55	0.00	0.00	0.09	0.02	0.00	0.00	0.00	0.15	0.00	0.01	2.10	0.81	0.43	0.20
PO_4*	0.090	0.739	0.089	0.247	0.778	0.281	0.111	0.108	0.025	0.046	0.162	0.129	0.001	0.095	0.346	0.036
SiO_2*	44.0	50.1	106.3	120.6	43.1	60.3	77.3	37.5	25.3	41.6	20.1	19.7	44.5	27.7	28.7	41.9
F*	2.6	2.7	3.4	5.5	2.0	2.3	4.1	3.5	2.0	11.3	0.4	0.1	2.1	0.8	2.0	2.1
Sr*	16.4	3.1				0.5						0.2	0.3	26.6	10.7	
H_3BO_3*	34.7	23.2				4.2						1.1	72.2	24.9	16.2	
As**	53.5	637.5	5.2	8.8	735.2	287.9	3.1	21.1	5.9	19.7	347.5	0.9	0.9	19.4	132.3	98.4
Ba**	82.3	140.2	75.6	74.3	172.9	57.1	56.5	6.7	63.2	21.6	29.7	21.1	4.3	1594.3	88.9	95.9
Cr**	1.1	1.4	0.0	0.0	0.3	0.0	0.0	0.3	1.3	0.0	0.6	0.2	0.0	4.5	1.3	2.3
Cu**	10.3	4.5	1.3	2.5	1.7	2.5	2.3	0.5	1.0	2.9	0.5	0.9	1.3	5.5	6.3	27.1
Ga**	2.7	4.3	2.6	2.5	5.5	1.9	1.9	1.0	2.2	0.9	1.0	0.7	0.5	48.5	2.9	3.1
Li**	1768.5	1119.8	1023.9	1051.5	694.3	182.6	382.7	60.6	63.7	102.1	119.2	3.6	25.6	2339.6	754.7	433.0
Mn**	824.9	39.6	289.6	270.8	145.3	30.4	0.2	0.2	0.5	25.5	0.4	0.2	0.4	19.9	0.2	79.8
Pb**	0.2	0.0	0.1	0.1	0.1	0.2	0.1	0.5	0.1	0.1	0.1	0.1	0.0	0.0	0.0	0.1
Rb**	650.0	700.8	212.6	222.3	291.9	27.3	193.7	19.8	7.1	43.9	28.2	0.2	2.1	526.9	177.1	128.2
Se**	9.5	5.8	0.3	0.3	3.5	0.2	0.3	0.4	0.6	0.6	0.0	1.0	2.4	12.7	16.6	33.4
U**	0.0	0.2	0.2	0.2	0.4	0.1	0.3	1.6	4.1	0.1	6.5	0.6	0.0	0.0	0.0	5.2
V**	19.9	8.2	0.8	1.2	4.3	0.7	1.0	1.6	4.7	0.8	2.6	0.6	0.7	23.1	22.1	48.1
Zn**	34.2	15.4	12.2	13.9	7.9	20.4	9.6	4.9	16.2	50.9	4.2	3.6	5.1	4.3	24.1	93.4
Mo**	0.0	0.1	1.6	0.1	0.0	0.1	0.1	2.7	1.6	6.3	3.7	0.5	0.2	0.0	0.0	1.0

*Concentration in mg/L, **Concentration in μg/L, ***Error in ±%, ****Concentration in kBq/m³.

Table 3.2b.   Chemical analyses of thermal waters. EC is given in mS/cm and Eh in mV.

	Back-arc regions (continued)			Arc regions—eastern Greece							Western Greece					
	Aidipsos	Polychnitos	Sea	Methana	Loutraki	Santorini	Melos	Nisyros	Kos	Sea	Kavassila	Sykies	Kyllini	Kaiafas	Vromoneri	Sea
T°C	63.0	55.8	26.1	32.4	32.3	34.0	50.0	45.0	45.0		30.6	20.6	27.6	32.3	22.5	28.9
EC	54.90	1.81	51.70	58.10	3.43						3.06	4.804	3.81	22.50	3.18	48.90
Eh	62	44	50	−298	95						−347	−112	−275	−352	−250	17
pH	6.80	6.81	7.95	6.20	7.05	7.10	7.2	6.5	6.6	7.90	7.33	7.48	7.3	6.70	6.46	8.1
CO_2*	240	48	100	848	70	95.0	98.0	89.0	98.0		64	202	60	124	134	
S^{2-}*	0	0	0	2.54	0.003	t race	t race	t race	t race	0	4190	0	12	23.4	0.003	0
$\overline{Rn}$****	0.11	80.00	0.43	4.40	68.00						6.60		1.90	7.10	6.20	
Error***	28	2	11	4.9	1.1						3.4		4.7	3.3	3.9	
$\overline{Tn}$****	0.43	3.60	0.05	0.56	1.50						0.24		<4.5	0.94	0.36	
Error***	20	9.2	30	12	11						25		22	10	15	
HCO_3*	536.8	237.9	140.3	1018.7	391.9	5148	1314.3			238.0	250.1	420.9	551.44	246.4	330.6	150.0
Cl*	17800	6780	18350	20600	786	1258	355	22720	22543	21804	430	1371	887	7420	695	20400
SO_4*	1595	443	2955	3005	132			3120	3408	2613	172	108	203	2510	260	3900
NO_3*	4	9	2	15	1						0	1	4	1	1	11
Na*	10299.4	3930.0	11999.0	11800.0	387.6	3036.0	711.0	11700.0	11700.0	11955.0	222.5	790.0	826.0	4542.1	216.8	12040.0
K*	351.0	36.0	460.0	563.0	34.7	86.0	164.0	714.0	651.0	438.0	15.8	41.0	14.2	130.4	15.0	680.0
Mg*	276.0	122.0	1344.0	1170.0	111.6	264.0	24.0	887.0	1118.0	1458.0	29.9	83.0	36.88	253.3	107.4	1333.3
Ca*	1656.0	568.0	492.0	1730.0	153.0	320.0	300.0	1900.0	1840.0	421.0	112.0	154.8	64.0	621.3	248.0	427.2
NO_2*	0.024	1.898	0.008	0.010	0.016	0.040	1.870	5.650	0.180	0.180	0.010	0.009	0.023	0.017	0.014	0.0152
NH_4*	1.140	8.000	0.000	0.090	0.000						0.880	7.500		2.350	0.090	0.000
PO_4*	0.058	0.230	0.008	0.276	0.040						0.211	0.014	0.033	0.040	0.065	0.007
SiO_2*	48.0	65.1	0.3	36.2	13.7	102.0	254.0	173.0	31.0	5.0	23.5	19.2	13.4	17.7	25.1	
F*	3.9		2.2	2.1	0.4						2.1	0.6	0.0	3.6	0.4	
Sr*		22.7				4.0	4.4	20.6	5.7	9.5		2.3	2.4	20.0	2.9	6.0
H_3BO_3*		19.9		0.0		6.2	69.3	8.1	38.4	25.7		2.3	6.0	10.5	2.3	26.3
As**	107.9	113.0	83.8	166.6	11.0			356		49.0	23.3	10.3	5.0	32.7	8.2	57.7
Ba**	244.2	279.0	8.3	76.2	16.7					9.0	60.3	230.2	70.0	25.8	72.7	6.2
Cr**	0.4	3.0	0.6	2.2	4.5					0.5	0.2	80.0	28.0	4.7	0.5	27.3
Cu**	43.9	75.1	49.1	126.8	1.3					20.5	0.5	27.0	21.0	37.0	1.4	57.8
Ga**	7.4	9.4	0.5	2.6	0.6					0.0	1.9	12.0	3.0	1.0	2.4	0.9
Li**	1808.9	7673.1	214.9	2872.2	21.3	111.0	1500.0		240.0	220.6	537.1	70.0	29.0	131.6	29.4	140.3
Mn**	73.3	31.9	2.3	3873.0	0.0		3940			0.0	3.2	683.0	64.0	24.7	53.6	0.7
Pb**	0.6	0.0	0.2	0.0	0.0					0.0	0.0	52.0	1.0	0.0	0.1	5.3
Rb**	228.1	884.2	95.8	1337.5	8.6	90.0	478.0		100.0	100.0	46.4	12.0	14.0	37.3	6.4	109.8
Se**	106.9	19.3	77.8	87.6	3.3		72.8			15.8	0.4	16.0	0.0	41.5	8.2	18.3
U**	0.0	0.1	2.7	1.3	0.4		0.3			2.0	0.0	0.0	0.0	0.3	0.3	3.0
V**	51.8	42.6	57.1	84.8	4.7		48.3			10.0	2.5	23.0	19.0	37.2	6.6	66.7
Zn**	97.0	170.5	166.1	216.8	6.9		105				10.8	1812.0	53.0	124.1	31.5	158.5
Mo**	0.1	2.6	8.9	0.2	0.2						0.0	3.0	0.0	0.0	0.4	

*Concentration in mg/L, **Concentration in μg/L, ***Error in ± %, ****Concentration in kBq/m³.

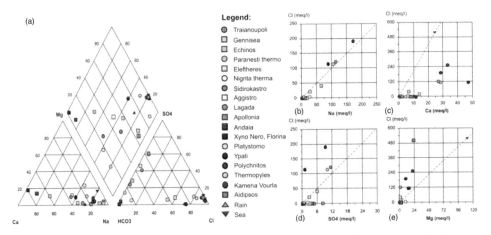

Figure 3.8.   Back-arc geothermal field classification according to the Piper diagram and major elements scatterplots with the seawater dilution line.

3.7.2.1.1   *Sodium-bicarbonate water Na–HCO₃*

In this group, Na^+ is the major cation and HCO_3^- the major anion. This group includes water samples from the Tertiary basins of Macedonia. The prevalence of sodium can be attributed to the increased potential of water to dissolve the aluminosilicate minerals of basement rocks due to increased CO_2 concentration (Ellis and Mahon, 1977).

According to Barnes *et al.* (1986) and Minissale *et al.* (1989), the origin of CO_2 in the geothermal fields of Greece is due to metamorphic reactions (decarbonization, recrystallization) of carbonate rocks at great depths. Nuti *et al.* (1980) came to the same conclusion for other regions, as well as other researchers (Bailey, 1989; Ceron and Pulido-Bosch, 1999). D'Alessandro *et al.* (2010) suggest that some of the CO_2 might have a deeper origin, although an origin from sedimentary rocks cannot be absolutely excluded.

Sodium prevalence against the rest cations (Fig. 3.8c) could also be attributed in some cases to cation-exchange processes despite the increased calcium concentrations. Samples from Florina, Aridaia, and Echinos are an exception to the above observation. Calcium is the major cation in thermal waters from these regions. According to Giggenbach (1991), this thermal water type can be formed either due to the dilution of CO_2 at low temperatures or due to the mixing with freshwater.

3.7.2.1.2   *Sodium-chloride water Na–Cl*

Sodium is the major cation, as in the former group of samples, with chloride being the major anion. Water samples from Alexandroupolis, Sperchios, Euboea, and Mytilini geothermal fields are classified in this hydrochemical type. For water samples from this group, Na/Cl ratio is equal or higher to 0.83, which is equivalent to the regular seawater ratio. It is assumed that high salt concentrations are the result of mixing between thermal fluids from greater depths and connate colder saline waters trapped in the shallow post-orogenic sediments (Minissale *et al.*, 1989). In the case of the Sperchios basin at the Thermopyles-Anthili area, the high salt concentrations are attributed to intruding seawater that becomes warm (Metaxas *et al.*, 2010). This is also the case at the geothermal fields of Euboea and Mytilini (Lambrakis and Kallergis, 2005; Lambrakis and Stamatis, 2008; Michelot *et al.*, 1993). According to Minissale *et al.* (1989), such saline contribution conceals the potential Na–HCO₃ composition of the deep rising waters of this category.

In all the above cases, during circulation and interaction with the host rocks, thermal waters acquire Ca and become depleted in Mg with respect to seawater (Fig. 3.8c and e). These changes

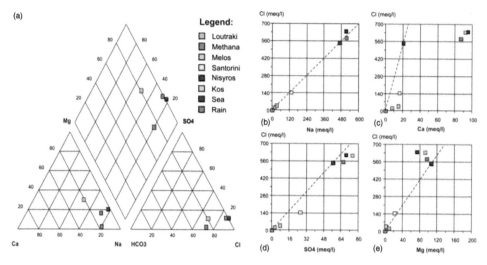

Figure 3.9.   Volcanic arc geothermal field classification according to the Piper diagram and major elements scatterplots with the seawater dilution line.

in water chemistry are probably related to hydrothermal weathering processes including the dissolution of primary minerals of the host rocks and precipitation of secondary (hydrothermal) solid phases. Carbon dioxide is probably involved to a considerable degree in these reactions as already discussed. Water samples from the geothermal fields of Traianoupolis, Eleftheres Kavalas, Apollonia, Ypati, and Thermopyles are rich in hydrogen sulfide and ammonia. Hydrogen sulfide is formed due to the reduction of sulfates deriving from the oxidation of sulfide minerals (Poutoukis and Dotsika, 1994) or directly from seawater (Duriez *et al.*, 2008).

3.7.2.2   *Hydrochemistry of Aegean volcanic arc geothermal fields*
The water samples from the geothermal fields along the active volcanic arc of the Aegean display a Na–Cl water type according to the Piper diagram (Fig. 3.9a).

Sodium is the major cation in this group of thermal waters, whereas the major anion is chloride, which is enriched in the fields of Nisyros and Kos as shown by the relative position of water samples from these two fields with respect to the seawater dilution line (Fig. 3.9b). Samples from the rest geothermal fields of this group are plotted along the line, as their Na/Cl ratio is similar to seawater, indicating its significant influence on the chemical composition of the thermal fluids. Many studies, some of them including oxygen and hydrogen stable isotopes, suggest that seawater is the major component of thermal fluids from the volcanic arc (Brombach *et al.*, 2003; Chiodini *et al.*, 1993; Dotsika *et al.*, 2009; Kavouridis *et al.*, 1999). Furthermore, Dotsika *et al.* (2009) suggest that the parent hydrothermal liquids of Nisyros and Melos are produced through mixing of seawater and arc-type magmatic water, with negligible to nil contribution of local groundwater. The contribution of magmatic water is considered to reach 70% in both cases. The same researchers investigated the contribution of magmatic water in other regions along the arc and concluded that in Santorini, magmatic water also contributes at the same proportion, whereas in Methana, magmatic water only accounts for 19% of the thermal water. From Figure 3.9c and d it is concluded that calcium and sulfate concentrations are increased, compared to seawater. On the contrary, Mg (Fig. 3.9e) is depleted. The increased Ca concentration can be attributed to the thermal water-rock interaction under increased CO_2 concentration and pressure (Tables 3.2a and b) (Dotsika *et al.*, 2009; Minissale *et al.*, 1997). According to Minissale *et al.* (1997), increased concentrations of K, Li, SiO_2, B, and other elements are due to the same process, which is enhanced by the presence of hot steam.

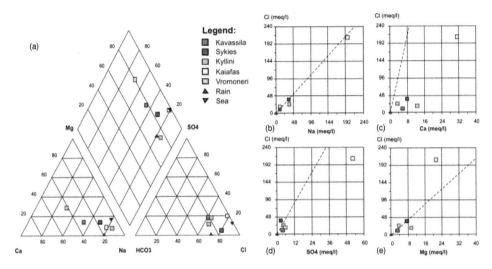

Figure 3.10. Volcanic arc geothermal field classification according to the Piper diagram and major elements scatterplots with the seawater dilution line.

High sulfate concentrations are possibly the result of hydrogen sulfide oxidation. Dotsika *et al.* (2009) suggest that high sulfate concentrations in thermal fluids from Kos and Santorini derive from the leaching of sulfur from the volcanic rocks along the arc. The Mg depletion compared to seawater is attributed to ion exchange taking place between Mg and Ca ions in Ca–Mg–silicate minerals under high temperatures (Giggenbach, 1988).

3.7.2.3 *Hydrochemistry of geothermal fields from western Greece*

According to the Piper diagram (Fig. 3.10a), the majority of samples from this group reveal a Na–Cl water type.

This group of water is characterized by the low radon and thoron concentrations compared to those from the back-arc fields. Hydrogen sulfide originates from the sulfate reduction by organic carbon. The sulfate come from the dissolution of the Ionian zone evaporite deposits and the reduction process is assisted by the action of sulfur-reducing bacteria (Kallergis and Lambrakis, 1992; 1993).

The influence of seawater on the chemical composition of thermal water from western Greece is evident, similarly to the other two groups of thermal water. Cl/Na ratio is more or less similar to this type of seawater, whereas the salinity is significantly increased compared to the local freshwater (Table 3.2b). The inclination of the ionic ratios among chloride ions and calcium, sulfate and magnesium ions from seawater (Fig. 3.10c–e) suggests that the primary composition of seawater was altered due to the interaction with the reservoir rocks. Thus, calcium and sulfate ions are enriched in the thermal water. The excess of sulfate may also be attributed to hydrogen sulfide condensation in the liquid phase (Ellis and Machon, 1977), whereas the depletion of magnesium is caused due to ion-exchange processes with calcium-magnesium silicate minerals (Ellis, 1971; Giggenbach, 1988).

3.7.2.4 *Minor and trace elements of Greek thermal waters*

The concentrations of minor and trace elements in Greek thermal waters are similar to those from other regions, e.g., central Turkey (Pasvanoğlu and Chandrasekharam, 2011) and Etna, Sicily (Giammanco *et al.*, 1997). Except for Li, B, Ba, Sr, and Rb, all trace elements display low concentrations. Some of them, such as Cr, Cu, Ga, Pb, Se, U, V, and Mo are below the maximum acceptable limits for drinking water standards. Others such as Cd, Co, and Li are even below detection limits so their values are not reported.

Strontium and barium are released from carbonate minerals and also from K-feldspars during incongruent dissolution processes. Lithium and rubidium are released mainly from feldspars and some clay minerals. These elements have been shown to be reliable indicators for groundwater residence time and their concentration increases with increased temperature, thus they have been used to discriminate between groundwater of thermal and shallow origin (Edmunds and Smedley, 2001).

Minor and trace element concentrations in thermal waters strongly depend on the redox potential in the reservoir. A change from oxidizing to reducing conditions in the same aquifer is accompanied by notable changes in the concentration of redox sensitive trace elements (Edmunds *et al.*, 2002). In Greek geothermal fields, several redox sensitive elements are investigated, including Mn, U, Cr, As, Se, Ni, Co, and Mo. Chromium is abundant in basaltic rocks and under oxidizing conditions is usually encountered in its oxyanion form (CrO_3^-). In a similar way, uranium is also encountered under oxidizing conditions in the $(UO_2) (CO_3)_2^{2-}$ ion form. Arsenic can be enriched under oxidizing conditions in the form of As(V), whereas selenium usually exhibits the same behavior. High Fe and Mn concentrations, as well as their cognate elements Co and Ni, are favored under reducing conditions. Molybdenum, although usually stable under oxidizing conditions in its oxyanion form, seems to be favored by dissolution reactions and long residence time (Edmunds and Smedley, 2001). Cu, Pb, and Zn solubility does not usually depend on the redox potential of the geothermal fluids.

Arsenic concentration exceeds $10\,\mu g/L$, which is the upper limit for drinking water, in the majority of the samples (Tables 3.2a and b). The origin of this element is usually attributed to geological factors although in modern era anthropogenic sources are also significant. Arsenic is usually contained in volcanic gases and geothermal fluids, whereas it is also a major element in certain sulfide minerals, such as arsenopyrite or orpiment and a secondary element in others, such as pyrite. The interaction between geothermal fluids and the leaching of sulfide-rich rocks, accelerated in the case of acidic water, results in the enrichment of As in the groundwater of geothermal fields. According to Casentini *et al.* (2011), as concentration in soils over geothermal fields in N. Greece ranges from 20 to 513 mg/kg, whereas at non-geothermal regions it varies from 5 to 66 mg/kg. At the geothermal fields of Kassandra, Sidirokastro, Thermopyles, Aidipsos, Kaiafas, Methana, and Nisyros, the measured As concentrations are remarkably high, exceeding $100\,\mu g/L$ (D'Alessandro *et al.*, 2008).

3.8 CONCLUSIONS

The thermal and mineral waters of Greece are an outcome of the country's geothermal fields and develop in regions with recent volcanic activity and active tectonics. They are also the carrier of geothermal energy for domestic central heating, but mostly for fish farming, greenhouse heating, and healing tourism.

The Greek geothermal fields are distinguished into back-arc, volcanic, arc and fault-induced geothermal fields of western Greece. The back-arc regions, which include the Tertiary basins of Central and East Macedonia Prefecture and Thrace Prefecture, north Aegean islands, Euboea, and Sperchios basin, are characterized by intense tectonic crust thinning. These regions host low- and medium-enthalpy fields. High-enthalpy fields are located along the Aegean volcanic arc, where recent volcanism exists. The low-enthalpy geothermal fields of western Greece formed due to the tectonic activity.

Reductive conditions favoring the formation of hydrogen sulfide and reduced nitrogen forms are the most common feature of the Greek geothermal fields. Thermal water from crystalline rocks in the back-arc geothermal fields revealed increased radon and thoron concentrations.

The water samples are classified into three major hydrochemical types, the Ca^{2+}, Mg^{2+}–HCO_3^- type, the Na^+–HCO_3^- type, and the Na^+–Cl^- type. Thermal water of the Ca^{2+},

Mg^{2+}–HCO_3^- type is generally considered to be of relative shallow origin and short circulation time (Minissale *et al.*, 1989; Lambrakis and Kallergis, 2004). Thermal water of the Na^+–HCO_3^- type is usually encountered in the Tertiary basins of northern Greece, where no influence of seawater exists. These waters are considered the most representative of back-arc areas. High concentrations of CO_2 accelerate the dissolution of silicate minerals from crystalline rocks and are thought to play a key role concerning the prevalence of sodium ions in such type of water. Most of the thermal waters belong to the Na^+–Cl^- type and originate from the mixing between seawater and freshwater and in a few cases (volcanic arc fields) magmatic water (Chiodini *et al.*, 1994). Seawater influence results in the differentiation of the major hydrochemical types, especially in the back-arc geothermal fields.

Trace element concentrations are typical for this type of water with As, B, Li, Mn, and Rb being enriched compared to local freshwater.

REFERENCES

Arvanitis, A.: Geothermal activities in Greece. *International Trade Fair Energy 2011 Renewable Energy Industry and Export Forum*, 4–8 April 2011, Hannover, Germany, 2011.

Arvanitis, A., Kavouridis, T. & Hatziyannis, G.: New perspectives for the development of geothermal energy in Greece. Lecture in the frames of the *13th National Energy Conference*, 10–14 November 2008, Athens, Greece, 2008.

Barnes, I., Leonis, C. & Papastamataki, A.: Stable isotope tracing of the origin of CO_2 discharges in Greece. In: A. Morfis & P. Paraskevopoulou (eds): *Proceedings of the 5th International Symposium on Underground Water Tracing*, Athens, 1986, pp. 25–42.

Bailey, D.K.: Carbonate melt from the mantle in the volcanoes of south-east Zambia. *Nature* 338 (1989), pp. 415–418.

Bencini, A., Duchi, V. & Martini, M.: Thermal waters from the islands of the Aegean arc (Greece). *Rendiconti Societa Italiana di Mineralogia e Petrologia* 37:2 (1981), pp. 921–928.

Brehm, H.: On the geology and the tectonics of the geothermal field of Loutraki – Sousaki Area. In: *Geothermics, thermal-mineral waters and hydrogeology* (in Greek). Theophrastus, 1985, Athens, Greece, pp. 47–56.

Brombach, T., Caliro, S., Chiodini, G., Fiebig, J., Hunziker, J.C. & Raco, B.: Geochemical evidence for mixing of magmatic fluids with seawater, Nisyros hydrothermal system, Greece. *Bull. Volcanol.* 65 (2003), pp. 505–516.

Casentini, B., Hug, S.J. & Nikolaidis, N.P.: Arsenic accumulation in irrigated agricultural soils in northern Greece. *Sci. Total Environ.* 409:22 (2011), pp. 4802–4810.

Ceron, J.C. & Pulido-Bosch, A.: Geochemistry of thermomineral waters in the overexploited Alto Guadalentin aquifer (south-east Spain). *Water Res.* 33:1 (1999), pp. 295–300.

Chiodini, G., Cioni, R., Leonis, C., Marini, L. & Raco, B.: Fluid geochemistry of Nisyros island, Dodecanese, Greece. *J. Volcanol. Geotherm. Res.* 56 (1993), pp. 95–112.

Chiodini, G., Cioni, R., Dotsika, E., Fytikas, M., Leonis, C., Magro, G., Marini, L., Michelot, J.L., Poutoukis, D. & Raco, B.: Fluid geochemistry for the volcanic surveillance of Thera island. *Proceedings International Workshop European Laboratory Volcanoes*, September 1994, Catania, Italy, 1994, pp. 193–196.

D'Alessandro, W., Bellomo, S., Brusca, L., Kyriakopoulos, K. & Karakazanis, S.: Arsenic in thermal waters of Greece. *26th European Conference-SEGH*, Athens, Greece, 2008.

D'Alessandro, W., Brusca, L., Martelli, M., Rizzo, A. & Kyriakopoulos, K.: Geochemical characterization of natural gas manifestations in Greece. *Bull. Geol. Soc. Greece, Proceedings of the 12th International Congress*, Patras, May, XLIII (5), 2010, pp. 2327–2237.

Dotsika, E., Poutoukis, D. & Dalampakis, P.: Geochemical-geothermal study of the Tychero-Fylakto thermal water (Evros Prefecture). *Proceedings of the 4th Hydrogeological Congress*, Thessaloniki, Greece, 1997, pp. 352–365.

Dotsika, E., Poutoukis, D., Michelot, J.L. & Raco, B.: Natural tracers for identifying the origin of the thermal fluids emerging along the Aegean Volcanic Arc (Greece): evidence of Arc-Type Magmatic Water (ATMW) participation. *J. Volcanol. Geoth. Res.* 179 (2009), pp. 19–32.

Doutsos, T. & Kokkalas, S.: Stress and deformation patterns in the Aegean region. *J. Struct. Geol.* 23 (2001), pp. 455–472.

Duriez, A., Marlin, C., Dotsika, E., Massault, M., Noret, A. & Morel, J.L.: Geochemical evidence of seawater intrusion into a coastal geothermal field of central Greece: example of the Thermopyles system. *Environ. Geol.* 54 (2008), pp. 551–564.

Ellis, A.J.: Magnesium ion concentration in the presence of Mg chloride, calcite, carbon dioxide and quartz. *Am. J. Sci.* 271 (1971), pp. 481–489.

Ellis A.J. & Mahon, W.A.: *Chemistry and geothermal systems.* Academic Press, New York, 1977.

Edmunds, W.M. & Smedley, P.L.: Residence time indicators in groundwater: the East Midlands Triassic sandstone aquifer. *Appl. Geochem.* 15 (2001), pp. 737–752.

Edmunds, W.M., Carrillo-Rivera, J.J. & Cardona, A.: Geochemical evolution of groundwater beneath Mexico City. *J. Hydrol.* 258 (2002), pp. 1–24.

Fytikas, M.: *Geological and geothermal study of Milos island* (in Greek). PhD Thesis, University of Athens, Athens, Greece, 1977.

Fytikas, M. & Andritsos, N.: *Geothermics.* Tziolas Publications, Thessaloniki, Greece, 2004 (in Greek).

Fytikas, M. & Kavouridis, T.: Geothermal area of Sousaki-Loutraki. In: *Geothermics, thermal-mineral waters and hydrogeology.* Theophrastus Publishing & Proprietary C.o. S.A., Athens, Greece, 1985, pp. 19–34.

Fytikas, M., Innocenti, F., Manetti, P., Mazzuoli, R., Peccerilio, A. & Villari, L.: Tertiary to Quaternary evolution of volcanism in the Aegean region, *Geological Society, London, Special Publications* (1984), v.17 pp. 687–699.

Fytikas, M., Kavouridis, T., Leonis, K. & Marini, L.: Geochemical exploration of the three most significant geothermal areas of Lesbos island, Greece. *Geothermics* 18:3 (1989), pp. 465–475.

Fytikas, M., Andritsos, N., Dalabakis, P. & Kolios, N.: Greek geothermal update 2000–2004. *Proceedings World Geothermal Congress 2005* Antalya, Turkey, 2005, Paper 0172.

Garagounis, K.: Feasibility study for Greek geothermal energy. *Mining and Metallurgical Annals* (Greece) 29–30 (1976), pp.17–41.

Georgalas, G. & Papakis, N.: Observations sur les sources ordinaries et thermominerales radioactives de la region karstique de Kammena Vourla (Grece Centrale). In: N. Knjiga (ed): *Proc. Int. Assoc Hydrogeologists.* Athens, Greece, 1966, pp. 221–227.

Giammanco, S., Ottaviani, M.V., Veschetti, E., Principio, E., Giammanco, G. & Pignato, S.: Major and trace elements geochemistry in the groundwaters of a volcanic area, Mount Etna (Sicily, Italy). *Water Res.* 32:1 (1997), pp. 19–30.

Giggenbach, W.F.: Geothermal solute equilibria, derivation of Na–K–Mg–Ca geoindicators. *Geochim. Cosmochim. Acta* 52 (1988), pp. 2749–2765.

Giggenbach, W.F.: Chemical techniques in geothermal exploration. In: F. D' Amore (ed): *Application of geochemistry in geothermal reservoir development.* UNITAR/UNDP, Rome, 1991, pp. 119–142.

Gioni-Stavropoulou, G.: Inventory of thermal and mineral springs of Greece, Part I, Aegean Sea (in Greek). Hydrological and Hydrogeological Investigation Report No 39, IGME, Athens, Greece, 1983.

Goguel, J.: Le regime termique de l'eau souterraine. *Ann. Mines* 10 (1953), pp. 1–29.

Gramann, F. & Kockel, F.: Das Neogen im Strimonbecken. *Geologisches Jahrbuch* 87, 1969, pp. 445–484.

Hatziyannis, E.G.: Update of the geothermal situation in Greece. *Proceedings European Geothermal Congress,* Unterhaching, Germany, 2007, pp. 1–4.

Hatziyannis, G.: The geothermal energy of Epirus: fields and and perspectives exploitation. *EGE Workshop "The geological research as a lever of development of Epirus"* July, 2011, Ioannina, Greece, 2011.

Innocenti, F., Kolios, N., Manetti, P., Mazzuoli, R., Peccerillo, G., Rita, F. & Villari, L.: Evolution and geodynamic significance of the tertiary orogenic volcanism in northeastern Greece. *Bull. Volcanol.* 47:1 (1984), pp. 25–37.

Kallergis, G. & Lambrakis, N.: Contribution a l'etude des sources thermominarales de Greece: les sources thermominerales de Kaiafa. *Hydrogeologie* 3 (1992), pp. 127–136.

Kallergis, G. & Lambrakis, N.: Contribution a l'etude des sources thermominarales de Greece: les sources thermominerales de Kyllini par rapport au regime hydrothermal du Peloponese accidental. *Steir. Beitr. Hydrogeol.* 44 (1993), pp. 207–220.

Karydakis, G.: Geothermal investigations of low enthalpy in the area of Therma Nigritas. Internal report, IGME, Athens, Greece, 1983.

Karydakis, G. & Kavouridis, T.: Geothermal investigations in the area of Lithotopos Heraklias Serron. Internal report, IGME, Athens, Greece, 1983 (in Greek).

Karydakis, G. & Kavouridis, T.: Geothermal investigations of low enthalpy in the area of Sidirokastron Serron. Internal report, IGME, Athens, Greece, 1989 (in Greek).

Katsanou, K., Stratikopoulos, K., Zagana, E. & Lambrakis, N.: Radon changes along main faults in the broader area of Aigion region, NW Peloponnese. *Bull. Geol. Soc. Greece, Proceedings of the 12th International Congress*, May 2010, Patras, 2010, pp. 1726–1736.

Katsanou, K., Siavalas, G. & Lambrakis, N.: Geological applications of 222Rn anomalies in groundwater and soil-gas. In: Z. Li & C. Feng (eds): *Handbook of radon properties, applications and health*. Nova Publishers, 2012, pp. 155–178.

Kavouridis, T., Karydakis, G., Kolios, N., Kouris, D. & Fytikas, M.: Geothermal investigations in Thira island. Internal report, IGME, Athens, Greece, 1982 (in Greek).

Kavouridis, T., Kuris, D., Leonis, C., Liberopoulou, V., Leontiadis, J., Panichi, C., La Ruffa, G. & Caprai, A.: Isotope and chemical studies for a geothermal assessment of the island of Nisyros (Greece). *Geothermics* 28 (1999), pp. 219–239.

King, C.Y.: Gas geochemistry applied to earthquake prediction: an overview. *J. Geophys. Res.* 91:12 (1986), pp. 269–281.

King, C.Y.: Gas-geochemistry approaches to earthquake prediction. In: L.Tommassino, G. Furlan, H.A. Khan & M. Monin (eds): *Proceedings of International Workshop on Radon Monitoring in Radioprotection, Environmental Radioactivity and Earth Sciences*, ITCP, 13–14 April 1989, Trieste, Italy, World Press, Singapore, 1990, pp. 244–274.

Kockel, F., Mollat, H. & Walther, H.: Geologie des Serbo-Mazedonischen Massivs und seines mesozoischen Rahmens (Nordgriechenland). *Geologische Jahrbuch* 83, 1971, pp. 529–551.

Kolios, N.: Geothermal investigations in the area of Potamias, S. Kessanis, Xanthi. Internal report, IGME, Athens, Greece, 1985 (in Greek).

Kolios, N.: Geothermal investigations in the area of the eastern basin of Nestos. Internal report, IGME, Athens, Greece, 1986 (in Greek).

Lambrakis, N. & Kallergis, G.: Contribution to the study of Greek thermal springs: hydrogeological and hydrochemical characteristics and origin of the thermal waters. *Hydrogeol. J.* 13 (2005), pp. 506–521.

Lambrakis, I.N. & Stamatis, N.G.: Contribution to the study of thermal waters in Greece: chemical patterns and origin of thermal water in the thermal springs of Lesvos. *Hydrol. Process.* 22 (2008), pp. 171–180.

Makris, J.: Geophysical investigations of the Hellenides. *Hamb. Geophys. Einzelschr.* (1977), 34, 124pp.

Makris, J., Papoulia, J. & Drakatos, G.: Tectonic deformation and microseismicity of the Saronikos Gulf, Greece. *Bull. Seismol. Soc. Am.* 94:3 (2004), pp. 920–929.

Marinos, P., Frangopoulos, J. & Stournaras, G.: The thermomineral spring of Hypati (Central Greece): hydrogeological, hydrodynamical, geochemical and geotechnical study of the spring and the surrounding area. *Ann. Geol. Pays Hellen.* 1:25 (1973), pp. 105–214 (in Greek).

Martin, P.S.: Paleoclimatology and a tropical pollen profile. Report of the *VIth International Congress on the Quaternary*, Warsaw, 1961. Volume 2, Paleoclimatological Section, Lodz, Poland, 1964, pp. 319–323.

Metaxas, A., Varvarousis, G., Karydakis, Gr., Dotsika, E. & Papanikolaou, G.: Geothermic status of Thermopyles – Anthili area in Fthiotida Prefecture. *Bull. Geol. Soc. Greece* XLIII: 5 (2010), pp. 2265–2273.

Michelot, J.L., Dotsika, E. & Fytikas, M.: A hydrochemical and isotopic study of thermal waters on Lesbos island (Greece). *Geothermics* 22:2 (1993), pp. 91–99.

Minissale, A., Duchi, V., Kolios N. & Totaro, G.: Geochemical characteristics of Greek thermal springs. *J. Volcanol. Geoth. Res.* 39 (1989), pp. 1–16.

Minissale, A., Duchi, V., Kolios, N., Nocenti, M. & Verrucchi, C.: Chemical patterns of thermal aquifers in the volcanic islands of the Aegean Arc, Greece. *Geothermics* 26:4 (1997), pp. 501–518.

Mountrakis, D.: *Geology of Greece*. University Studio Press, Thessaloniki, Greece, 1985 (In Greek).

Mountrakis, D. & Kilias, A.: Geological mapping of sections of the Serbopmacedonian Massif and Strimonas Basin; Regions of Kerkini and Lithotopos, Krousia Mountain, Exploration of Serrer Prefecture geothermal fields, Prefecture of Serres. Aristotle University of Thessaloniki, Department of Geology, 1992, pp. 10–20.

Nuti, S., Noto, P. & Ferrara, G.: The system H_2O–CO_2–CH_4–H_2 at Travale, Italy: tentative interpretation. *Geothermics* 9 (1980), pp. 287-295.

Orfanos, G.: Inventory of thermal and mineral springs of Greece, Peloponnesus, Zakynthos, Kythira. Hydrological and Hydrogeological Investigation Report No. 39, IGME, Athens, Greece, 1985 (in Greek).

Papanikolaou, D., Lekkas, E. & Syskakis, D.: Tectonics of the Melos geothermal field. *Bull. Geol. Soc. Greece* XXIV (1990), pp. 27–46 (in Greek).

Papanikolaou, D., Lekkas, E. & Sakellariou, T. D.: Geological structure and evolution of Nisyros volcano. *Bull. Geol. Soc. Greece* XXV:1 (1991), pp. 406–419.

Pasvanoğlu, S. & Chandrasekharam, D.: Hydrogeochemical and isotopic study of thermal and mineralized waters from the Nevşehir (Kozakli) area, Central Turkey. *J. Volcanol. Geoth. Res.* 202 (2011), pp. 241–250.

Pe-Piper, G. & Piper, D.J.W.: *The igneous rocks of Greece: the anatomy of an orogeny.* Gebrüder Borntraeger, Berlin, 2002.

Poutoukis, D. & Dotsika, E.: Hydrochemical and isotopical research of the Lagada-Volvi region. *Proceedings 2nd National Hydrogeological Congress*, 24–28 November 1993, Patras, Greece, B, 1994, pp. 679–689.

Richon, P., Bernard, P., Labed, V., Sabroux, J.C., Beneïto, A., Lucius, D., Abbad, S. & Robe, M.C.: Results of monitoring [222]Rn in soil gas of the Gulf of Corinth region, Greece. *Radiat. Meas.* 42:1 (2007), pp. 87–93.

Robertson, A.H.F.: Origin and emplacement of an inferred late Jurassic subduction-accretion complex, Euboea, eastern Greece. *Geol. Mag.* 128 (1991), pp. 27–41.

Sfetsos, S.C.: Inventory of thermal and mineral springs of Greece, III, Continental, Greece. Hydrological and Hydrogeological Investigation Report No, 39, IGME, Athens, Greece, 1988.

Tataris, A.: The Eocene in the semimetamorphosed basement of Thera island. *Bull. Geol. Soc. Greece* III:1 (1964), pp. 232–238.

Traganos, G.: The present state and results of the second phase of geothermal investigation in Mygdonia. Geological and geophysical research report, IGME, Athens, Greece, 1991 (in Greek).

Tranos, M.: Tectonic-stratigraphical study of the Fylakto-Tychero. *Technical report. Geothermiki Ellados,* 1995, pp. 74

Volti, T.K.: Magnetotelluric measurements on the Methana peninsula (Greece): modelling and interpretation. *Tectonophysics* 301 (1999), pp. 111–132.

Vriniotis, D. & Papadopoulou, K.: The role of Louros and Arachtos rivers in the development of sediments of the Arta's plain with the contribution of geochemical parameters. *Bull. Geol. Soc. Greece* XXXVI (2004), pp. 150–157.

Vrouzi, F.: Research and development of geothermal resources in Greece: present status and future prospects. *Geothermics* 14:2–3 (1985), pp. 213–227.

CHAPTER 4

Geological setting, geothermal conditions and hydrochemistry of south and southeastern Aegean geothermal systems

Maria Papachristou, Konstantinos Voudouris, Stylianos Karakatsanis, Walter D'Alessandro & Konstantinos Kyriakopoulos

4.1 INTRODUCTION

Greece is particularly rich in geothermal resources due to favorable geologic conditions. The intense tectonic and, in some cases, magmatic activity as well as increased heat flow have created extended thermal anomalies with geothermal gradients reaching up to 100°C/km.

In the southern Aegean, geothermal activity is often associated with volcanism and the presence of shallow magma chambers, especially along the South Aegean Active Volcanic Arc (SAAVA) (Megalovassilis, 2005), which is characterized by high heat flow and several high, medium, and low enthalpy geothermal fields (Fytikas, 1977). The active fault systems favor the circulation of deep hot fluids and the development of thermal springs. High-enthalpy fluids have been discovered at the volcanic islands of Milos and Nisyros, whereas low-enthalpy geothermal systems have been found in Milos (shallow reservoirs), Kimolos, Santorini, Chios, Kos, and Ikaria islands (Fig. 4.1b).

Numerous investigations were carried out during the last 30 years in order to investigate the origin and composition of thermal waters in Greece (Chiodini *et al.*, 1993; Dotsika *et al.*, 2009; Lambrakis and Kallergis, 2005; Minissale *et al.*, 1997). This chapter summarizes the geological/geothermal setting and investigates the water chemistry of the SAAVA, in particular the Chios and Ikaria geothermal systems, using conventional hydrochemical techniques and statistical analysis (cluster and factor analysis).

4.2 GENERAL GEOLOGICAL SETTING

The Aegean area (Fig. 4.1a) constitutes one of the most rapidly deforming segments of the Alpine-Himalayan belt (Francalanci *et al.*, 2005). It displays the highest deformation rates in the whole Africa–Europe collision zone (McKenzie, 1972; Papazachos and Comninakis, 1971), as is manifested by the extremely high number of seismic events (more than 60% of the European seismicity) occurring in this area (Papazachos, 2002). The deformation is associated with the subduction of the eastern Mediterranean lithosphere under the Aegean plate along the Hellenic Arc and the westward motion of Anatolia along the North Anatolian Fault (Baba *et al.*, 2009; Le Pichon and Angelier, 1979; Papazachos and Comninakis, 1969; Papazachos and Delibasis, 1969).

The geodynamic and tectonic regime of the Aegean area during the most recent geologic period and the formation of the SAAVA (Fig. 4.1a) have contributed to the increase of the heat flow and the development of several important geothermal reservoirs at relatively shallow depths (e.g., Milos, Kimolos, Nisyros, Santorini). The volcanic zone (Fig. 4.1a) extends from the Saronikos Gulf as far as Nisyros and the south-western edge of Kos, including the active centers of Methana, Milos, Santorini, and Nisyros (Fytikas, 1980).

The older formations of Milos, Kimolos, and Santorini, as well as the island of Ikaria, belong to the Attic-Cycladic massif (Fig. 4.2), a metamorphic complex and heterogeneous zone that consists of several formation units, tectonically related to each other (Mountrakis, 2010; Ring,

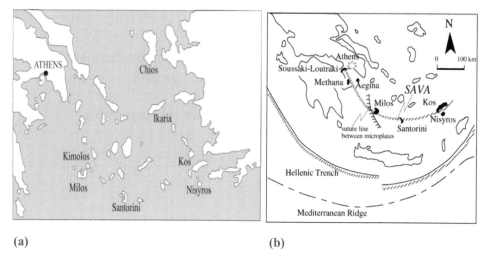

(a)   **(b)**

Figure 4.1.   (a) Studied geothermal areas and (b) schematic structural map of south Greece. The dashed
area indicates the South Aegean Active Volcanic Arc (SAAVA) and the main volcanic centers
(source: La Ruffa *et al.*, 1999).

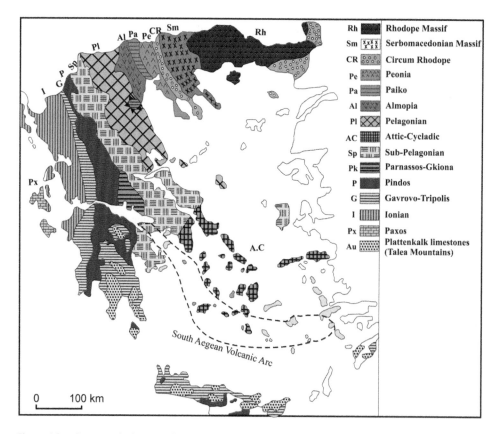

Figure 4.2.   Geotectonical zones of Greece (Mountrakis, 2010).

2010). The zone is affected by two main metamorphic events: high pressure eclogitic (HP/HT) to glaucophanitic (HP/LT) metamorphism, due to the convergence of the Apulian and European plates, and retrogressive low-pressure green-schist (LP/LT) metamorphism, during the orogenic uplift.

The islands of Kos and Chios belong to the Sub-Pelagonian geotectonic zone (Fig. 4.2), which paleogeographically represents the western slope of Pelagonian zone towards the Axios (Vardar) through (Mountrakis, 2010). It is regarded as the ophiolitic suture of the oceanic area that extends to the west of the Pelagonian continental block, constituting a unified oceanic space with the Axios zone.

4.2.1 *South Aegean Active Volcanic Arc*

Cenozoic volcanism in the Aegean area took place in successive phases from the Upper Eocene–Lower Oligocene until the present day (Fytikas *et al.*, 1984). During the Lower Pliocene, the rate of convergence between the two continental margins increased and initiated the volcanism and the development of several hydrothermal systems at the SAAVA.

The volcanoes of the SAAVA lie on a continental crust of 32–34 km thickness at the western and eastern edges and 23–26 km at the central part (Innocenti *et al.*, 1981; Makris, 1977). Their position is controlled by large tectonic lineaments, trending E–W to NW–SE for the western part and NE–SW for the central and eastern part of the arc (Francalanci *et al.*, 2005).

Milos, Nisyros, and Santorini belong to the high heat flow zone of the Aegean, which is attributed not only to the presence of large solidified magma chambers at relatively shallow depths but also to the intrusions of molten magma (3–5 km below surface) due to the strong extensional tectonic regime and several volcano-tectonic regime (Fytikas, 1989; Fytikas and Andritsos, 2004).

4.3 REGIONAL GEOLOGICAL AND GEOTHERMAL SETTING

4.3.1 *Milos*

Milos is situated in the central part of the SAAVA and is the main island of the so-called "Milos archipelago." The island is almost entirely composed of volcanic formations, which lie on the metamorphic and sedimentary pre-volcanic basement (Fig. 4.3).

The oldest geological formation on Milos is the crystalline metamorphic basement, which belongs to the Attic-Cycladic massif (Fytikas, 1989; Fytikas and Marinelli, 1976). It outcrops only sporadically at the southern part of island. The predominant rocks are schists, whereas the presence of marbles, which in other Cycladic islands, are common, extensive and indicated only by the occurrence of a few xenolithes brought to the surface by volcanic eruptions and phreatic explosions (Fytikas, 1989).

The Neogene sediments were deposited during the Miocene/Pliocene transgression phase. They outcrop in the southern part of the island, partially covering the metamorphic basement; however, they were found in the northern part of Milos, in depths between 500 and 800 m (Fytikas, 1989). They consist of basal conglomerate and alternations of limestones, marls, sandstones, and clays. Their rich marine microfauna is characteristic of Upper Miocene–Lower Pliocene period (Fytikas, 1977; 1989).

The volcanic activity on Milos started 3.5 Ma ago and continued up to very recent historic times with hydrothermal (phreatic) explosions (Francalanci *et al.*, 2005; Fytikas *et al.*, 1984, 1986; Fytikas and Vougioukalakis, 1993; Traineau and Dalambakis, 1989). The volcanic formations consist of lava and mainly pyroclastic formations of andesitic, dacitic, rhyolitic, and rarely basaltic andesitic composition (Fytikas, 1989). They were almost entirely covered by the products of the strong phreatomagmatic activity, which were deposited as a "chaotic" formation consisting of metamorphic (schists) and volcanic fragments and a fine-grained mass, originally named as

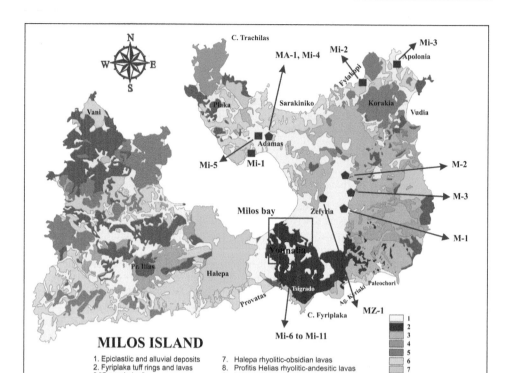

Figure 4.3.   Milos island geological map with the location of deep geothermal boreholes in Zephyria and Adamas (modified from Fytikas, 1977; 1989).

the "Green lahar" (Fytikas and Marinelli, 1976). The hydrothermal alteration resulted in the transformation of the volcanics into clay minerals and the creation of several important mineral deposits on the island (e.g., kaolinite, bentonite).

The youngest deposits in Milos consist of non-volcanic material (alluvial, fluvial–torrential terraces, etc.) with limited extent, mainly due to the recent age of volcanism and the morphology of the island. The older formations have been affected by Alpine tectonic processes (Delimbasis and Drakopoulos, 1993; Fytikas, 1989). The positive geothermal anomaly is definitely associated with the more recent (Pliocene–Quaternary) tectonic activity, which continues up to present times (Delimbasis and Drakopoulos, 1993; Drakopoulos and Delimbasis, 1973; Papadopoulos, 1992). The NE–SW extensional stress field during the Pliocene created NW–SE normal faults, while the most recent ENE–WSW faults were formed by intense Quaternary tectonic activity. The intersection of the above faults favored greatly the circulation of geothermal fluids (Fytikas, 1989) and the development of the geothermal reservoirs. Additional information and detailed studies on the tectonics and neotectonics of Milos can be found in Angelier *et al.* (1977), Fytikas (1977), Simeakis (1985), Jarrige (1978), Ochmann *et al.* (1989), and Papanikolaou *et al.* (1990a).

The high-enthalpy geothermal field of Milos is regarded as Greece's most important resource due to its great geothermal potential and very high temperatures. The numerous active thermal manifestations (fumaroles, hot springs, and hot grounds with maximum temperatures between 76 and 101°C) that can be observed on the island, as well as in the geochemical data, the hydrothermal craters, and the very high thermal gradients, were strong indications for the great geothermal anomaly (Fytikas, 1989). The geothermal exploration identified and confirmed two different geothermal systems. The first regards the deep high-enthalpy geothermal reservoir, which is mainly developed inside the metamorphic basement. The second is formed inside the rhyolitic lavas, at depths less than 150 m below surface, with temperatures between 50 and 100°C (low-enthalpy system). These shallow aquifers are conductively heated by the bentonitic and kaolinitic deposits that constitute the cap rock of the deeper high-enthalpy reservoir.

The geothermal gradient on the island has proven particularly high in the central and eastern parts, with values up to more than 80°C/100 m (Fytikas, 1977; 1989), delimitating two interesting regions, Zephyria plain and Adamas area. Less important anomalies were discovered at the western part of Milos, whereas a strong geothermal anomaly was found southwards of Zephyria (Fytikas *et al.*, 1976).

Deep boreholes drilled at the Zephyria and Adamas areas have confirmed the existence of the high-enthalpy system. The boreholes MZ-1 (Zephyria, 1101 m) and MA-1 (Adamas, 1163 m), reached the metamorphic basement and produced two-phase geothermal fluids from the high-enthalpy reservoir. The productive zone in MZ-1 started at the depth of 809 m, inside the fractured quartzites and calc-schists (Fytikas *et al.*, 1976). The temperature at 932 m was 310°C. In the MA-1 borehole, the top of the geothermal reservoir was encountered a few meters above the Neogene limestones, inside the volcanic formations (lava-breccias). The productive zone included the limestones and continued inside the crystalline basement, with temperatures of 250°C at 1095 m (Fytikas *et al.*, 1976). The boreholes M-1, −2, and −3, also drilled in the area of Zephyria, reached the depth of 1180, 1381, and 1017 m, respectively. The production zones started at a depth of 800–900 m, inside the metamorphic basement. During the production tests, each borehole produced 50–120 t/h of two-phase fluids. The enthalpy measurements in M-2 borehole (>1700 kJ/kg) indicated reservoir temperatures between 300 and 320°C.

The low-enthalpy (shallow) geothermal reservoir of central Milos (Vounalia area) is located inside the extremely permeable recent volcanics. Its thickness is approximately 100–150 m (Fytikas *et al.*, 2005) and, according to the borehole logs, it consists of three main layers, from the top to the bottom: (i) the rhyolitic lavas, (ii) the lahar formation, and (iii) the hydrothermally altered tuffs (kaoline, bentonite, etc.). The temperature measurements suggest two different productive zones: (i) the upper one, located at the top of the reservoir, at approximately sea-level elevation, with thickness and temperature that vary between 10 and 20 m and 53 and 95°C, respectively, (ii) the lower zone that lies at depths greater than 20 m below sea level, with temperatures around 80–110°C at the inland boreholes and 40–50°C at those drilled close to the coast. The most productive boreholes produced more than 100 m^3/h of water each, with well-head temperatures ranging between 80 and 100°C (Fytikas *et al.*, 2005).

The geothermal conditions and the exploration surveys have been described thoroughly by several authors, such as Dominco and Papastamataki (1975), Fytikas and Marinelli (1976), Fytikas (1977, 1989), Fytikas *et al.* (1976, 1988, 1989, 2005), ENEL and PPC (1981a; 1981b; 1982), Cataldi *et al.* (1982), Thanassoulas (1983), Vrouzi (1985), Mendrinos (1988, 1991), Koutroupis (1992), Tsokas (1985), Karytsas *et al.* (2003), and Kyriakopoulos (2010).

4.3.2 *Kimolos*

Kimolos is a very small (35.7 km^2) island to the northeast of Milos, made up mainly by volcanic products (Fytikas and Vougioukalakis, 1993). It was built up at the period between 3.5 and 1.6 Ma ago (Francalanci *et al.*, 2007) and has partly emerged since the Middle Pliocene (Francalanci *et al.*, 1994; 2003; Fytikas and Vougioukalakis, 1993; Fytikas *et al.*, 1986).

The pre-volcanic basement occurs in two very limited outcrops at the western part of the island (Fig. 4.4) and consists of mica and quartz schists, as well as Neogene clastic marine sediments, which have been faulted and uplifted to the surface as a consequence of the tectonic activity and igneous intrusions (Tsokas *et al.*, 1995). The granitic intrusion is evidenced in three small outcrops in the central part of the island, covering a very small area around the Petali Heights (Tsokas *et al.*, 1995) and inside the large pyroclastic flow in the Kastro area (Fytikas and Vougioukalakis, 1993).

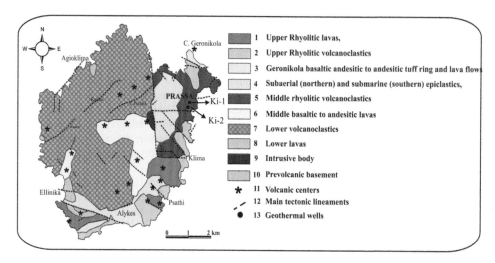

Figure 4.4.   Kimolos island geological sketch map (Francalanci *et al.*, 2005).

The granite occurrence is related to the igneous volcanic activity on the island (Fytikas and Vougioukalakis, 1993), 3.15 Ma ago (Ferrara *et al.*, 1978). As suggested by the geophysical study of Kimolos (Tsokas *et al.*, 1995), the granitic and metamorphic units are strongly fractured, favoring the circulation of hot fluids.

The explosive volcanic activity resulted in the deposition of two massive ignimbritic formations (Kastro and Prassa areas) and the production of other volcanic materials (Fig. 4.4) that belong to the calc-alkaline and high K calc-alkaline series (Fytikas *et al.*, 1986), ranging from basalts to rhyolites (Francalanci *et al.*, 2007). The tectonic regime of Kimolos is dominated by four impressive fault systems: trending N–S, NE–SW, E–W, and NW–SE, the primary being the N–S system (Tsokas *et al.*, 1995).

Several hot springs are located in NW coast of Kimolos, along the NE–SW tectonic lineaments, with significant flow rates and temperatures up to 56°C (Tsokas *et al.*, 1995; Karytsas *et al.*, 2002). Hot springs have also been found in the NE coastal area with temperatures up to 46°C. The various thermal manifestations and the temperature (up to 61.5°C) measured in ten shallow (50 m) and two deeper (188 and 238 m) boreholes (in the Prassa region) have revealed the geothermal interest of the area.

The low-enthalpy geothermal reservoir is located inside the porous volcanoclastic formations, where ascending hot seawater has been entrapped (Fytikas, 1995). Up to six different productive layers separated by impermeable pyroclastic material have been identified in the area of the deeper drilling exploration (Karytsas *et al.*, 2002).

Temperature progressively increases from 40–45°C at the depth of 20–30 m to 59°C at 50 m and 61°C at 200 m. The chemical analyses performed in samples from hot springs and boreholes showed remarkable similarity with a seawater composition (Karytsas *et al.*, 2002), thus confirming the geothermal model suggested by Fytikas (1995). The SiO_2, Na/K, and Na–K–Ca geothermometers indicated 64–78, 108–114, and 169–228°C reservoir temperatures, respectively (Karytsas *et al.*, 2002).

4.3.3   *Santorini*

Santorini (or Thera) is the main island in a group consisting also of Therasia, Aspronissi, Palea Kameni, and Nea Kameni islands (Fig. 4.5). It covers an area of 75.8 km^2 and is located at the southernmost part of the SAAVA, comprising one of its six volcanic centers. The island is almost

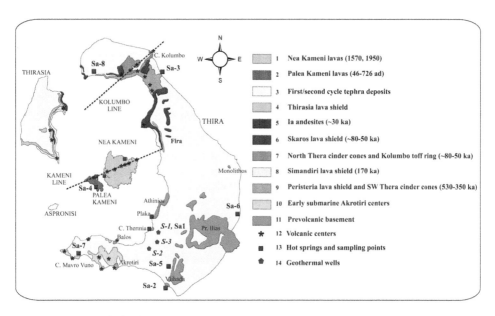

Figure 4.5.   Santorini island group geological sketch map (modified from Francalanci *et al.*, 2005).

entirely made up of Pliocene to recent pyroclastics and lavas (Skarpelis and Liati, 1990) as a result of a complex history of volcanic eruptions over about 0.7 Ma (Fig. 4.5). Santorini constitutes one of the most violent caldera-forming volcanoes of the world. The last collapse of the caldera was linked to the catastrophic Minoan explosive eruption about 3600 years ago, whereas the islands of Palea and Nea Kameni were formed during post-caldera volcanic activity that became subaerial around 197 BC (Francalanci *et al.*, 2007).

The geologic evolution of the island has been the subject of several detailed studies, since the end of the 19th century: Fouqoué (1879), Washington (1926), Ktenas (1927), Reck (1936), Georgalas (1953), Georgalas and Papastamatiou (1953), Nicholls (1971), Bond and Sparks (1976), Watkins *et al.* (1978), Druitt and Sparks (1982), Druitt (1983), Druitt *et al.* (1989), Fytikas *et al.* (1990a; 1990b), Heiken and McCoy (1990), Papadopoulos (1990), Davis *et al.* (1998), Cioni *et al.* (2000), Vougioukalakis and Fytikas (2005), Francalanci *et al.* (2005; 2007), Dimitriadis *et al.* (2005; 2009) and Stiros *et al.* (2010).

The volcanics are deposited over the metamorphic basement that belongs to the Attico-Cycladic massif (Fytikas *et al.*, 1990a; 1990b). The non-volcanic formations outcrop in the SE part of Santorini and consist of a phyllitic epimetamorphic series of Upper Jurassic–Lower Cretaceous age (Fytikas *et al.*, 1990b; Tataris, 1964) and a thick layer of Upper Triassic–Upper Cretaceous recrystallized limestones that have overthrusted the phyllites (Blake *et al.*, 1981; Fytikas *et al.*, 1990b; Papastamatiou, 1958). Low-grade metamorphosed conglomerates and sandstones of Lower Tertiary age have also been recognized (Skarpelis and Liati, 1990). The glaucophanitic schists outcropping in the Athinios and Thermia areas have been partly affected by contact metamorphism due to a young (7 Ma) granitic intrusion (Skarpelis and Liati, 1990).

The post-alpine volcanism started during the Early Quaternary in the area of Akrotiri Peninsula and probably at the location of the Christiania islands (20 km SW of Santorini). It can be divided into six main stages (Druitt *et al.*, 1989), with the most recent phase dating from 1613 BC (Minoan explosion) up to the present. The products of the first activity (0.7 Ma ago) outcrop at the south-western part of the island and were produced by submarine volcanic centers (Fytikas *et al.*, 1990b). The last paroxysmal event covered the entire island with pumice, flows, and surge deposits (Fytikas *et al.*, 1990b). Post-Minoan activity is almost entirely confined inside the caldera

(Fytikas *et al.*, 1990b; Arriaga *et al.*, 2008), with the exception of the 1650 AD activity at the Kolumbo submarine volcanic center.

The tectonic setting of Santorini is complicated and reflects many caldera collapses (Fytikas *et al.*, 1990b). The predominant NE–SW faults follow the still active extensional regime of the central Aegean and offset earlier faults that have an E–W and N–S trend (Perissoratis, 1995). The caldera of Santorini, the Kolumbo's crater, and the Christiana islands are aligned in a NE–SW direction, along fracture zones, named "Kameni" and "Kolumbo" lines, also marked by deep seismic activity (Perissoratis, 1995 and references therein). The volcano-tectonic system of Kolumbo (Kolumbo line) is the most active tectonic feature of Santorini (Arriaga *et al.*, 2008). Some NNE faults at the southern part of Santorini have been identified and mapped by Vougioukalakis (2007).

The geotectonic and magmatic conditions of Santorini have created a favorable environment for geothermal and hydrothermal activity. This is evidenced by the existence of shallow magma chambers and thermal springs on the island (Fytikas *et al.*, 1990a). Numerous hot springs and fumaroles are also present at Palea and Nea Kameni islands, with maximum temperatures at 97°C, whereas the temperature of the fluids that outflow from the bottom of Kolumbo caldera reaches 230°C. The majority of the hot springs of Santorini appears in the internal part of the caldera along the faults, fractures, and fissures of the pre-volcanic basement. The main springs are Plaka (33.6°C), Athermi Christou (56°C), and Vlihada (32°C). The scarcity of spectacular thermal manifestations is probably attributed to the existence of permeable superficial "fresh" volcanics that allow seawater to intrude diluting the uprising hot fluids.

Preliminary geothermal exploration has indicated an area of an anomalous geothermal gradient at the southern part of Santorini inside the formations between Akrotiri and the pre-volcanic basement outcrop. On the contrary, no significant temperature gradient was detected in northern Santorini (Fytikas *et al.*, 1990a). The maximum measured temperature in three deeper boreholes was 64.7°C at the depth of 240 m (S-1 borehole), whereas in S-2 and S-3 the temperature did not exceed 52.3°C at 440 m and 51.2°C at 260 m, respectively (Fytikas *et al.*, 1990a), thus indicating a low-enthalpy geothermal system. However, a deeper geothermal reservoir, with at least medium enthalpy fluids, is probably located at greater depths. The K–Na–Mg geothermometer, applied by Mendrinos *et al.* (2010) to the shallow aquifer waters, suggested deep equilibrium temperatures in the range of 110–175°C. The reservoir depth was estimated at 800–1000 m below surface.

4.3.4   *Nisyros*

Nisyros is a small (42 km^2), almost circular island, located at the eastern edge of the SAAVA. It is entirely volcanic (Fig. 4.6) formed during the Late Quaternary, with a truncated cone of 8 km basal diameter and a central caldera depression of 4 km diameter. It constitutes the youngest volcano of the SAAVA (Francalanci *et al.*, 2005; Vougioukalakis and Fytikas, 2005 and references therein).

The geothermal field of Nisyros has been characterized as the most dynamic and hottest field of the world in a plate collision environment, due to the presence of magma chamber at extremely shallow depths (www.igme.gr). The volcano-seismological studies carried out in the area (Papadopoulos, 1984; Papadopoulos *et al.*, 1998) suggested a magma chamber at a depth of approximately 2 km, which was confirmed by the findings of geothermal drilling. The confirmed high-enthalpy geothermal field reaches temperatures above 400°C at depths between 1500 and 1800 m.

The pre-volcanic basement consist of Mesozoic carbonates and Neogene sediments (Barberi *et al.*, 1988; Varekamp, 1992) that lies at depths of 600 and 1000 m below sea level at the NW and southern part of the caldera, respectively (Geotermica Italiana, 1983; 1984). The infilling of the caldera is made of lacustrine, alluvial, and ash deposits (Caliro *et al.*, 2005; Tibaldi *et al.*, 2008b). Diorites with associated metamorphic rocks were encountered at the depth of 1816 m below sea level. The oldest volcanic formations have an age of less than 160 ka (Vougioukalakis and Fytikas, 2005) and are represented by a few submarine basaltic andesites outcropping at the NW part of the island (Innocenti *et al.*, 1981). The post-caldera activity includes the emplacement

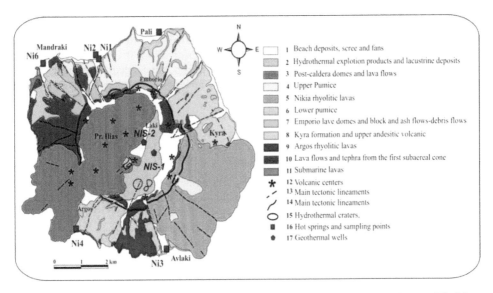

Figure 4.6.   Nisyros island geological sketch map and deep geothermal borehole location (modified from Francalanci *et al.*, 2005).

of dacitic domes (NE–SW direction), pumice falls, lava flows of dacitic-rhyolitic composition, and surge deposits (Caliro *et al.*, 2005; Davis, 1967; Di Paola, 1974; Innoccenti *et al.*, 1981).

The southern sector of Nisyros caldera and the NE–SW domes have been affected by several hydrothermal eruptions such as those occurred in the years 1871–1873 and 1888 that took place in historic times and formed the craters of Polyvotis Megalos, Phlegethon, and Polyvotis Mikros (Caliro *et al.*, 2005; Marini *et al.*, 1993).

Five major fault systems have been identified in Nisyros, trending NE–SW, NW–SE, E–W, N–S, and ESE–WNW, the latter characterizing the entire vicinity of the island (Zouzias *et al.*, 2011 and references therein). Some of the active faults were reactivated during seismic crises of the last century (Vougioukalakis *et al.*, 1998).

The high geothermal anomaly of Nisyros is manifested at surface by the existence of several hot springs and fumaroles. The major fumarolic vents are located in the craters of Kaminakia, Polybotes Megalos, Polybotes Mikros, and Stephanos (www.igme.gr). The temperature of the gases and the ground around the fumaroles is about 100°C. Several thermal springs discharging hot seawater (T_{max}: 54°C) are located along the northern and southern coasts.

Extensive surface geothermal exploration studies have indicated the presence of a high-enthalpy reservoir at moderate depths (1000–1500 m) and a shallow medium enthalpy system at depths between 100 and 500 m. The deep drilling exploration project confirmed the excellent geothermal potential of the island (Koutroupis, 1992; Vrouzi, 1985). The temperature gradient in the first 300 m below surface is approximately 35°C/100 m (Vrouzi, 1985). The deep boreholes were drilled in the area of Lakki, inside the caldera. The first borehole (NIS-1) encountered the high-enthalpy geothermal reservoir at the depth of 1420–1816 m, within the thermo-metamorphic and hydrothermally altered crystalline rocks (marbles, schists, and volcanics), with very high temperatures (>400°C) and extreme (>45 atm) pressures (www.igme.gr). The second borehole (NIS-2) found the high-enthalpy liquid dominated reservoir at shallower depths, between 1000 and 1547 m (Mendrinos *et al.*, 2010; Ungemach, 2002) with temperatures exceeding 350°C. The productive zones were inside fractured limestones intercalated with subintrusive quartz-dioritic rocks (Mendrinos *et al.*, 2010).

A shallow aquifer with temperature around 100°C was encountered by the NIS-1 borehole inside the fractured, altered volcanics (Mendrinos *et al.*, 2010). NIS-2 penetrated a medium- to high-enthalpy aquifer (T 150°C) at depths between 370 and 470 m within andesitic lavas and breccias (Geothermica Italiana, 1984; Mendrinos *et al.*, 2010). According to Kavouridis *et al.* (1999), the fluids of the shallow aquifers with temperatures of 120–170°C are probably present beneath the entire island, whereas the more concentrated and hot brines occur beneath the caldera. Low-temperature fluids (45–75°C, in some cases >95°C) in very shallow depths can be found in several sites on the island (www.igme.gr).

4.3.5   *Kos*

Kos island is located at the easternmost part of SAAVA, a few kilometers off the Turkish coast. It is characterized by a complex geological structure, volcanism, and superficial thermal manifestations (Bardintzeff *et al.*, 1989; Dalabakis, 1987; Papanikolaou and Lekkas, 1990), that relate to the extensional tectonics and the geodynamic setting of the broader area.

The geological structure of Kos is illustrated in the maps of Figure 4.7, with emphasis on the SW part of the island, where the geothermal exploration was carried out. The tectonic evolution of the island is controlled by the dominant WNW–ESE and the NE–SW faults, which are related to extensional processes and volcanic activity during the Pleistocene and Pliocene (Lagios *et al.*, 1998; Papanikolaou and Lekkas, 1990).

The volcanic products of western Kos (Fig. 4.7b) derive from two periods of intense volcanic activity (La Ruffa *et al.*, 1999). The oldest volcanics (trachytic welded ignimbrite) were deposited during the Upper Miocene and are related to a pre-Aegean Volcanic Arc subduction zone (Allen *et al.*, 2009; Keller, 1982). The Pliocene activity associated with the SAAVA volcanism started about 3.4 Ma ago (Allen *et al.*, 2009; Dalabakis, 1987; Keller *et al.*, 1990) with calc-alkaline deposits (dacites), followed by rhyolitic domes (2.7–1.6 Ma ago), the pyroclastics of Kefalos tuff ring, and the Zini perlitic obsidian dome (Allen *et al.*, 2009; Francalanci *et al.*, 2005). The later activity caused a small caldera collapse in the Kamari Bay area (Dalabakis and Vougioukalakis, 1993; Francalanci *et al.*, 2005).

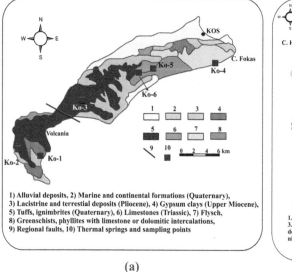

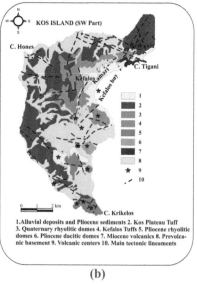

(a)   (b)

Figure 4.7.   (a) Simplified geological map of Kos island (La Ruffa *et al.*, 1999 with modifications) and (b) SW Kos detailed geological map (Francalanci *et al.*, 2005).

Two interesting geothermal areas are distinguished in Kos: the area of Volcania (SW Kos) and the area around Agios Fokas thermal spring (T 45°C) in the NE part of the island, which is the only hot spring emerging at Kos.

The Volcania area is a 1 km diameter basin with 14 small circular hydrothermal zones arranged in two intersecting lineaments (Bardintzeff *et al.*, 1989). The temperatures measured in the two exploration boreholes drilled in Volcania were as low as 24°C, revealing a subnormal geothermal gradient at the depth of 400–450 m. This could be possibly attributed to the intense alteration processes and self-sealing phenomena that prevent the deep hot fluids rising toward the surface (La Ruffa *et al.*, 1999). The hydrothermally altered marls and marly limestones, which are found at the depth of 300–550 m, probably constitute the impermeable cover of the main geothermal reservoir, which is expected at greater depths (>1 km below surface, Lagios *et al.*, 1998). The solute geothermometry performed by La Ruffa *et al.* (1999), for several samples from Kos, indicate temperatures up to 110°C for the reservoir feeding the Kefalos waters, whereas all the other waters seem to be related to thermal systems with lower temperatures (up to a maximum of 60°C).

4.3.6 *Ikaria island*

Ikaria is located in the east-central part of the Aegean area and it is well known for its numerous thermal and radioactive springs, which are classified among the most radioactive springs in the world. According to several authors (Photiades, 2002 and references therein), Ikaria occupies a transitional geotectonic position between the Attic-Cycladic massif and the Pelagonian zone. It consists of metamorphic rocks and granites (Fig. 4.8) that belong to two major lithotectonic units (Ktenas, 1969; Pe-Piper and Photiades, 2006; Photiades, 2002; 2004):

(i) The Lower Unit, which is characterized by a gneissic basement (Ikaria unit) that is overlain by a marble sequence and passes upward into a sequence of intercalated schists and marbles (Messaria unit). The intrusion of two major granitic bodies of Miocene age (Baltatzis *et al.*, 2009; Photiades, 2002; Ring, 2007) that outcrop at the western (Raches) and eastern (Xylosyrtis) parts of the island resulted in the green-schist to mid-amphibolitic and a green-schist to blue-schist retrogressive metamorphism that affected the entire unit (Photiades, 2002).

(ii) The Upper Unit (Photiades, 2002) occurs at the central part of the island, in the region of Kefala (Kefala unit according to Papanikolaou, 1978; Baltatzis *et al.*, 2009) and in the area of Faros (NE Ikaria). The formations in Kefala consist of an ophiolitic mélange, tectonically

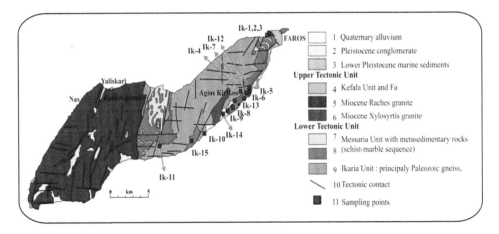

Figure 4.8. Geological map of Ikaria (modified from Photiades, 2004).

overlain by Late Triassic crystalline limestones and marbles (Papanikolaou, 1978; Pe-Piper and Photiades, 2006; Photiades, 2002). At Faros area, a coarse-grained terrestrial conglomerate (Faros Mollase unit) directly overlies the detachment fault over the Lower Unit and consists of ophiolitic clasts and Triassic limestones, which are uncomformably overlain by transgressive Early Pliocene sediments (Pe-Piper and Photiades, 2006; Photiades, 2002;).

Very rare Pliocene rhyolitic dykes have been discovered in the NW part of Ikaria (Nas area), cutting across the granitic rocks. They are considered as the result of volcanism in the Aegean back-arc region and are described and discussed in detail by Baltatzis *et al.* (2009). The major faults and tectonic features of Ikaria are illustrated in the geologic map of Ikaria (Photiades, 2004). Details on the tectonic structure and evolution of the island can be found in Papanikolaou (1978), Papanikolaou *et al.* (1991), Photiades (2002), Kumerics *et al.* (2005), Ring (2007) and included references.

The hot springs that spout all over the island are related to the presence of active fracture zones. Their temperatures vary between 32.5 and 58.4°C. The radioactivity of the water is attributed to the presence of radon that enriches the water as the latter comes in contact with rocks (e.g., granite, granodiorite) with radioactive minerals.

The geothermal conditions in Ikaria have not been systematically explored; however, the island is regarded as a very promising geothermal area, since the thermal springs provide clues to only a small portion of the actual geothermal potential that is "trapped" underground.

4.3.7 *Chios island*

Chios is situated at the central-eastern part of Aegean Sea, just a few kilometers away from the Turkish coast. It is an island with many interesting geothermal areas, including the confirmed geothermal fields of Nenita and Thymiana.

Geologically, it belongs to the Sub-Pelagonian geotectonic zone (Mountrakis, 2010). The extremely thick (3–4 km) Paleozoic basement (Fig. 4.9) outcrops at the NW sector of the island and consists of pre-alpine clastic and recrystallized formations (marls, limestones greywackes, sandstones, and shales), which are disrupted by volcanic rocks of the Lower Carboniferous (Dotsika *et al.*, 2006). The Mesozoic formations are mainly limestones and dolomites (Kavouridis *et al.*, 2009) that cover the largest part of Chios. The Triassic system consists of conglomerates, marls, sandstones, dolomites, limestones, and tuffs (Dotsika *et al.*, 2006). The Jurassic is represented by conglomeratic sandstones and carbonates. The total thickness of the Jurassic carbonates exceeds 300 m.

The SE part of Chios constitutes a sedimentary basin that is filled by Neogene formations. The Lower Unit consists of thin-layered marls, ferrous sandstones, and a sandy–clayey series which contains a horizon of a white pumice tuff (Kavouridis, 2009). The ferrous sandstones are in direct contact with the Mesozoic carbonates toward the west. The upper unit is composed of fluvial-lacustrine sediments (carbonate mudstones) with a thickness of more than 250 m.

The Lower–Middle Miocene (15 Ma) volcanic rocks in Chios are represented by lava extrusions of acid to basic composition and volcanic centers in the north and south part of the island (Kavouridis *et al.*, 2000; 2009; Dotsika *et al.*, 2006). The composition of the lava extrusions in northern Chios is silicic, whereas in the southern part (near Nenita) it is silicic to mafic and submafic (Dotsika *et al.*, 2006). The volcanic activity preceded the Miocene sedimentation and is probably associated with the initiation of the extensional tectonism that started in that period and resulted in the formation of long and deep faults that facilitated the magma uprise.

The most recent sediments cover the plain areas of Chios and consist of Quaternary thin fluvial-terrestrial and coastal deposits.

The SE region of Chios is characterized by NE–SW and NW–SE fault structures as well as by a less significant NNE–SSW fault system (Kavouridis, 2009). The Neogene sediments are faulted and have been influenced by the post-alpine compressional and extensional tectonic activity. The

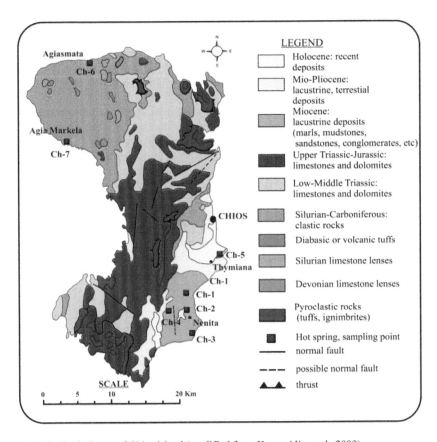

Figure 4.9. Geological map of Chios island (modified from Kavouridis *et al.*, 2009).

most recent extensional tectonism, which started in the Pleistocene and is still active, created normal faults that explain the upflow and circulation of geothermal fluids.

The systematic geothermal exploration of Chios was focused on the SE part of the island, where 13 geothermal boreholes were drilled and two interesting areas were identified: Thymiana and Nenita. The thermal gradient was estimated to be four times the mean value in Thymiana. The geothermal reservoir of Nenita was encountered inside the limestones and consists of two productive zones: one at depths between 25 and 53 m and temperature of 27°C, which is probably influenced by seawater, and the second at depths >70 m with temperature of 32°C. The confirmed geothermal field covers 6.5 km^2. In Nenita, the geothermal gradient is about seven times higher than the mean value and the temperatures, up to the depth of 500 m, vary between 36 and 83°C. The confirmed geothermal field covers 6 km^2 but the unexplored area of interest is evidently much larger. The main reservoir is located inside the fractured Mesozoic black or dark-colored carbonates at depths between 300 and 500 m with temperatures ranging between 78°C and 83°C. The high geothermal anomaly is associated to the N–S trending uplift of the Mesozoic limestones (Kavouridis *et al.*, 2009).

In northern Chios, the saline thermal springs of Agiasmata and Agia Markella, with temperatures of 54 and 35°C, respectively, indicate the existence of another hot aquifer system. However, no systematic exploration has been carried out in this sector of the island. According to Mendrinos *et al.* (2010), the geothermometer K–Na–Mg suggests a deep resource with temperatures of 150–210°C in the Nenita area, while similar equilibrium temperatures are implied for the hot springs of Agia Markella and Agiasmata.

4.4   SAMPLING AND DATA ANALYSIS

Fifty-five (55) water samples were collected from springs during the years 2005–2007. The samples were collected from the following islands: Chios (7), Ikaria (15), Kos (6), Milos (11), Santorini (8), Nisyros (6), and Kimolos (2). The water sampling locations are shown in Figures 4.3–4.9.

The temperature (T), the electrical conductivity (EC), pH, and redox potential (E_h) of water samples were measured *in situ* by portable instruments. The samples were filtered in situ, with 0.45 μm filters (millipore) and were stored in High Density Polyethylene (HDPE) bottles. The aliquots for major and trace metal analyses were preserved by acidification with HNO_3.

All chemical analyses were performed at INGV-Pa (Istituto Nazionale di Geofisica e Vulcanologia—Sezione di Palermo, Italy). The following chemical parameters were determined: Na^+, K^+, Mg^{2+}, Ca^{2+}, Cl^-, NO_3^-, SO_4^{2-}, F^- *via* ionic chromatography, Li, Be, B, Al, Mn, Co, Ni, Cu, Zn, Rb, Sr, Ag, Cd, Sb, Ba, Hg, Tl, Pb, Bi, U, V, Cr, As, Fe, Se, Mo, Cs, and SiO_2 by using inductively coupled plasma techniques (ICP-MS or ICP-OES), while HCO_3^- was determined by titration (HCl 0.1N). Chemical analyses from previous studies were also used for comparison (Chiodini *et al.*, 1993; Dotsika *et al.*, 2006; 2009; Fytikas *et al.*, 2005; Karytsas *et al.*, 2002; Koutinas, 1989; La Ruffa *et al.*, 1999; Minissale *et al.*, 1989). The results of chemical analyses are listed in Table 4.1.

4.5   RESULTS

4.5.1   *Major elements composition*

The data in Table 4.1 show a wide dispersion for the majority of parameters (coefficient of variation generally greater than 100%). A significant relationship is observed between the major ions Na–Cl, Ca–Na, Ca–SO_4, K–Cl, Ca–Mg (Table 4.2). High correlation (>0.80) and significant at p-level 0.05 have been obtained between: Na^+–Cl^- (0.99), Na^+–SO_4^{2-}(0.81), Na^+–HCO_3^-(0.86), Cl^-–SO_4^{2-}(0.82), and Cl^-–HCO_3^- (0.85). Strong linear correlations between ions suggest that they have a common origin or result from a common process.

The highest values of Na^+ and Cl^- concentrations were observed in the spring-water of islands that was affected by seawater intrusion. However, it should be noticed that the salinity of rainfall in the Cyclades archipelago is high, due to airborne sea spray and this has a strong effect on the chemical composition of the groundwater of the islands even when no direct seawater intrusion is observed (Dazy *et al.*, 1997).

Sulfur is widely distributed in a reduced form in both igneous and sedimentary rocks, as metallic sulfides (Ellis and Mahon, 1977). The high sulfate concentrations in water could be attributed to the dissolution and oxidation of the mineral pyrite (FeS_2). Sculpture in reduced or oxidized form may be volatilized and released in large amounts in volcanic regions and can also be present in geothermal water (Hem, 1985).

The maximum value of SiO_2 is observed in Milos and can be associated with the high temperature dissolution of igneous rocks. The deep aquifer in marbles has low SiO_2 content and the shallow aquifer in basement gneiss shows high SiO_2 content. The maximum value of K^+ concentration recorded in Nisyros is related to K-feldspar dissolution.

High nitrate (NO_3^-) concentrations are recorded in some sites (Milos, Santorini) indicating the contribution of fresh water. The high CO_2 flux in the Southern Aegean Volcanic Arc derives from deep geogenic sources either directly from the mantle or from thermo-metamorphism of marine limestones (Barnes *et al.*, 1986; D'Alessandro *et al.*, 2008; 2010).

4.5.2   *Hydrochemistry and water types*

Based on the results of chemical analyses, the following hydrochemical types of waters can be identified in each island.

Table 4.1. Chemical analyses for groundwater and thermal water samples.

Sample		T (°C)	pH	EC (μS/cm)	Na (mg/L)	K (mg/L)	Mg (mg/L)	Ca (mg/L)	Cl (mg/L)	NO$_3$ (mg/L)	SO$_4$ (mg/L)	HCO$_3$ (mg/L)	SiO$_2$ (mg/L)
1	Ik1	36.5	6.90	22030	4535	166	452	521	8230	0.0	1053	238	15
2	Ik2	21.0	7.42	2936	377	11	44	182	807	9.3	119	281	17
3	Ik3	28.1	7.23	33100	7213	264	743	646	12789	0.0	1697	250	14
4	Ik4	42.8	6.65	46800	11192	390	889	1332	19941	0.0	2716	162	24
5	Ik5	43.2	6.51	36600	8317	294	624	1017	14403	0.0	2061	201	32
6	Ik6	53.0	6.52	49700	11990	419	907	1420	21819	0.0	2939	162	25
7	Ik7	41.7	6.94	46900	11088	391	849	1331	20349	0.0	2736	159	21
8	Ik8	18.3	6.97	246	29	2	4	12	41	0.0	11	56	36
9	Ik9	21.0	7.45	232	21	2	4	18	29	0.0	14	67	26
10	Ik10	19.5	7.40	438	25	4	8	56	46	0.0	25	177	19
11	Ik11	16.9	7.93	342	28	4	9	24	47	22.3	29	64	11
12	Ik12	53.1	6.65	50600	12321	425	931	1536	22362	0.0	3105	143	20
13	Ik13	58.3	6.64	50000	12354	434	928	1520	22273	0.0	3140	125	21
14	Ik14	18.0	7.06	280	25	0	6	22	39	1.1	16	79	23
15	Ik15	18.0	7.72	329	24	4	5	35	41	0.6	28	101	16
16	Mi1	15.8	8.76	3300	351	32	80	253	1040	18.6	169	171	95
17	Mi2	22.2	6.90	3850	453	29.3	94.5	259.3	1135.8	96.7	252	162	76
18	Mi3	23.2	6.85	3910	591	31	90	150	1013	38.4	73	668	96
19	Mi4	43.7	6.53	4610	673	123	59	264	1141	9.9	319	729	219
20	Mi5	32.7	6.00	9370	724	88	186	1324	2979	58.3	1229	171	146
21	Mi6	97.0	6.86	55100	12200	2620	192	1515	25125		310	56	158
22	Mi7	89.0	6.36	43900	9400	1800	120	1350	19900		275	59	202
23	Mi8	85.0	6.69	25200	5280	1150	85	725	11200		170	82	162
24	Mi9	55.0	7.75	32800	7600	1370	94	850	13700		1350	85	104
25	Mi10	97.0	7.20	57100	12600	2700	348	1520	23900		2180	90	138
26	Mi11	96.0	7.50	69000	15100	3780	1560	2120	30300		850	86	142
27	Sa1	50.0	6.96		979	38	56	98	1182	120.9	711	256	118
28	Sa2	23.0	8.02	4890	884	29	109	131	1548	55.2	316	168	46
29	Sa3	21.2	7.90	2680	413	6	44	89	703	17.4	101	220	50
30	Sa4	31.0	7.60		1130	38	115	110	1990		350	100	42
31	Sa5	37.0	5.70		12530	630	1500	540	22700	6.0	2700	450	83
32	Sa6	23.0	7.80		225	8	40	90	310	5.0	210	220	59
33	Sa7	23.0	6.90		3580	120	465	480	7400	35.0	1030	130	57
34	Sa8	24.0	7.40		820	12	190	260	1960	18.0	310	210	
35	Ko1	19.3	7.30	674	95	8	14.46	23	162	1.2	37	79	72
36	Ko2	19.8	7.33	1371	160	9	43.01	92	260	6.2	78	360	53
37	Ko3	14.7	7.42	1373	98	8	44.23	171	187	6.8	261	342	43
38	Ko4	42.0	6.00	51100	11948	415	1156	1578	21809	0.0	3555	253	21
39	Ko5	15.6	7.30	616	23	3	21	81	37	5.0	50	278	15
40	Ko6	18.0	7.42	571	22	3	21	91	36	0.0	3	61	19
41	Ni1	27.5	7.07	38600	8150	622	487	1450	15698	22.3	1081	997	148
42	Ni2	20.5	8.72	941	138	24	8	55	140	16.7	41	293	26
43	Ni3		7.05		1380	70	47	18	2110		296	62	50
44	Ni4		6.70		11380	821	773	1180	21370		1610	145	177
45	Ni5		6.60		10520	626	895	802	18990		2130	398	133
46	Ni6		7.00		4760	337	447	770	9500		870	212	105
47	Ch1	22.0	6.80	1030	67	3	78	75	119		54	488	
48	Ch2	21.0	6.90	1500	104	36	78	139	173		91	549	30
49	Ch3	30.0	7.30	3250	421	34	102	93	770		114	518	
50	Ch4	30.0	7.50	1610	135	6	92	87	196		75	671	40
51	Ch5	26.0	6.90	20000	10300	406	1143	667	19063		2390	335	12
52	Ch6	54.0	6.70	15600	2270	263	19	221	4083		46	396	60
53	Ch7	35.0	5.80	20000	13730	1125	614	1600	22250		1315	403	58
54	Ki1	61.5	6.80	47986	10345	391	1240	1066	19217		3564	109	35
55	Ki2	49.2	6.80	45965	10115	391	1230	1042	18579		3569	109	28

Mi6–Mi11: Fytikas *et al.* (2005); Sa4, Sa8: Dotsika *et al.* (2009); Ko4, Ko6: La Ruffa *et al.* (1999); Ch1–Ch7: Dotsika *et al.* (2006); Ki1, Ki2: Karytsas *et al.* (2002).

Table 4.2.   Correlation coefficients between major ions in groundwater and thermal springs.

	Na	K	Mg	Ca	Cl	SO$_4$	NO$_3$	HCO$_3$	SiO$_2$
Na	1	0.63	0.16	0.97	0.99	0.81	−0.1	0.86	−0.19
K		1	0.44	0.66	0.67	0.75	0.22	0.69	−0.11
Mg			1	0.27	0.19	0.57	0.15	0.31	0.24
Ca				1	0.96	0.86	−0.13	0.85	−0.2
Cl					1	0.82	−0.09	0.85	−0.19
SO$_4$						1	0.05	0.73	−0.07
NO$_3$							1	−0.15	−0.09
HCO$_3$								1	−0.1
SiO$_2$									1

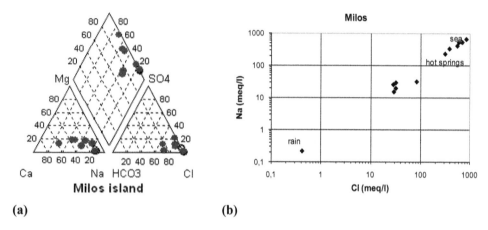

(a)   (b)

Figure 4.10.   (a) Piper diagram of the analyzed water samples in Milos island and (b) logarithmic plot of Na concentration (meq/L) *vs* Cl (meq/L) in Milos island.

4.5.2.1   *Milos island*

The majority of the samples are plotted in the center of the right side of the diamond-shaped field (Na–Cl type) of Piper diagram (Fig. 4.10). The other water types representing mixed waters are: Na–Ca–Cl–HCO$_3$, Na–Cl–HCO$_3$, and Ca–Na–Cl–SO$_4$. The composition of these waters should be attributed to local petrology, and their physical and chemical characteristics are temperature values ranging between 23.2°C and 43.7°C (thermal waters), pH values are less than 7, EC between 2370 and 3910 μS/cm, and TDS values between 2750 and 6906 mg/L.

Cold waters (Mi1, Mi2) belong to Na–Ca–Cl type. The temperature varies from 15.8 to 22.2°C, EC ranges from 3300 to 3850 μS/cm, pH values are greater than 7, and TDS values range between 2210 and 2558 mg/L. As shown in Figure 4.10b, all the samples from Milos are very close to the rainwater–seawater mixing line. High Cl$^-$ concentrations of thermal waters arise from the contribution of seawater and/or a Na–Cl geothermal fluid.

The Cl$^-$ concentration of groundwater from Vounalia ranges between 11,200 mg/L (sample Mi8), which is approximately half of the chloride concentration of local seawater and 30,300 mg/L (sample Mi11), which is 40% higher than the local seawater (Fytikas *et al.*, 2005). The chemistry of the groundwater is the result of the mixing of three types of waters: deep geothermal water, seawater, and meteoric water.

Valsami-Jones *et al.* (2005) group the hydrothermal fluids into two types: (i) low chloride waters containing low concentration of alkalis (Na, K, Li) and calcium and low concentration of silica and sulfate, and (ii) high chloride waters containing high concentration of alkalis and

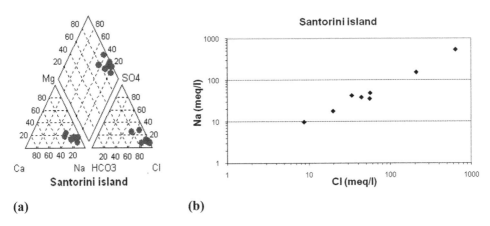

Figure 4.11. (a) Piper diagram of the analyzed water samples in Santorini island and (b) logarithmic plot of Na concentration (meq/L) *vs* Cl (meq/L) in Santorini island.

calcium and lower concentration of silica and sulfate. Both fluid types show low magnesium and high metal concentrations. Especially, the first type has the highest cobalt, nickel, aluminum, iron, and chromium concentrations, and the second one has the highest zinc, cadmium, manganese, and lead concentrations.

Dotsika *et al.* (2009) suggest that all the Br/Cl and B/Cl ratios of all the thermal springs are lower than those of local seawater, indicating that the high-salinity end-member cannot be the marine water. It could be either marine water modified by water-rock interaction at high temperature or a seawater/arc type magmatic water mixture. At Milos and Nisyros islands, the silica (SiO$_2$) content of the thermal spring is higher than the other islands, indicating a higher thermal gradient and the ascent of high enthalpy geothermal fluids to the surface along fractures (Minissale *et al.*, 1997).

4.5.2.2 *Kimolos island*
Based on chemical analyses from two boreholes (depth 188 and 238 m) drilled in Kimolos island, Karytsas *et al.* (2002) concluded that the geothermal water has a composition very similar to that of seawater (electrical conductivity 47,000 μS/cm) with a slightly increased content of Ca (1050 mg/L) and SO$_4$ (3500 mg/L) and decreased content of Na (10,230 mg/L) and Cl (18,900 mg/L). The temperature of groundwater ranges between 49.2 and 61.5°C.

4.5.2.3 *Santorini island*
Based on results of chemical analyses from eight (8) samples, it is concluded that the Na$^+$ is the most abundant cation, while Cl$^-$ is the most abundant anion. The Na$^+$ concentration is between 225 and 12,530 mg/L, while Cl$^-$ concentration is from 310 to 22,700 mg/L. The Ca^{2+} and Mg^{2+} concentrations range from 89 to 540 mg/L and from 40 to 1500 mg/L, respectively. The concentrations of HCO$_3^-$ and SO$_4^{2-}$ vary from 100 to 450 mg/L and 210 to 2700 mg/L, respectively. High major ion concentrations are recorded at Palea Kameni island. The temperature of water samples ranges from 21.5 to 50°C.

The majority of the samples are plotted on the right side of the diamond-shaped field of Piper diagram (Fig. 4.11a). The main hydrochemical type is Na–Cl, except for two samples (Sa1, Sa6), which belong to Na–Cl–SO$_4$ hydrochemical type. The Na–Cl waters are highly mineralized with TDS values above 1580 mg/L. The Na/Cl ratios in meq/L are between 0.7 and 1.2 (Fig. 4.11b).

In general, the water from thermal springs of Santorini (Plaka, Vlychada) island is close to the mixing line between seawater and cold groundwater suggesting that these thermal waters originate from shallow aquifer systems. On the other hand water from thermal spring (temperature 37.0°C) of Palea Kameni island, which is made up entirely from volcanic products, has similar chemical composition to local seawater (Dotsika *et al.*, 2009).

According to Chiodini *et al.* (1993), the high values of electrical conductivity (50,000 μS/cm) and TDS may be related to the presence of mantle fluids in Santorini island. The high HCO_3^- concentration in the aqueous solution could be associated with the addition of CO_2 gas to the seawater infiltrating inland and its titration to HCO_3^- ion through interaction with rocks (Dotsika *et al.*, 2009).

4.5.2.4 *Nisyros island*

Based on the results of chemical analyses performed in samples from Nisyros, the following conclusions can be drawn (Fig. 4.12a,b):

One sample belongs to Na–Ca–HCO_3–Cl hydrochemical type with low temperature 20.5°C, pH 8.72, Cl^- 140 mg/L, Na^+ 138 mg/L, low values of EC (941 μS/cm), and TDS (741 mg/L), representing cold water affected by meteoric water of the area. Cold waters are characterized by Cl^- concentration ranging from 200 to 5000 mg/L and temperature between 17 and 23°C. These are seawater and groundwater mixtures (Chiodini *et al.*, 1993; Dotsika and Michelot, 1993).

Samples from hot springs belong to Na–Cl water type with temperature ranging from 37 to 52°C, pH values from 6.60 to 7.07, and high TDS values ranging between 28,650 and 34,360 mg/L. Based on data analyzed by Kavouridis *et al.* (1999), it is concluded that the thermal springs of Nisyros have Cl^- and Na^+ concentrations very close to local seawater and are enriched in Ca^{2+} and HCO_3^- and slightly depleted in SO_4^{2-} and Mg^{2+}, indicating high temperature seawater-rock interaction. Shallow thermal waters characterized by high Cl^- and Na^+ concentrations represent the mixing of three water types: geothermal water, seawater, and meteoric water. The Nisyros springs are often affected by NH4 and B-rich steam condensate (Chiodini *et al.*, 1993). Minissale *et al.* (1997) suggest that this process is quite likely to take place also at other islands, for example, Milos and Santorini where fumaroles are also present.

The geothermal brine from geothermal borehole of IGME shows similar concentration to those of seawater, but it shows low concentration of SO_4, HCO_3, and Mg and high concentration of Ca^{2+} with respect to local seawater. The decrease in SO_4 is probably due to the precipitation of sulfate minerals and/or reduction to sulfide (Dotsika *et al.*, 2009). The depletion of Mg^{2+} could be associated to the formation of Mg-rich secondary minerals within the hydrothermal system.

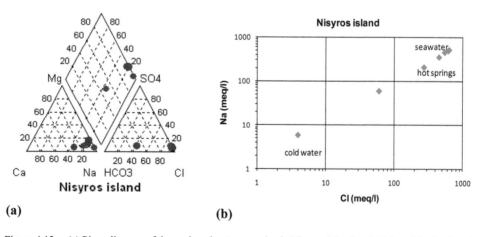

(a)   **(b)**

Figure 4.12.   (a) Piper diagram of the analyzed water samples in Nisyros island and (b) logarithmic plot of Na concentration (meq/L) *vs* Cl (meq/L) in Nisyros island.

4.5.2.5   *Kos island*

The maximum values are recorded in sample from the hot spring near Fokas Cape (eastern part), emerging from sedimentary rocks: temperature 44°C, EC 51,100 μS/cm, and TDS 40,736 mg/L. The water type of this hot spring is classified as sodium type (Na: 11,948 mg/L) in the cations triangular plot, and as chloride type (Cl: 21,809 mg/L) in that of anions (Na–Cl hydrochemical water type), consisting of seawater modified by interaction with sedimentary rocks and by dissolution/precipitation of the calcite-gypsum mineral pair (La Ruffa *et al.*, 1999). Aghios Fokas spring (Ko4) is the main thermal manifestation on the island and is enriched in Ca^{2+}, K^+, and SO_4^{2-}.

Other hydrochemical types are: Na–Ca–Mg–Cl–HCO$_3$, Ca–Na–Mg–HCO$_3$–SO$_4$–Cl, Ca–Mg–HCO$_3$, and Ca–Mg–Na–Cl–HCO$_3$ (Fig. 4.13a). These types correspond to relatively cold waters with temperature ranging between 14.7 and 19.8°C. The EC varies from 571 to 1373 μS/cm, pH values from 7.30 to 7.42, and TDS between 257 and 1161 mg/L. High Ca concentrations are recorded at Kos island, indicating their origin from Mesozoic marbles and dolomite breccia.

Based on data analyzed by La Ruffa *et al.* (1999), the groundwater from sedimentary aquifers (eastern part) and volcanic environments (western part, Volcania area) shows TDS values less than 500 mg/L. Water form 500 m depth (IGME borehole) in Volcania area (temperature 20°C) exhibits TDS values of 5500 mg/L and HCO$_3$ of 3900 mg/L. This is probably the result of the addition of CO_2-rich gas to the groundwater, resulting in the conversion of dissolved CO_2 to HCO$_3$ (La Ruffa *et al.*, 1999).

4.5.2.6   *Ikaria island*

About 53% of the water samples (Ik1, Ik2, Ik3, Ik4, Ik5, Ik6, Ik7, Ik12, Ik13) belongs to the Na–Cl or Na–Ca–Cl hydrochemical types and are classified as primary geothermal fluids (Nicholson, 1993). The temperatures range between 28.1°C and 58.3°C. The mean pH value is less than 7 indicating the slightly acidic character of waters with a maximum value of 7.23. EC varies from 22,030 to 50,600 μS/cm and is indicative of solutions close to marine composition.

The second dominant type Na–Ca–Cl–HCO$_3$ (33.3% of the water samples; Ik8, Ik9, Ik10, Ik14, Ik15) represents cold waters with high concentrations of Ca and HCO$_3$, indicating mixing processes with water of meteoric origin. Their temperature values are moderate ranging between 18°C and 21°C, pH values are greater than 7, indicating the slightly alkaline character of waters. These waters are characterized by low EC (232–438 μS/cm) and TDS (182–360 mg/L) values, respectively. The other hydrochemical water type is Na–Ca–Mg–Cl–HCO$_3$ (sample Ik11), which is probably related to local petrology of the island.

The chemical composition of water is shown in a Piper diagram (Fig. 4.14a). Hydrochemical facies can be classified on the basis of the dominant ions. The samples from cold springs are

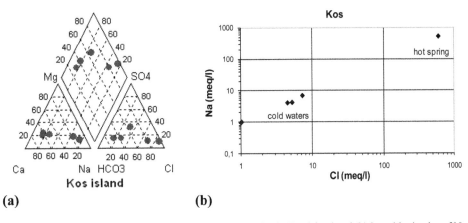

(a)   **(b)**

Figure 4.13.   (a) Piper diagram of the analyzed water samples in Kos island and (b) logarithmic plot of Na concentration (meq/L) *vs* Cl (meq/L) in Kos island.

plotted in the center of the diamond-shaped field (Ca–HCO$_3$ type), while the thermal waters are projected on the right side (Na–Cl hydrochemical water type). Furthermore, all the samples from Ikaria island are very close to the rainwater–seawater mixing line Na–Cl (Fig. 4.14b) indicating the common marine origin of these ions.

4.5.2.7   Chios island

Groundwater (samples Ch1, Ch2, Ch4) from the southern part of the island (Nenita area) belongs to sodium-bicarbonate type (Na–HCO$_3$). The temperature ranges between 21 and 30°C and pH values range between 6.8 and 7.5. EC varies between 1030 and 1610 μS/cm. These waters are characterized by relatively low concentrations of Cl$^-$ (119–196 mg/L), excluding the mixing of seawater with groundwater from the shallow aquifers.

Water from thermal springs (samples Ch5, Ch6, Ch7) and sample Ch3 from borehole in the Nenita area belongs to the sodium-chloride (Na–Cl) hydrochemical type (Fig. 4.15a,b). The temperature ranges between 26 and 54°C and EC varies from 15,600 μS/cm to greater than 20,000 μS/cm. This type of water is rich in lithium (Li$^+$) and is representative of thermal waters of Chios island, related to interactions processes between water and rocks (Dotsika *et al.*, 2006; Kavouridis *et al.*, 2001). Furthermore, the relatively high concentration of HCO$_3$ can be associated with the absorption of CO$_2$-bearing gases or the condensation of CO$_2$ geothermal steam.

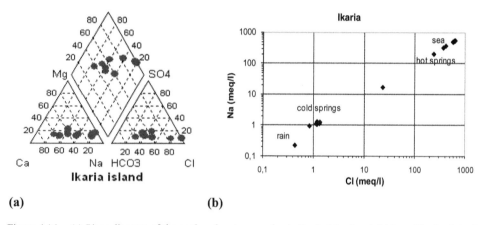

(a)   **(b)**

Figure 4.14.   (a) Piper diagram of the analyzed water samples in Ikaria island and (b) logarithmic plot of Na concentration (meq/L) *vs* Cl (meq/L) in Ikaria island.

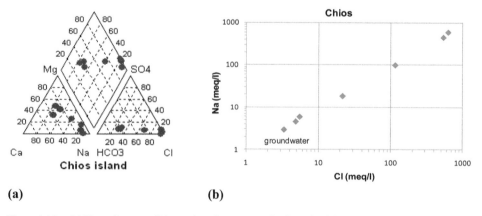

(a)   **(b)**

Figure 4.15.   (a) Piper diagram of the analyzed water samples in Chios island and (b) logarithmic plot of Na concentration (meq/L) *vs* Cl (meq/L) in Chios island.

On the other hand, the thermal waters of springs, which are rich in Cl$^-$, are mixed waters between local groundwater and seawater. Based on isotopic geothermometer data, Dotsika *et al.* (2006) suggest the probable existence of a deep geothermal reservoir of high enthalpy (220°C) in the northern part (Agiasmata and Agia Markella) of the island and another reservoir of low enthalpy (80°C) in the southeastern part (Nenita–Thymiana area) of the island.

4.5.3 *Trace elements composition*

The thermal waters show high concentrations of arsenic (As), strontium (Sr), boron (B), and fluoride (F). High concentration of Sr observed in the majority of samples (Ikaria, Nisyros, and Kos islands) is related to carbonate minerals. The highest As concentration is recorded at Nisyros island and can be associated with the high-enthalpy geothermal fluids. The presence of Li can be associated with the increased temperatures of the geothermal fluids. It is well known that high temperatures increase its mobility and release into waters (Appelo and Postma, 1994; D'Amore and Panichi, 1987).

The iron (Fe) concentration derived either from mafic mineral dissolution or from sulfide minerals oxidation is locally high. Elevated manganese (Mn) concentrations are related to granite of Ikaria (Ikaritis), rich in Mn (Mitropoulos and Tarney, 1992; Varnavas and Cronan, 1988). The high boron (B) concentration (mean value 2.1 mg/L) is attributed to geothermal fluids and intensive water-rock interactions.

High concentrations of elements like Ba, Zn, and Ni occur at Milos and can be associated with the presence of metamorphic graywacke and crystalline basement of schists (Kyriakopoulos, 1998). It is pointed out that the concentration of the tracer elements is not normally distributed. Minissale *et al.* (1997) suggest that apart from water-rock interaction and rock dissolution, higher SiO_2, B, Li, and NH_4 concentrations in hot spring waters than in seawater and cold aquifers is likely to be caused by steam loss, due to subsurface boiling.

4.6 STATISTICAL ANALYSIS

4.6.1 *Cluster analysis*

Cluster analysis is a simple approach of classifying groundwater quality (Hussein, 2004). The Euclidean distance is used as a measure of similarity between every pair of groundwater samples. The distance between two samples (d_{ij}) is given by:

$$d_{ij} = \left[\sum_{k=1}^{m} (X_{ik} - X_{jk})^2 \right]^{1/2}$$

where X_{ik} denotes the k-th parameter measured on sample i and X_{jk} is the k-th parameter measured on sample j.

The samples were grouped into distinct clusters (hydrochemical groups). Table 4.3 shows the cluster centers for each parameter of water in each group. It is concluded that the data were classified into two groups, corresponding to main hydrochemical types. All the samples collected

Table 4.3. Final cluster centers.

Cluster	pH	EC	NO$_3$	SO$_4$	HCO$_3$	Cl	SiO$_2$	Ca	K	Na	Mg
1	6.73	44812	0.1	2627	171	19134	23	1257	373	10630	838
2	7.36	3354	4.5	162	133	950	20	108	24	633	66

Parameters in mg/L, EC in μS/cm.

from the thermal coastal springs belong to cluster 1; most of the samples collected from the cold groundwater were found to be in cluster 2. A few samples seem to do fall in either group.

4.6.2   Factor analysis

In order to examine the relationships between the hydrochemical parameters, R-mode factor analysis was applied (Karakatsanis et al., 2011). Because the different parameters use different units, e.g., mg/L and μS/cm, the initial data must be standardized with a mean value of 0 and a standard deviation of 1 (Stamatis et al., 2001; Voudouris et al., 1997; 2000). Using factor analysis, the parameter distribution in the samples is explained in terms of five (5) factors.

Factor I: It accounts for 51.5% of total variance with high loadings in the parameters Cl (0.97), Na (0.97), SO_4 (0.975), K (0.95), Mg (0.94), Ca (0.93), Sr (0.87), and B (0.72). This factor can be associated with the intrusion of seawater into aquifers, which increases the concentration of the previous parameters.

Factor II: The second factor represents 17.3% of the total variation and has positive loading in the parameters Mn (0.86), Rb (0.80), Li (0.77), Fe (0.73), and B (0.54). The high loading of Mn and Fe is related to reduced environment and presence of igneous rocks. The high loading of parameters should be attributed to granite dissolution, e.g., at Ikaritis, which is rich in these elements.

Factor III: This factor is dominated by HCO_3 (0.89), Al (0.67), and SiO_2 (0.52), and accounts for 8.35% of the total variation. HCO_3 ions derive from the dissolution of geogenic CO_2 originating from the recent volcanic activity (D'Alessandro et al., 2010). The high loadings of the Al and SiO_2 can be associated with the dissolution of silicate minerals enhanced by the low pH values induced by CO_2 dissolution.

Factor IV: It accounts for 6.8% of the total variation and exhibits high loadings with respect to NO_3 (0.84) and SiO_2 (0.66). Nitrate is the most abundant nutrient in water and can be attributed to different sources, the most important of which are the excessive application of fertilizers and the use of septic tanks in conjunction with the disposal of untreated domestic effluent into abandoned wells that are currently used as septic tanks. The high positive loading of SiO_2 reflects the dissolution of silica minerals and high temperatures.

4.7   DISCUSSION AND CONCLUSION

The aim of this chapter is to summarize the geological context and determine the water chemistry of southern Aegean geothermal systems, which are related to the recent volcanic activity. The Aegean region represents an active convergent zone, where continental microplates exhibit a complex interaction under the influence of the overall N–S convergence between the African and the Eurasian plates.

From a geological point of view, most of the islands belong to the Attic-Cycladic massif and are composed of metamorphic rocks such as schists, phyllites, marbles, and granite. Milos, Kimolos, and Nisyros are almost entirely made up of (or covered by) recent volcanic rocks. They are characterized by the presence of several surface thermal manifestations such as hot springs with relatively high temperatures. The geothermal potential of the area was confirmed in most cases by the drilling exploration that has revealed very high temperatures, especially in Milos and Nisyros islands.

The volcanic arc is characterized by the presence of high CO_2 flux derived from deep geogenic sources either directly from the mantle or from the thermo-metamorphism of marine limestones (D'Alessandro et al., 2008; 2010).

The majority of aquifers, which discharge through springs in islands of the southern Aegean, are associated with the seawater intrusion and the mixing of meteoric waters and deep thermal water. Depending on its residence time in the aquifer, seawater may modify the chemical composition as shown by losing Mg^{2+} and SO_4^{2-} ions and acquiring SiO_2, Ca^{2+}, HCO_3^-, NH_4^+, and B after water-rock interaction at variable depth and temperature. High silica concentrations are recorded in Milos and Nisyros thermal springs, due to the presence of active hot hydrothermal system at shallow depth (Minissale *et al.*, 1997).

The predominant hydrochemical water type is Na–Cl. This type is recorded in samples from island geothermal springs and from coastal aquifers and could be associated with seawater mixed with meteoric water. Most of these samples are characterized by a Na/Cl ratio very close to that of seawater (0.84). The presence of trace elements is attributed to dissolution of minerals and hydrothermal solutions of the volcanic arc formations.

ACKNOWLEDGMENTS

The authors would like to thank the Istituto Nazionale di Geofisica e Vulcanologia—Sezione di Palermo (Italy) for analyzing the water samples and Emeritus Professor Michael Fytikas for his useful and valuable comments during the preparation of this chapter.

REFERENCES

Allen, S.R., Vougioukalakis, G.E., Schnyder, C., Bachmann, O. & Dalabakis, P.: Comments on: On magma fragmentation by conduit shear stress: evidence from the Kos plateau tuff, Aegean Volcanic Arc, by Palladino, Simei and Jyriakopoulos (JVGR, 178, 807–817). *J. Volcanol. Geoth. Res.* 184 (2009), pp. 487–490.

Angelier, J., Cantagrel, J.M. & Vilmont, J.C.: Néotectonique cassante et volcanisme plio-quaternaire dans l'arc égéen interne: l'ile de Milos (Grèce). *Bull. Soc. Geol. France* 19 (1977), pp. 119–124. Appelo, C. & Postma, D.: *Geochemistry, groundwater and pollution*. Balkema, The Netherlands, 1994.

Arriaga, M.-C.S., Tsompanakis, Y. & Samaniego, F.: Geothermal manifestations and earthquakes in the caldera of Santorini, Greece: an historical perspective. *Proceedings of the 33rd Workshop on Geothermal Reservoir Engineering*, 28–30 January 2008, Stanford University, Stanford, CA, 2008.

Baba, A., Yuce, G., Deniz, O. & Ugurluoglu, D.Y.: Hydrochemical and isotopic composition of Tuzla geothermal field and its environmental impacts. *Environ. Forensics* 10 (2009), pp. 144–161.

Baltatzis, E., Kostopoulos, D., Godelitsas, A., Zachariadis, P. & Papanikolaou D.: Pliocene tourmaline rhyolite dykes from Ikaria island in the Aegean back-arc region: geodynamic implications. *Geodin. Acta* 22:4 (2009), pp. 189–199.

Barberi, F., Navarro, J.M., Rosi, M., Santacroce, R. & Sbrana, A.: Explosive interaction of magma with ground water: insights from xenoliths and geothermal drillings. *Rend. Soc. It. Mineral. Petrol.* 43 (1988), pp. 901–926.

Bardintzeff, J.M., Dalabakis, P., Traineau, M. & Brousse, R.: Recent explosive volcanic episodes on the island of Kos, Greece: associated hydrothermal paragenesis and geothermal area of Volcania. *Terra Nova* 1:1 (1989), pp. 75–78.

Barnes, I., Leonis, C. & Papastamataki, A.: Stable isotope tracing of the origin of CO_2 discharges in Greece. *Proceedings of the 5th International Symposium on Tracer Methods in Hydrology*, Athens, Greece, 1986, pp. 29–43.

Blake, M.C., Bonneau, M., Geussant, J., Kienast, J.R., Lepvier, C., Maluski, H. & Papanikolaou, D.: A geologic reconnaissance of the Cycladic blueschists belt, Greece. *Geol. Soc. Am. Bull.* 92 (1981), pp. 247–254.

Bond, A. & Sparks, R.S.J.: The Minoan eruption of Santorini, Greece. *Geol. Soc. London J.* 132 (1976), pp. 1–16.

Caliro, S., Chiodini, G., Galluzzo, D., Granieri, D., La Rocca, M., Saccorotti, G. & Ventura, G.: Recent activity of Nisyros volcano (Greece) inferred from structural, geochemical and seismological data. *Bull. Volcanol.* 67 (2005), pp. 358–369.

Cataldi, R., del Gaudio, P., Nevi, G., Rossi, U., Toneatti, R., Koutroupi, N., Michalakis, C. & Vondicakis, M.: Milos geothermal project. *Proceedings International Conference on Geothermal Energy*, 11–14 May 1982, Florence, Italy, 1982, pp. 97–111.

Chiodini, G., Cioni, R., Leonis, C., Marini, L. & Racco, B.: Fluid geochemistry of Nisyros island, Dodecanese, Greece. *J. Volcanol. Geoth. Res.* 56 (1993), pp. 95–112.

Cioni, R., Gurioli, L., Sbrana, A. & Vougioukalakis, G.E.: Precursory phenomena and destructive events related to the Late Bronze Age Minoan (Thera, Greece) and AD 79 (Vesuvius, Italy) Plinian eruptions; inferences from the stratigraphy in the archaeological areas. In: The archaeology of geological catastrophes. *Geol. Soc. London, Spec. Publ.* 171, 2000, pp. 123–141.

Dalabakis, P.: *Le volcanisme récent de l'Isle de Kos.* PhD Thesis, Univ. Paris Sud Orsay, Paris, France, 1987.

Dalabakis, P. & Vougioukalakis, G.: The Kefalos tuff ring (W Kos): depositional mechanisms, vent position and model of the evolution of the eruptive activity. *Bull. Geol. Soc. Greece,* XXVIII:2 (1993), pp. 259–273 (in Greek with English abstract).

D'Alessandro, W., Brusca, L., Kyriakopoulos, K., Michas, G. & Papadakis, G.: Methana, the westernmost active volcanic system of the South Aegean Arc (Greece): insight from fluids geochemistry. *J. Volcanol. Geoth. Res.* 178 (2008), pp. 818–828.

D'Alessandro, W., Brusca L., Martelli, M., Rizzo, A. & Kyriakopoulos, K.: Geochemical characterization of natural gas manifestations in Greece. *Proceedings of the 12th International Congress of the Geological Society of Greece,* Patras, Greece, May, 2010, *Bull. Geol. Soc. Greece* 43:5 (2010), pp. 2327–2337.

D'Amore, F. & Panichi, C.: Geochemistry in geothermal exploration. In: M. Economides & P. Ungemach (eds): *Applied geothermics.* Wiley & Sons, New York, 1987.

Dando, R.P., Aliani, S., Arab, H., Bianchi, C.N., Brehmer, M., Cocito, S., Fowlers, S.W., Gundersen, J., Hooper, L.E., Kölbh, R., Kuevere, J., Linke, P., Makropoulosr, K.C. Meloni, R., Miquel, J.-C., Morri, C., Müller, S., Robinson, C., Schlesner, H., Sieverts, S. *et al.*: Hydrothermal studies in the Aegean Sea. *Phys. Chem. Earth* B 25:1 (2000), pp. 1–8.

Davis, E.: Zur Geologie und Petrologie der Inseln Nisyros und Jali (Dodecanes). *Proceedings of University of Athens* 42, 1967, pp. 235–252.

Davis, E., Gartzos, E. & Dietrich V.J.: Magmatic evolution of the Pleistocene Akrotiri volcanoes. In: R. Casale, M. Fytikas, G. Sigvaldasson & G.E. Vougioukalakis (eds): *The European Laboratory Volcanoes, Proceedings of the 2nd Workshop,* 2–4 May 1996, Santorini, Greece, EUR 18161 EN, European Commission, Luxembourg, 1998, pp. 49–67.

Dazy, J., Drogue, C., Charmanidis, Ph. & Darlet, Ch.: The influence of marine inflows on the chemical composition of groundwater in small islands: the example of the Cyclades (Greece). *Environ. Geol.* 31:3–4 (1997), pp. 133–141.

Delimbasis, N.D. & Drakopoulos, J.C.: The Milos island earthquake of March 20, 1992 and its tectonic significance. *PAGEOPH* 141:1 (1993), pp. 44–58.

Dimitriadis, I.M., Panagiotopoulos, D.G., Papazachos, C.B. & Hatzidimitriou, P.M.: Recent seismic activity (1994–2002) of the Santorini volcano using data from a local seismological network. In: M. Fytikas & G.E. Vougioukalakis (eds): The South Aegean Active Volcanic Arc, present knowledge and future perspectives. *Developments in Volcanology,* volume 7, 2005, pp. 185–203.

Dimitriadis, I., Karagianni, E., Panagiotopoulos, D., Papazachos, C., Hatzidimitriou, P., Bohnhoff, M., Rische, M. & Meier, T.: Seismicity and active tectonics at Coloumbo reef (Aegean Sea, Greece): monitoring an active volcano at Santorini volcanic center using a temporary seismic network. *Tectonophysics* 465 (2009), pp. 136–149.

Di Paola, G.M.: Volcanology and petrology of Nisyros island (Dodecanese, Greece). *Bull. Volcanol.* 38 (1974), pp. 944–987.

Dominco, E. & Papastamataki, A.: Characteristics of Greek geothermal waters. *Proceedings of the 2nd U.N. Symposium on the Development and Use of Geothermal Resources,* San Francisco, CA, 1975, pp. 109–121.

Dotsika, E. & Michelot, J.L.: Hydrochemistry, isotope contents and origin of geothermal fluids at Nisyros (Dodecanese). *Bull. Geol. Soc. Greece* XXVIII:2, 1993, pp. 293–304.

Dotsika, E., Leontiadis, I., Poutoukis, D., Cioni, R. & Raco, B.: Fluid geochemistry of the Chios geothermal area, Chios island Greece. *J. Volcanol. Geoth. Res.* 154 (2006), pp. 237–250.

Dotsika, E., Poutoukis, D., Michelot, J.L. & Raco, B.: Natural tracers for identifying the origin of the thermal fluids emerging along the Aegean Volcanic Arc (Greece): evidence of arc-type magmatic water (ATMW) participation. *J. Volcanol. Geoth. Res.* 179 (2009), pp. 19–32.

Drakopoulos, J.C. & Delibasis, N.D.: Volcanic-type microearthquake activity in Milos, Greece. *Ann. Geofis.* 26 (1973), pp. 131–153.

Druitt, T.H.: *Explosive volcanism on Santorini, Greece.* PhD Thesis, University of Cambridge, UK, 1983.

Druitt, T.H. & Sparks, R.S.J.: A proximal ignimbrite breccia facies on Santorini, Greece. *J. Volcanol. Geoth. Res.* 13 (1982), pp. 147–171.

Druitt, T.H., Mellors, R.A., Pyle, D.M. & Sparks, R.S.J.: Explosive volcanism on Santorini, Greece. *Geol. Mag.* 126 (1989), pp. 95–126.

Ellis, A.J. & Mahon, W.A.: *Chemistry and geothermal systems*. Academic Press Inc., New York, 1977.

ENEL & PPC: Milos Geothermal Project, Report on borehole MILOS-1. GR/MI-9, May 1981, 1981a.

ENEL & PPC: Milos Geothermal Project. Report on borehole MILOS-2. GR/MI-11, July 1981, 1981b.

ENEL & PPC: Milos Geothermal Project. Report on borehole MILOS-3. GR/MI-14, January 1982, 1982.

Ferrara, G., Fytikas, M., Giuliani, O. & Marinelli, G.: Age of the formation of the Aegean Volcanic Arc. In: C. Doumas (ed): *Thera and the Aegean world II*. The Thera Foundation, London, 1978, pp. 37–41.

Fouqué, F.: *Santorin et ses éruptions*. Masson et cie, Paris, France, 1879.

Francalanci, L., Fytikas, M. & Vougioukalakis, G.E.: Volcanological and geochemical evolution of Kimolos and Polyegos centres, Milos island group, Greece. *Abstracts of International Association of Volcanology and Chemistry of the Earth's Interior (IAVCEI) Congress*, Ankara, Turkey, 1994.

Francalanci, L., Fytikas, M. & Vougioukalakis, G.E.: Kimolos and Polyegos volcanoes, South Aegean Arc, Greece: volcanological and magmatological evolution based on stratigraphic and geochemical data. *International Conference, The South Aegean Active Volcanic Arc: Present Knowledge and Future perspectives*, 17–20 September 2003, Milos, Greece, Abstracts, 2003, pp. 25–26.

Francalanci L., Vougioukalakis, G.E., Perini G. & Manetti, P.: A west-east traverse along the magmatism of the South Aegean Volcanic Arc in the light of volcanological, chemical and isotope data. In: M. Fytikas & G.E. Vougioukalakis (eds): The South Aegean Active Volcanic Arc, present knowledge and future perspectives. *Developments in Volcanology*, Elsevier, Amsterdam, 2005.

Francalanci, L., Vougioukalakis, G.E. & Fytikas, M.: Petrology and volcanology of Kimolos and Polyegos volcanoes within the context of the South Aegean Arc, Greece. *Geol. Soc. Am., Spec. Paper* 418, 2008, pp. 33–65.

Fytikas M.: *Geological and geothermal investigation research in the island of Milos*. PhD Thesis, Aristotle University of Thessaloniki, Thessaloniki, Greece, 1977 (in Greek).

Fytikas, M.: Geothermal exploration in Greece. In: A. Strub & P. Ungemach (eds): *Proceedings of the 2nd International Seminar on Results of EC Geothermal Research*, 4–6 March 1980, Strasbourg, France. D. Reidel Publ. Co., Dordrecht, The Netherlands, 1980, pp. 213–237.

Fytikas, M.: Geothermal situation in Greece. *Geothermics* 17:2–3 (1988), pp. 549–556.

Fytikas, M.: Updating of the geological and geothermal research on Milos island. *Geothermics* 14 (1989), pp. 485–496.

Fytikas, M.: Geological-geothermal survey of the region of Prassa Kimolos island — Selection of the geothermal productive borehole drilling sites. Report for CRES, THERMIE GE.438.94.HE, 1995 (in Greek).

Fytikas M. & Andritsos, N.: *Geothermal energy*. Ed. Tziolas, Thessaloniki, Greece, 2004 (in Greek).

Fytikas, M. & Marinelli, G.: Geology and geothermics of the island of Milos (Greece). *Proceedings of the International Congress on Thermal Waters, Geothermal Energy and Volcanism of the Mediterranean Area*, IGME, 1976.

Fytikas, M. & Vougioukalakis, G.E.: Volcanic structure and evolution of Kimolos and Polyegos (Milos island group). *Greek Geol. Soc. Bull.* XXVIII:2 (1993), pp. 221–237 (in Greek with English summary).

Fytikas, M., Kouris, D., Marinelli, G. & Surcin, J.: Preliminary geological data from the first two productive geothermal wells drilled at the island of Milos (Greece). *Proceedings of the International Congress on Thermal Waters, Geothermal Energy and Volcanism of the Mediterranean Area*, 1, 1976, pp. 511–515.

Fytikas, M., Innocenti, F., Manetti, P., Mazuoli, R., Peccerilo A. & Vilari, L.: Tertiary to Quaternary evolution of the volcanism in the Aegean region. In: J.E. Dixon & A.H.F. Robertson (eds): The geological evolution of the eastern Mediterranean. *Geol. Soc. London, Spec. Publ.* 17, 1984, pp. 687–699.

Fytikas, M., Innocenti, F., Kolios, N., Manetti, P., Mazzuoli, R., Poli, C., Rita, F. & Villari, L.: Volcanology and petrology of volcanic products from the island of Milos and neighboring islets. *J. Volcanol. Geoth. Res.* 28 (1986), pp. 297–317.

Fytikas, M., Garnish, J.D., Hutton, V.R.S., Stratoste, E. & Wohlenberg, J.: An integrated model for the geothermal field of Milos. *Geothermics* 18:4 (1989), pp. 611–621.

Fytikas, M., Karydakis, G., Kavouridis, Th., Kolios, N. & Vougioukalakis, G.: Geothermal research on Santorini. In: D.A. Hardy, J. Keller, V.P. Galanopoulos, N.C. Fleming & T.H. Druitt (eds): *Thera and the Aegean world III*. The Thera Foundation, London, 2, 1990a, pp. 241–249.

Fytikas, M., Kolios, N. & Vougioukalakis, G.E.: Post-Minoan volcanic activity of the Santorini volcano. Volcanic hazard and risk. Forecasting possibilities. In: D.A. Hardy, J. Keller, V.P. Galanopoulos, N.C. Fleming & T.H. Druitt (eds): *Thera and the Aegean world III*. The Thera Foundation, London, 2, 1990b, pp. 183–198.

Fytikas, M., Radoglou, G., Karytsas, C., Mendrinos, D., Vasalakis, A. & Andritsos, N.: Geothermal research in Vounalia area, Milos island (Greece), for seawater desalination and power production. *Proceedings of World Geothermal Congress 2005*, 24–29 April 2005, Antalya, Turkey, 2005.

Georgalas, G.: L' éruption du volcan de Santorin en 1950. *Bull. Volcanol.* 13 (1953), pp. 39–55.

Georgalas, G. & Papastamatiou, J.: L' eruption du volcan du Santorin en 1939–41. L'éruption du dome Fouqué. *Bull. Volcanol.* 13 (1953), pp. 3–18.

Geotermica Italiana: Nisyros 1 geothermal borehole, PPC-EEC report, 1983. Geotermica Italiana: Nisyros 2 geothermal borehole, PPC-EEC report, 1984.

Heiken, G. & McCoy, F.: Precursory activity to the Minoan eruption, Thera, Greece. In: D.A. Hardy, J. Keller, V.P. Galanopoulos, N.C. Fleming & T.H. Druitt (eds): *Thera and the Aegean world III*. The Thera Foundation, London, 2, 1990, pp. 79–88.

Hem, J.D.: Study and interpretation chemical characteristics of natural water. US Geol. Survey, Water Supply, 1985, Paper No 2254.

Hussein, M.T.: Hydrochemical evaluation of groundwater in the Blue Nile basin, eastern Sudan, using conventional and multivariate techniques. *Hydrogeology J.* 12 (2004), pp. 144–158.

IGME: Geological Map of Milos island, 1:25.000, 1977.

Innocenti, F., Manetti, P., Peccerilo, A. & Poli, G.: South Aegean Volcanic Arc: geochemical variations and geotectonic implications. *Bull. Volcan.* 44:3 (1981), pp. 371–391.

Jarigge, J.: *Etudes néotectoniques dans l'arc volcanique égéen. Les iles de Kos, Santorini, Milos*. Thesis, Univ. Paris, XI, Paris, France, 1978.

Karakatsanis, S., D'Alessandro, W., Kyriakopoulos, K. & Voudouris, K.: Chemical characterization of the thermal springs along the South Aegean Volcanic Arc and Ikaria island. In: N. Lambrakis, G. Stournaras & K. Katsanou (eds): *Advances in the research of aquatic environment*. Springer, 2, 2011, pp. 239–247.

Karytsas, C., Alexandou, V. & Boukis, I.: The Kimolos geothermal desalination project. *Proceedings of International Geothermal Days "Greece 2002" — International Workshop on the possibilities of geothermal Development on the Aegean islands Region*, 5–8 September 2002, Milos, 2002.

Karytsas, C., Mendrinos, D. & Radoglou, G.: Geothermal field of Milos island in Greece: current geothermal exploration and development. CD-ROM Proceedings of *Start in eine neue Energiezukunft: 1 Fachkongress Geothermischer Strom*, 12–13 November 2003, Neustadt Glewe, Germany, 2003.

Kavouridis, T., Kuris, D., Leonis, C., Liberopoulou, V., Leontiadis, J., Panichi, C. & La Ruffa, G.: Isotope and chemical studies for a geothermal assessment of the island of Nisyros, Greece. *Geothermics* 28 (1999), pp. 219–239.

Kavouridis, T., Vrellis, G., Vougioukalakis, G. & Chatzis, M.: Exploration and evaluation of the geothermal potential in Chios island. IGME Report, Athens, Greece, 2000.

Kavouridis, T., La Rouffa, G. & Panichi, C.: Isotope and chemical studies for a geothermal assessment of Chios island, Greece. In: R. Cidu (ed): *Water-rock interactions 2001*. Swets & Zeitlinger, Lisse, The Netherlands, 2001, pp. 867–870.

Kavouridis, T., Vrellis, G., Vakalopoulos, P. & Xenakis, M.: Geothermal potential in SE Chios. CD-ROM *Proceedings of GES 2009*, 11–12 December 2009, Thessaloniki, Greece, 2009.

Keller, J.: Mediterranean island Arc. In: R.S. Thorpe (ed): *Andesites*. John Wiley, New York, 1982, pp. 307–325.

Keller, J., Rehren, T. & Stadlbauer, E.: Explosive volcanism in the Hellenic Arc: a summary and review. In: D.A. Hardy, J. Keller, V.P. Galanopoulos, N.C. Fleming & T.H. Druitt (eds): *Thera and the Aegean world III*. The Thera Foundation, London, 2, 1990, pp. 13–26.

Koutinas, G.: High salinity fluid handling in Milos geothermal field. *Geothermics* 18 (1980), pp. 175–182.

Koutroupis, N.: Update of geothermal energy development in Greece. *Geothermics* 21 (1992), pp. 881–890.

Ktenas, C.: L' eruption du volcan des Kammenis (Santorin) en 1925, II. *Bull. Volcanol.* 4 (1927), pp. 7–46.

Ktenas, C.A.: La géologie de l'ile de Nikaria (Rédigée des restes de l'auteur par G. Marinos). *Geological and Geophysical Researches*, Athens, 13, 1969, pp. 57–85.

Kumerics, C., Rong, U., Brichau, S., Glodny, J. & Monie, P.: The extensional Messaria zone and associated faults, Aegean Sea, Greece. *J. Geol. Soc. London* 162 (2005), pp. 701–721.

Kyriakopoulos, C.: Natural degassing of carbon dioxide and hydrogen sulphide ant its environmental impact at Milos island, Greece. *Proceedings of the 12th International Congress of the Geological Society of Greece*, Patras, Greece, 2010, Patras, 2010.

Kyriakopoulos, G.K.: K-Ar & Rb-Sr isotopic data of white micas from Milos island geothermal boreholes field. *Anal. Geol. Pays Hell.* 38 (1998), pp. 37–48.

Lagios, E., Galanopoulos, D., Hobbs, B.A. & Dawes. G.J.K.: Two dimensional magnetotelluric modelling of the Kos island geothermal region (Greece). *Tectonophysics* 287 (1998), pp. 157–172.

Lambrakis, N. & Kallergis, G.: Contribution to the study of Greek thermal springs: hydrogeological and hydrochemical characteristics and origin of thermal waters. *Hydrogeology J.* 13 (2005), pp. 506–521.

La Ruffa, G., Panichi, C., Kavouridis, T., Liberopoulou, V., Leontiadis, J. & Caprai, A.: Isotope and chemical assessment of geothermal potential of Kos island, Greece. *Geothermics* 28 (1999), pp. 205–217.

Le Pichon, X. & Angelier, J.: The Hellenic arc and trench system: a key to the neotectonic evolution of the eastern Mediterranean area. *Tectonophysics* 60 (1979), pp. 1–42.

Makris, J.: Geophysical investigations of the Hellenides. *Hamburger Geophysikalische Einzelschriften* 33 (1977), p. 128.

Marini, L., Principe, C., Chiodini, G., Cioni, R., Fytikas, M. & Marinelli, G.: Hydrothermal eruptions of Nisyros (Dodecanese, Greece). Past events and present hazards. *J. Volcanol. Geoth. Res.* 56 (1993), pp. 71–95.

McKenzie, D.: Active tectonics of the Mediterranean region. *Geophys. J.R. Astr. Soc.* 30 (1972), pp. 109–185.

Megalovasilis, P.: *Geochemical investigation of sediments of bottom and regions of Aegean Volcanic Arc.* PhD Thesis, University of Athens, Athens, Greece, 2005.

Mendrinos, D.: *Modeling of geothermal fields in Greece.* Master of Engineering Thesis, University of Auckland, New Zealand, 1988.

Mendrinos, D.: Calculation of the geothermal potential of Milos, sensitivity analysis. Renewable Energy Sources Seminar, C.R.E.S., Pikermi, Greece, 1991, p. 22 (in Greek).

Mendrinos, D., Choropanitis, I., Polyzou, O. & Karytsas, C.: Exploring for geothermal resources in Greece. *Geothermics* 39 (2010), pp. 124–137.

Minissale, A., Duchi, V., Kolios, N., Nocenti, M. & Verucchi, C.: Chemical patterns of thermal aquifers in the volcanic islands of the Aegean Arc, Greece. *Geothermics* 26:4 (1997), pp. 501–518.

Mitropoulos, P. & Tarney, J.: Significance of mineral composition variations in the Aegean islands arc. *J. Volcanol. Geoth. Res.* 15 (1992), pp. 283–303.

Mountrakis, D.: *Geology and geotectonic evolution of Greece.* University Studio Press, Thessaloniki, Greece, 2010

Nicholls, I.A.: Santorini volcano, Greece: tectonic and petrochemical relationships with volcanics of the Aegean region. *Tectonophysics* 11:5 (1971), pp. 377–385.

Nicholson, K.: *Geothermal fluids-chemistry and exploration techniques.* Springer Verlag, Berlin, Gemany, 1993. Ochmann, N., Hollnack, D. & Wohlenberg, J.: Seismological exploration of the Milos geothermal reservoir, Greece. *Geothermics* 18:4 (1989), pp. 563–577.

Papadopoulos, G.A.: Seismic properties in the eastern part of the South Aegean Volcanic Arc. *Bull. Volcanol.* 47 (1984), pp. 143–152.

Papadopoulos, G.A.: Deterministic and stochastic models of the seismic and volcanic events in the Santorini volcano. In: D.A. Hardy, J. Keller, V.P. Galanopoulos, N.C. Fleming & T.H. Druitt (eds): *Thera and the Aegean world III.* The Thera Foundation, London, 2, 1990, pp. 151–159.

Papadopoulos, G.A.: The 20th March 1992 South Aegean, Greece, earthquake (Ms=5.3): possible anomalous effects. *Terra Nova* 5:4 (1992), pp. 399–404.

Papadopoulos, G.A., Sachpazi, M., Panopoulou, G. & Stavrakakis, G.: The volcanoseismic crisis of 1996–97 in Nisyros, SE Aegean Sea, Greece. *Terra Nova* 10 (1998), pp. 151–154.

Papanikolaou, D.: Contribution to the geology of Ikaria island, Aegean Sea. *Annales Géologiques de Pays Helléniques* 29 (1978), pp. 1–28.

Papanikolaou, D. & Lekkas, E.: Miocene tectonism in Kos, Dodecanese islands. *IESCA 1990*, Abstract, 1990, pp. 179–180.

Papanikolaou, D., Lekkas, E. & Syskakis, D.: Structural analysis of Milos island geothermal field. *Bull. Geol. Soc. Greek* XXIV, Athens, Greece (1990), pp. 27–46.

Papanikolaou, D., Sakellariou, D. & Leventis, A.: Microstructural observations on the granite of Ikaria island, Aegean Sea. *Bull. Geol. Soc. Greek* XXV (1991), pp. 421–437.

Papastamatiou, J.: Sur l'âge des calcaires cristallines del' ile de Thera (Santorin). *Bull. Geol. Soc. Greece* 3 (1958), pp. 104–113.

Papazachos, B.C. & Comninakis, P.E.: Geophysical features of the Greek island arc and the Eastern Mediterranean Ridge. *Proceedings of C. R. Seances de la Conference Reunie a Madrid*, 16, 1969, pp. 74–75.

Papazachos, B.C. & Delibasis, N.D.: Tectonic stress field and seismic faulting in the area of Greece. *Tectonophysics* 7 (1969), pp. 231–255.

Papazachos, B.C. & Comninakis, P.E.: Geophysical and tectonic features of the Aegean Arc. *J. Geophys. Res.* 76 (1971), pp. 8517–8533.

Papazachos, C.B.: The active crustal deformation field of the Aegean area inferred from seismicity and GPS data. *CD-Proceedings of the 11th General Assembly of the WEGENER Project*, 12–14 June 2002, Athens, Greece, 2002.

Pe-Piper, G. & Photiades, A.: Geochemical characteristics of the Cretaceous ophiolitic rocks of Ikaria island, Greece. *Geol. Mag.* 143:4 (2006), pp. 417–429.

Perissoratis, C.: The Santorini volcanic complex and its relation to the stratigraphy and structure of the Aegean Arc, Greece. *Marine Geol.* 128 (1995), pp. 37–58.

Photiades, A.: The ophiolitic mollase unit of Ikaria island (Greece). *Turkish J. Earth Sci.* 11 (2002), pp. 27–38.

Photiades, A.: Geological map of Ikaria in scale 1:50.000, IGME, Athens, Greece, 2004.

Reck, H.: *Santorini — Der Werdergang eines Inselvulkans und sein Ausbruch 1925–1928*. Dietrich Reimer, Berlin, Germany, 3 volumes, 1936.

Ring, U.: The Geology of Ikaria island: the Messaria extensional shear zone, granites and the exotic Ikaria nappe. In: G.Lister, M. Forster & U. Ring (eds): Inside the Aegean metamorphic core complexes. *J. Virtual Explorer*, electronic edition 27, 2007, paper 3.

Ring, U., Glodny, J., Will, T. & Thomson, S.N.: The Hellenic subduction system: high-pressure metamorphism, exhumation, normal faulting, and large-scale extension. *Annu. Rev. Earth Planet. Sci.* 38 (2010), pp. 45–76.

Simeakis, K.: Neotectonic evolution of Milos island. IGME report, Athens, Greece, 1985 (in Greek).

Skarpelis, N. & Liati, A.: The prevolcanic basement of Thera at Athinios: metamorphism, plutonism and mineralization. In: D.A. Hardy, J. Keller, V.P. Galanopoulos, N.C. Fleming & T.H. Druitt (eds): *Thera and the Aegean world III*. The Thera Foundation, London, 2, 1990, pp. 172–182.

Stamatis, G., Voudouris, K. & Karefilakis, Th.: Groundwater pollution by heavy metals in historical and mineral area of Lavrio (Attica). *Water Air Soil Poll.* 128 (2001), pp. 61–83.

Stiros, S., Psimoulis, P., Vougioukalakis, G. & Fytikas, M.: Geodetic evidence and modeling of a slow, small-scale inflation episode in the Thera (Santorini) volcano caldera, Aegean Sea. *Tectonophysics* 494 (2010), pp. 180–190.

Tataris, A.: The iron and manganese in the sands of Thera coasts. *Bull. Geol. Soc. Greece* 6:1 (1964), pp. 65–83.

Thanassoulas, C.: Technical report on the execution of gravimetric measurements in Milos island. IGME report 3868, Athens, Greece, 1983 (in Greek).

Tibaldi, A., Pasquarè, F.A., Papanikolaou, D. & Nomikou, P.: Tectonics of the Nisyros island, Greece, by field and offshore data and analogue modeling. *J. Struct. Geol.* 30 (2008), pp. 1489–1506.

Traineau, H. & Dalabakis, P.: Mise en evidence d'une éruption phreatique historique sur l'ile de Milos (Grece). *C.R. Acad. Sci. Paris*, 308, Serie II, 1989, pp. 247–252.

Tsokas, G.: *Geophysical sounding on Milos and Kimolos islands*. PhD Thesis, Aristotle University of Thessaloniki, Thessaloniki, Greece, 1985 (in Greek).

Tsokas, G.N., Hansen, R.O., Fytikas, M., Vassilellis, G.D. & Thanassoulas, C.: Geological and geophysical study of the island of Kimolos (Greece) and geothermal implications. *Geothermics* 24:5 (1995), pp. 679–693.

Ungemach, P.: Energy development problematic in the Mediterranean. The Aeolian & Aegean islands. The geothermal energy case. *Proceedings International Workshop on the Possibilities of Geothermal Development of the Aegean islands Region*, 5–8 September 2002, Milos, Greece, 2002.

Valsami-Jones, E., Baltatzis, E., Bailey, E.H., Boyce, A.J., Alexander, J.L., Magganas, A., Anderson, L., Waldron, S. & Ragnarsdottir, K.V.: The geochemistry of fluids from an active shallow submarine hydrothermal system: Milos island, Hellenic Volcanic Arc. *J. Volcanol. Geoth. Res.* 148 (2005), pp. 130–151.

Varekamp, J.C.: Some remarks on volcanic vent evolution during plinian eruptions. *J. Volcanol. Geoth. Res.* 54 (1992), pp. 309–318.

Varnavas, S. & Cronan, D.S.: Arsenic, antimony and bismuth in sediments and waters from the Santorini hydrothermal field, Greece. *Chem. Geol.* 67 (1988), pp. 295–305.

Voudouris, K., Lambrakis, N., Papatheodorou, G. & Daskalaki, P.: An application of factor analysis for the study of the hydrogeological conditions in Plio-Pleistocene aquifers of NW Achaia (NW Peloponnesus, Greece). *Math. Geol.* 29:1 (1997), pp. 43–59.

Voudouris, K., Panagopoulos, A. & Koumantakis, J.: Multivariate statistical analysis in the assessment of hydrochemistry of the northern Korinthia Prefecture alluvial aquifer system (Peloponnesus, Greece). *Nat. Resour. Res.* 9:2 (2000), pp. 135–146.

Vougioukalakis, G.E.: Santorini guide to volcano. Institute for the Study and Monitoring of the Santorini Volcano, Thera, Greece, 2007, 82 p.

Vougioukalakis, G., Sachpazi, M., Perissoratis, C. & Lyberopoulou, Th.: The 1995–1997 seismic crisis and ground deformation on Nisyros volcano, Greece: a volcanic unrest? *Abstracts of the Colima Volcano: 6th International Meeting*, 26–31 January 1998, Colima, Mexico, 1998.

Vougioukalakis, G.E. & Fytikas, M.: Volcanic hazards in the Aegean area, relative risk, evaluation, monitoring and present state of the active volcanic centers. In: M. Fytikas & G.E. Vougioukalakis (eds): The South Aegean Active Volcanic Arc, present knowledge and future perspectives. *Developments in Volcanology*, Elsevier, Amsterdam, The Netherlands, 2005, pp. 161–163.

Vrouzi, F.: Research and development of geothermal resources in Greece: present statues and future prospects. *Geothermics* 14 (1985), pp. 213–227.

Washington, H.S.: Santorini eruption of 1925. *Bull. Geol. Soc. Am.* 37 (1926), pp. 349–384.

Watkins, N.D., Sparks, R.S.J., Sigurdsson, H., Huand, T.C., Federman, A., Carey S. & Ninkovich, D.: Volume and extent of the Minoan tephra from Santorini volcano: new evidence from deep-sea cores. *Nature* 271 (1978), pp. 122–126.

Zouzias, D., Miliaresis, G.C. & Seymour, K.St.: Interpretation of Nisyros volcanic terrain using land surface parameters generated from the ASTER Global Digital Elevation Model. *J. Volcanol. Geoth. Res.* 200 (2011), pp. 159–170.

CHAPTER 5

Application of hydrogeochemical techniques in geothermal systems; examples from the eastern Mediterranean region

Ayşen Davraz

5.1 INTRODUCTION

Geothermal systems are found throughout the world in a range of geological settings and are increasingly being developed as an energy source. Hot springs, geysers, mud pots, and fumaroles are dynamic surface features that represent interacting subterranean system of water, heat, and rocks. Three geological components are required for the formation of any geothermal water: water, heat, and reservoir rock. Each of the different types of geothermal system has distinct characteristics that are reflected in the chemistry of the geothermal fluids and their potential applications. Meteoric water that has gained depths of several kilometers through fractures and permeable horizons forms most of hydrothermal fluids. The source of heat is either magma, in the case of volcano-related geothermal systems, or high geothermal gradient due to decay of radioactive elements within the depths of the earth. Fractures in rocks often create permeability; however, in some systems interconnected pores or large cave systems allow fluids to flow (Heasler et al., 2009).

Geothermal systems are commonly classified by a series of descriptive terms. They are referred to as liquid or vapor dominated, low or high temperature, sedimentary or volcanic hosted, etc. When a geothermal region is identified, the next step is to use different available exploration techniques to localize the potential geothermal resource areas and identify suitable drilling targets for extraction of geothermal fluids or steam. To obtain all this information, it is necessary to employ a suite of exploration techniques. Important among these are: geological and hydrological, geochemical, geophysical, remote sensing techniques, and exploratory drilling. Because the geological settings of different geothermal resource areas vary widely, the exploration tools used differ as well does the order of investigation (Chandrasekharam and Bundschuh, 2008).

Hydrothermal systems can change through time because of local or distant events that alter the water source or flow path, the heat source, the thermal characteristics along the subsurface flow path, and/or the fractured rock that the water is flowing through. Major tectonic activities like earthquakes, volcanic eruptions can have a large influence on the physical and chemical processes and characteristics of geothermal systems. Such activities can change the circulation paths of the hydrothermal system, and can change the chemical composition of the fluids and gases. Hydrogeochemical methods are extensively used and play a major role in geothermal exploration and exploitation. Hydrothermal flow systems are complex (Henley and Ellis, 1983) and often water chemistry analysis is the best way to assess subsurface flow systems. Kharaka et al. (1991), summarizing their work and the work of Hem (1985) and Ellis and Mahon (1977), noted that chemical data, among other uses, could help to determine:

- original chemical signature of the thermal fluids;
- chemical evolution of the thermal fluids and circulation pattern of the geothermal system;
- reservoir temperature,;
- subsurface mixing of different fluids (e.g., deep hydrothermal and local cold water);
- water-rock interactions;
- hydraulic connections among widely spaced thermal sites;

- possible contributions of magmatic water to hydrothermal systems;
- the age of hydrothermal systems, through dating waters or hydrothermal deposits.

The goal of this chapter is to describe the hydrogeochemical techniques, which are generally used in the active geothermal system, and case studies that focus on the eastern Mediterranean provinces.

5.2   HYDROGEOCHEMICAL EVALUATION OF GEOTHERMAL FLUIDS

Geothermal fluids coming from the deep reservoir have diverse chemical compositions depending on the specific reservoir conditions. The waters circulating in high-enthalpy geothermal areas contain variable solute concentrations, which depend on temperature, gas content, heat source, rock type, permeability, age of the hydrothermal system, and fluid source (Barbier, 2002). Cations (e.g., sodium, potassium, lithium, calcium, magnesium, rubidium, cesium, manganese, and iron), anions (e.g., sulfate, chloride, bicarbonate, fluoride, bromide, and iodide), and neutral species (e.g., silica, ammonia, boron, and noble gases) are the most common species found in the waters. Accordingly, four types of waters have been proposed: sodium chloride water, acid-sulfate/chloride water, acid-sulfate water, and calcium bicarbonate water (Ellis and Mahon, 1977; Henley et al., 1984; Giggenbach, 1988). However, it must be emphasized that each of these waters may mix with each other giving rise to hybrid water types.

Sodium chloride water: The most common type of fluid found at depth in large water-dominated geothermal systems is near-neutral pH, with chloride as the dominant anion. It contains 1000–10,000 mg/kg of chloride and mainly carbon dioxide gas. When this type of water approaches the surface, it loses steam and carbon dioxide and consequently becomes slightly alkaline. It is believed that these waters are formed from the absorption of magmatic volatiles such as HCl, CO_2, SO_2, and H_2S into deeply circulating meteoric water (Ellis and Mahon, 1977).

Acid-sulfate/chloride water: These are superficial fluids formed by condensation of geothermal gases into near-surface, oxygenated groundwater. Oxidation of sulfide to bisulfate at deeper levels causes its acidity. As the waters rise from depth, they cool down and become acidic (Ellis and Mahon, 1977). They are found on the margins of a field at some distance from a major upflow area, in perched water tables and over boiling zones. Although usually found near the surface ($\sim$100 m), sulfate waters can penetrate to depth through faults into geothermal system. Here they are heated, take part in rock alteration reactions, and mix with the ascending chloride fluids (Nicholson, 1993).

Acid-sulfate water: Typically found in fumarolic areas above the upflow part of the geothermal systems, where steam rising from hot-water reservoirs condenses as it approaches the surface. The separated vapor may condense in shallow groundwaters or surface waters to form steam-heated waters. In this environment, atmospheric oxygen oxidizes H_2S to sulfuric acid producing acid-sulfate waters. These are characterized by low chloride contents and low pH values (0–3) and react quickly with host rocks to produce advanced argillic alteration, with kaolinite, halloysite, crystobalite, and alunite. Extensive leaching of surface lithologies by these waters, or acidic steam, can produce silica residue (Nicholson, 1993).

Bicarbonate water: Originates through either dissolution of CO_2-bearing gases or condensation of geothermal steam in relatively deep, oxygen-poor groundwaters. Because the absence of oxygen prevents oxidation of H_2S, the acidity of these aqueous solutions is due to dissociation of H_2CO_3 (Ellis and Mahon, 1977). Such fluids can occur in an umbrella-shaped perched condensate zone overlying the geothermal system, and can also result due to mixing of ascending thermal water with near-surface groundwater. Bicarbonate waters found in non-volcanogenic, high-temperature systems (e.g., Turkey) are of more problematic origin and may constitute the deep reservoir fluid (Nicholson, 1993).

5.2.1 *Collection of samples*

A hydrogeochemical investigation of geothermal fluids essentially involves four main steps. These are:

1. Collection of samples
2. *In situ* measurements
3. Chemical analyses
4. Data interpretation

Collection of samples for chemical analysis is the first step for evaluation of geothermal fluids. It is imperative that this step is properly carried out because all subsequent steps depend on it. This procedure must be implemented by well-trained personnel with insight into possible errors. The most common error made during sampling involves the use of improper containers, improper cleaning, and lack of or improper treatment for the preservation of samples. If the sampling is incorrect, chemical data interpretation becomes meaningless. The strategies of sampling and analysis of the natural waters circulating in the geothermal area to be investigated (i.e., the number of samples to be collected and the analytical routine procedure chiefly depend on the scale and state of advancement of each specific project, the available funds, and logistic constraints. In general, it is not advisable (i) to analyze a large number of chemical and isotopic parameters in a small number of samples or (ii) to determine a small number of chemical parameters in a huge number of samples. The best strategy is to collect samples from a reasonable number of thermal and non-thermal waters, distributed all over the investigated area. Sample locations should be selected according to conceptual hydrogeological model of the geothermal area and places of thermal springs. The hydrogeochemical survey carried out two seasons (before and after the rains) so that mixing phenomena can better be understood.

During collection, the sample has to be cooled. If pH, total carbonate, and hydrogen sulfide are going to be analyzed in a laboratory, the sample has to be cooled and stored in airtight bottles. Commonly samples for determination of volatile components are kept in plastic bottles without previous cooling. The bottles are then taken to laboratories and in spite of accurate analyses of the water in the bottle, the result does not represent the spring or the well. During storage, the volatiles degas and the recorded values cannot be used in a detailed evaluation of the sample. The samples collected for determination of cations must be filtrated and acidified. Acidification is needed to preserve cation contents of high-temperature waters, which become supersaturated upon cooling, and to prevent precipitation of trace metals from both high-temperature and low-temperature waters. Acidification is usually done by the addition of either HCl (e.g., 1 mL HCl 1:1 [$\sim$6 N] to 50 mL of sample) or HNO_3 (e.g., 0.5 ml HNO_3 1:3 [$\sim$4 N] to 50 mL of sample). In addition, samples for determination of anions can be collected without any treatment.

5.2.2 *In situ measurements*

Physical parameters must be measured in the field, but it may also be necessary to determine certain key physical and chemical components in the field as they may change on transportation of samples to the laboratory. When the sampling point has been selected, it should be described and *in situ* measurements carried out. This includes measurement of temperature, pH, electrical conductivity (EC), redox potential (Eh), flow rate, and geographical coordinates for the sampling site. Some determinations are best done in the field, such as dissolved oxygen (O_2), hydrogen sulfide (H_2S), and total carbonate analysis. Electrical conductivity is a measure of the ionized chemicals in water. Generally, high amounts of dissolved minerals and salt concentration in water yield high electrical conductivities. Measurements of electrical conductivity can provide estimates of the hydrothermal fluid's purity and contribute to understanding its flow path.

A measure of the hydrothermal fluid's acidity or alkalinity is its pH; a pH of 7 is neutral, less than 7 is acidic, and above 7 is basic. Changes in the pH may indicate changes in the flow path of the geothermal fluid, changes in the temperature of the geothermal fluid, loss of CO_2, or anthropogenic activities. It should be kept in mind that accurate determination of the pH is essential for all further calculations since it participates in almost all chemical reactions. Calcite scaling prediction is based on the measured pH. An inaccurate pH value may therefore lead to a wrong prediction.

5.2.3   Chemical analyses

Constituents dissolved in hydrothermal waters can include common anions, cations, trace elements, organic compounds, isotopes, and radionuclides. The quality of water analysis is usually checked computing the ionic balance; however, possible errors for minor constituents (e.g., Li, F, but also Mg and SO_4 in high-temperature geothermal liquids) or neutral species (e.g., SiO_2, H_3BO_3, NH_3) cannot be detected in this way. At best, ionic balance gives an indication on the analytical accuracy of major constituents. The anion-cation balance check is based on a percentage difference between the total positive charge and the total negative charge, defined as follows:

$$Charge\text{-}balance\ error\ (difference - \%) = 100(\Sigma cations - \Sigma anions)/(\Sigma cations + \Sigma anions)$$

where contributions to charge are in units of meq/L. A charge-balance error of the waters of less than 5% is within the limits of acceptability.

5.2.4   Data interpretation

Water composition is controlled by the rock composition with which the thermal water is in contact within the basin, and hydrologic characteristics such as the permeability of the formation and residence time of the circulating fluids below the surface. Determination of hydrogeochemical facies of geothermal waters and of chemical changes resulting from interaction with rocks is performed through the interpretation of the chemical data. For this purpose, different classification methods are used. Chemical classification of waters is essential for a correct utilization of geochemical techniques, which can be confidently applied only to particular kinds of fluids with limited ranges of composition, reflecting the provenance environment. Various diagrams have been developed for visualization of water chemistry. The most commonly used diagrams are summarized briefly below.

   One of the most popular diagrams in geothermal investigation is the use of ternary or binary variation diagrams such as the Piper (Piper, 1944) or Langelier and Lundwing diagrams, diagrams that discriminate different groups of waters based on relative concentration of major ions. Piper diagram is based on the relative amounts of Na+K, Mg, Ca, Cl, SO_4, and HCO_3+CO_3 in a fluid. These components are the major ions in thermal and non-thermal waters, and classifications based on them agree well with observations on the formation of various geothermal water types. In the Piper diagram, major ions are plotted in the two base triangles as cation and anion milliequivalent percentages. Total cations and total anions are each considered as 100%. The respective cation and anion locations for an analysis are projected into the diamond field, which represents the total ion relationship. The Piper diagram has been widely used to study the similarities and differences in the composition of waters and to classify waters into "hydrogeochemical facies." Water plotting in the upper half of both, the cation and anion, triangles would be referred to as magnesium sulfate-type water. Water plotting in the lower left-hand side of the cation triangle and the lower right-hand side of the anion triangle would be calcium chloride-type water. If both cation and anion compositions plot in the middle of the two triangles, then the waters would be referred to as mixed cation-mixed anion types. If a water plots near the middle of one of the edges of the triangles, then one might refer to, for example, magnesium-calcium sulfate water.

   Another diagram is the Langelier-Ludwig square diagram (1942). This diagram is a sort of square version of the more popular diamond-shaped Piper plot. The position of a sample in the Langelier-Ludwig plot is obtained by first calculating the sum of main anions $(Cl + SO_4 + HCO_3)$ and of main cations $(Ca + Mg + Na + K)$ that refer to the concentration of the component in meq%. Then the percentage of each cation and each anion is determined and plotted on the diagram. The advantages of this diagram are that mixing lines are straight lines, all available samples can be plotted; grouping and trends can be evaluated, the vertices display the composition of salts present in nature, e.g., calcite, anhydrite, and halite Dissolution or precipitation of these salts is suggested by trends moving toward or away from the pertinent vertex (Marini, 2000).

The Durov diagram is an alternative to the Piper diagram. The Durov diagram plots the major ions as percentages of milliequivalents in two base triangles. The total cations and the total anions are set equal to 100% and the data points in the two triangles are projected onto a square grid which lies perpendicular to the third axis in each triangle. The intersection of lines extended from the two sample points on the triangle to the central rectangle gives a point that represents the major-ion compositions on a percentage basis.

From this point, lines extending to the adjacent scaled rectangles provide for representations of the analyses in terms of two parameters selected from various possibilities, such as total major-ion concentrations, total dissolved solids, ionic strength, specific conductance, hardness, total dissolved inorganic carbon (DIC), or pH. This plot reveals useful properties and relationships for large sample groups. The main purpose of the Durov diagram is to show clustering of data points to indicate samples that have similar compositions.

The Chadha diagram is a somewhat modified version of the Piper diagram (Piper, 1944) and the expanded Durov diagram (Durov, 1948). In this diagram, the difference in milliequivalent percentage between alkaline earth (calcium plus magnesium) and alkali metals (sodium plus potassium), expressed as percentage values, is plotted on the X axis, and the difference in milliequivalent percentage between weak acidic anions (carbonate plus bicarbonate) and strong anions (chloride plus sulfate) is plotted on the Y axis (Chadha, 1999).

The Schoeller diagram is a semilogarithmic diagram, which is developed to represent major-ion analyses (SO_4, HCO_3, Cl, Mg, Ca, Na/K) in meq/L and to demonstrate different hydrochemical water types on the same diagram. This type of graphical representation is advantageous in that, unlike the trilinear diagrams, actual sample concentrations are displayed and compared. The diagram gives absolute concentration, but the line also gives the ratio between two ions in the same sample. If a line joining two points representing ionic concentrations in a single sample is parallel to another line joining a second set of concentrations from another sample, the ratio of those ions in those samples are equal.

The Cl–SO_4–HCO_3 diagram is used to classify geothermal fluids based on the major anion concentrations (Cl, SO_4, and HCO_3) (Giggenbach, 1988; Giggenbach and Goguel, 1989). Using this diagram, several types of thermal water can be distinguished: mature waters, peripheral waters, and volcanic and steam-heated waters. The diagram may provide an initial indication of a mixing relationship. The position of a data point in this plot is obtained by first calculating the sum Σ_{an} of the concentrations C_i (in mg/kg) of all three species involved:

$$\Sigma_{an} = C_{Cl} + C_{SO_4} + C_{HCO_3}$$

Then the percentages of chloride, %Cl, and bicarbonate, %HCO_3, are evaluated according to:

$$\%Cl = 100C_{Cl}/\Sigma_{an}$$

$$\%HCO_3 = 100C_{HCO_3}/\Sigma_{an}$$

In this plot are indicated the compositional ranges for the different kinds of waters typically found in geothermal areas, such as:

- mature NaCl waters of neutral pH, which are rich in Cl and plot near the Cl vertex; Na–HCO_3 waters, here indicated as peripheral waters
- volcanic and steam-heated waters, generated through absorption into groundwater of
- either high-temperature, HCl-bearing volcanic gases or lower temperature H_2S-bearing geothermal vapors.

The advantages of this diagram are:

- the three main anion are plotted separately on the three vertices of the plot;
- mixing lines are straight lines; all available samples can be plotted;
- groupings and trends can be evaluated.

Its limitations are:

- relative ratios between Cl, SO_4, and HCO_3 are displayed;
- the content of each species relative to water is obliterated in this plot;
- apparent correlations may be accidental;
- correlations have to be checked by means of additional independent data.

5.2.4.1   Case studies in the eastern Mediterranean region

The distribution of geothermal fields in Turkey closely follows the tectonic patterns. The eastern Mediterranean region is tectonically active due to the Isparta Angle. The Isparta Angle is a major geological lineament in the eastern Mediterranean area and is one of the most important geographical structures in SW Anatolia, Turkey, developed by a bending of the Taurides in the northern part of Antalya bay (Poisson et al., 2003; Yagmurlu et al., 1997). In this section, hydrogeochemical analyses results of thermal and mineral water springs in the Isparta, Burdur, Fethiye, Sandıklı, and Usak regions (Fig. 5.1) which are located in eastern Mediterranean region were interpreted using above mentioned diagrams.

Isparta region: The Arslandogmus thermal waters are discharged from the two locations (Arslandogmus and Ilidere, Fig. 5.2) with 22°C in the Sarkikaraagac district of Isparta province (Davraz et al., 2006; Örmeci, 2006). The Arslandogmus springs discharge 10 L/s from a fault zone located in between the Caltepe limestone consist of dolomitic limestone and the Sultandede formation which is composed of schist, quartzite, and sandstone (Davraz et al., 2006). In addition, the Icmeler mineral water is also located in the Sarkikaraagac district. It discharges from a contact between the Sultandede formation and the Bagkonak formation consists of sandstone, claystone, conglomerate, and mudstone. The discharging spring water has a temperature of 16°C (Davraz et al., 2006).

The Degirmendere mineral water is a typical example of acid-sulfate waters in the Keciborlu district (Isparta province). It is discharged from a natural pool with 8°C (Fig. 5.3). The total dissolved solids (TDS) of the Degirmendere mineral water is 5440 mg/L (Davraz, 2003). Hydrothermal alteration products are widely observed in surrounding of the Degirmendere springs. In addition, the region is the location of a former sulfur mine and volcanic gases (H_2S) are observed in this area. The Degirmendere mineral water is enriched in SO_4. Thermal waters emerging through the faults do not represent the geochemical signature of the geothermal reservoir but represent the rock with which they interacted during circulation. Chemical analyses of original serpentinite, tuff, and limestone rocks and alteration products of these rocks from the area surrounding the Degirmendere spring had been undertaken in previous investigations (Güneş, 1993; Sarıiz, 1985).

Analysis shows that SiO_2, Fe_2O_3, and Al_2O_3 contents of altered serpentinite increased and MgO decreased in reference to original serpentinite. Al_2O_3 and S contents of altered tuff rock increased at significant ratios and SiO_2 ratio decreased compared to the original tuff rock sample from this region. CaO and MgO ratios of altered limestone significantly decreased and SiO_2, Fe_2O_3, and Al_2O_3 contents increased compared to the original limestone sample (Güneş, 1993; Sarıiz, 1985). Fe, Al, Ca, Mg, Si, and Na contents of the Degirmendere spring are increased due to water-rock interaction between meteoric water and altered rocks. The major cations of the Degirmendere spring are $Ca^{2+} > Mg^{2+} > Na^+ > K^+$. Analysis of major anions SO_4 and Cl resulted in concentrations of 12,000 and 0.017 mg/L, respectively; chloride occurs in traces as is typical for acid-sulfate waters (Nicholson, 1993). The spring water does not include the HCO_3 anions as bicarbonate is typically either absent or at low concentrations, since in very acid waters the dissolved carbonate is usually lost from solution as carbon dioxide gas (Nicholson, 1993).

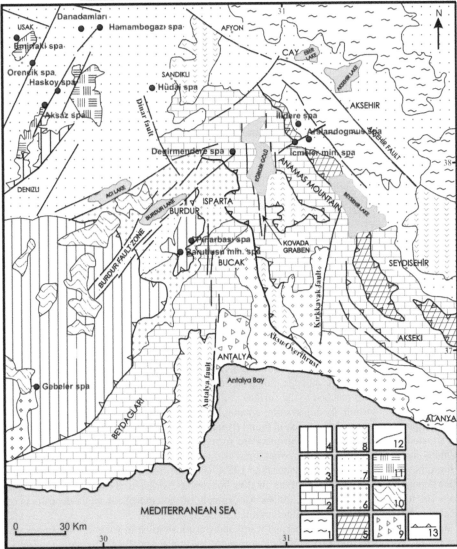

Figure 5.1.   Location of geothermal and mineral waters in the eastern Mediterranean region (modified by Ersoy *et al.*, 2011; Yagmurlu and Sentürk, 2005) 1. Paleozoic metamorphic rocks, 2. Mesozoic carbonated rocks, 3. Antalya nappes, 4. Lycian nappes, 5. Beyşehir-Hoyran nappes, 6. Marine Tertiary sediments, 7. Continental Neogene sediments, 8. Neogene volcanites, 9. Antalya travertine, 10. Ophiolitic nappes, 11. Menderes massif metamorphics, 12. Normal fault, and 13. Trust fault.

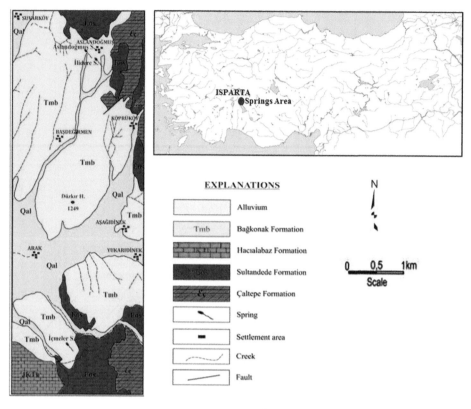

Figure 5.2.   Arslandogmus and Ilidere thermal waters and İcmeler mineral waters in Isparta/SW Turkey (Davraz *et al.*, 2006).

The SiO_2 concentration of the spring water is 179.85 mg/L. Values of Fe (1942.74 mg/L) and Al (938.69 mg/L) concentration were the highest in the Degirmendere thermal water.

The presence of hydrothermal alteration rocks, active volcanism, horst, and graben structures show that these regions are suitable for evolution of geothermal springs. Fluid heating with geothermal gradient is reached to the surface in the geothermal systems that is developed in the less hilly and flat terrain. In this system, a natural pool with acid-sulfate, alkali-chloride water, and precipitates of amorphous silica is found (Kocak, 2002). Natural amorphous silica deposits are situated in the surrounding Degirmendere spring. Deposits of siliceous sinter are common to many high-temperature hydrothermal areas and are indicated the presence of hydrothermal reservoirs with temperatures of greater than 175°C (Fournier and Rowe, 1966). The amorphous silica deposits occur with cooling of the alkaline hot springs, which have enough silica in solution from 100 to 50°C (Rodgers *et al.*, 2004). An example of the amorphous silica deposits is only located in New Zealand in the Tikitere region.

Burdur region: The Pınarbası thermal water and Barutlusu mineral water are the major water springs in Tefenni district of the Burdur province. Barutlusu and Pınarbası springs discharge from an overthrust zone, which is developed between the allochthonous Kızılcadag ophiolite and Dutdere limestone (Figs. 5.4 and 5.5). The discharge temperatures of Barutlusu and Pınarbası springs are 17.6 and 27°C, respectively. The reservoir rock of Pınarbası thermal water is Dutdere limestone and the cover rock is Pliocene aged Cameli formation consist of limestone, marl, sandstone, and conglomerate. The heat source is probably the high geothermal gradient resulting from the tectonic regime (Varol and Davraz, 2010).

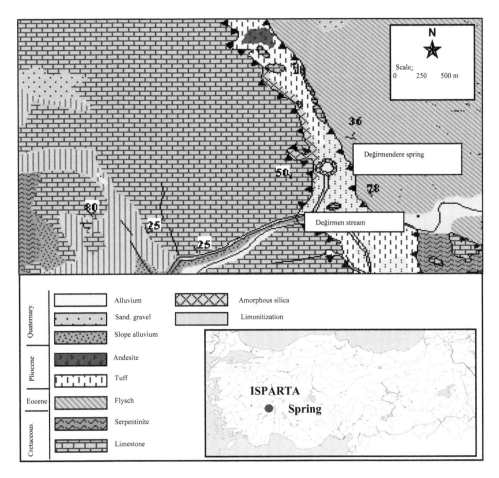

Figure 5.3.   Degirmendere spring (Isparta region; Davraz, 2003).

Fethiye region: Gebeler thermal water spring is situated in Fethiye district of Mugla province. Gebeler thermal water is discharged with 35°C within a natural pool in the Girmeler cave most probably associated with a NE-SW trending fault (Fig. 5.6). The geological structure of this region is complex with geological units belonging to Yesilbarak and Likya nappes. The reservoir rock of the Gebeler thermal water is the Beydagları carbonate platform, which tectonically overlies the Yesilbarak nappes (Davraz and Sener, 2006).

Usak region: The Usak province in the western part of Turkey has several thermal springs. Hamambogazı, Orencik, and Emirfakı thermal waters are discharged *via* drilled wells. Aksaz and Haskoy are natural thermal springs. Danadamları mineral water is discharged from a natural pool. The water temperatures range between 33 and 63°C. These waters hosted by the Menderes metamorphic rocks emerge along fault lineaments from two geothermal reservoirs present in the area (Fig. 5.7). The first reservoir consists of gneiss, schists, and marbles of the Menderes metamorphic rocks. Another reservoir is also Pliocene lacustrine limestone. Fractures and faults in these rocks facilitate circulation of water at depth where it is heated. The Neogene units are impermeable cap rocks of the Usak geothermal system (Davraz, 2008).

Sandıklı region: The Sandıklı region is an important geothermal field in the Afyon province. Sandıklı region is an area with intensive hydrothermal alteration and the presence of a geothermal system related to neotectonic tectonism and active volcanism in western Anatolia. The Hüdai fault,

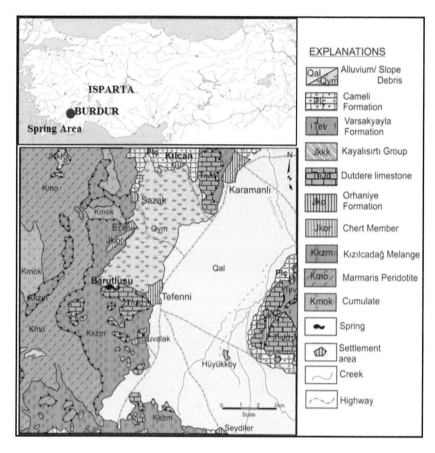

Figure 5.4.   Barutlusu (Burdur) mineral water (Varol and Davraz, 2010).

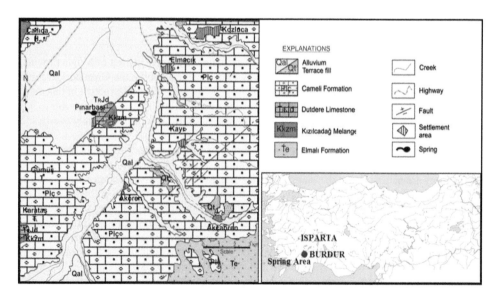

Figure 5.5.   Pınarbaşı (Burdur) thermal water (Varol and Davraz, 2010).

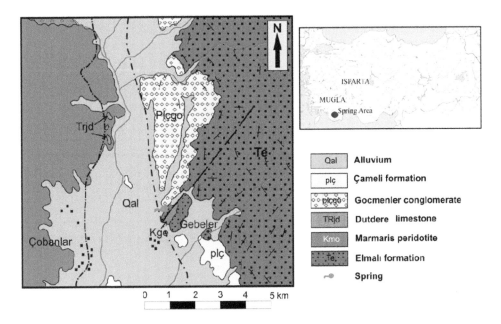

Figure 5.6.   Gebeler thermal water (Davraz and Sener, 2006).

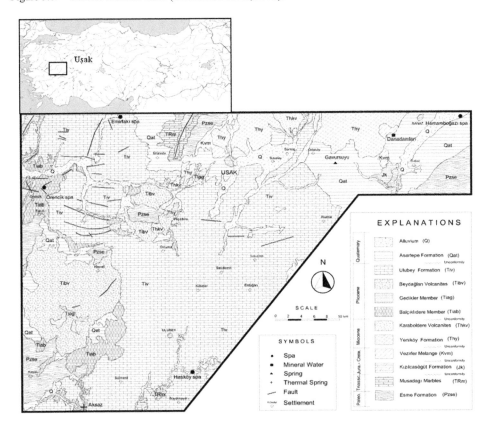

Figure 5.7.   Usak geothermal waters (Davraz, 2008).

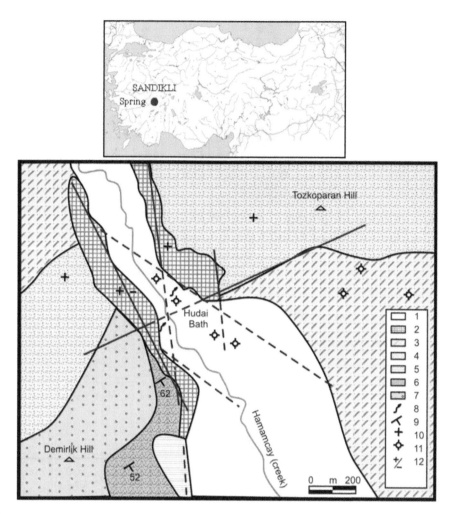

Figure 5.8.   Geological map of Sandıklı geothermal field (Afsin and Canik, 1996). 1. Alluvium, 2. Travertine, 3. Pliocene, 4. Late Miocene, 5. Late Cambrian–Early Ordovician, 6. Middle Cambrian,
7. Lower Cambrian, 8. Thermal–mineral waters (Hudai Bath), 9. Strike and dip of bed/foliation
10. Horizontal bed, 11. Drilling well, 12. Fault (possible) (Afsin *et al.*, 2012).

which is situated at the western edge of the Sandıklı depression area, is the youngest NNW-SSE
trending fault and it is developed in the Quaternary units. The Hüdai thermal water springs are the
discharges of the hottest water in this region. These springs discharged in the past with 62–68°,
but they do not exist nowadays due to exploitation of geothermal water in wells, which resulted
in decreasing groundwater level. The main reservoir rocks of the thermal water are quartzite
and limestone (Fig. 5.8; Afsin, 1991; 1997; Afşin and Canik, 1996; Afsin *et al.*, 2012; Aksever,
2011). Temperatures of Sandıklı geothermal system in wells are 54–82°C in between 53 and
1022 m depth. pH values are 6.5–7.47 and EC values range between 1360 and 2653 µS/cm (Afsin
et al., 2012).

The chemical data of above-mentioned thermal and mineral waters are presented in Table 5.1.
In Isparta region, the major cations of the spring are Ca > Mg > Na > K and Ca > Na > Mg > K,
and the major anions of the spring are HCO_3 > SO_4 > Cl and SO_4 > Cl > HCO_3. In Burdur region,
the major cations of the spring are Na > Ca > Mg > K, Ca > Mg > Na > K, and the major anions

Table 5.1. *In situ* measurements and chemical analyses results in Isparta, Burdur, Uşak, Fethiye, and Sandıklı regions (Fig. 5.1).

Location	pH	T (°C)	EC (μS/cm)	Ca (mg/L)	Mg (mg/L)	Na (mg/L)	K (mg/L)	HCO$_3$ (mg/L)	CO$_3$ (mg/L)	SO$_4$ (mg/L)	Cl (mg/L)
Arslandogmus (Isparta)	6.8	21.6	710	115.0	27.48	4.6	0.39	224.5	–	124.87	0.709
Ilidere(Isparta)	6.8	22.2	700	115.2	27.36	4.83	0.39	222.7	–	114.79	0.709
Icmeler (Isparta)	6.4	16.4	523	970.1	207.33	247.9	36.36	2700	–	1549.9	32.97
Degirmendere (Isparta)	1.7	8.0	10860	505.5	170.35	96	13.59	0.0	–	12000	0.017
Barutlusu (Burdur)	9.2	17.6	1066	16.76	1.25	35.25	0.33	39.5	1530	1	46.44
Pınarbasi (Burdur)	7.1	27	1372	46.04	27.47	17.06	1.33	215	–	0.03	10.64
Gebeler (Fethiye)	6.7	35	4310	290.1	89.10	506.8	11.65	500	–	675	835
Hamambogazı (Uşak)	6.2	63	4900	304.8	136.6	690.1	106.1	1675	–	1050	135
Aksaz (Uşak)	6.1	37	4370	397.5	108.5	631.4	91.7	1575	–	950	83
Haskoy (Uşak)	6.3	33	2950	282.7	72.6	171.7	23.6	1200	–	625	31
Orencik (Uşak)	6.4	38	4180	150.9	53.6	785.6	82.4	2150	–	250	87
Danadamları (Uşak)	6.3	24	3180	350.7	130.4	180.8	33.1	1425	–	650	31
Emirfakı (Uşak)	6.1	38	3840	271.8	71.1	253.3	27.7	925	–	775	42
Hüdai (Sandıklı)	6.5	67.4	3192	191.4	28.21	216.6	34.01	651.1	–	417.86	71.27

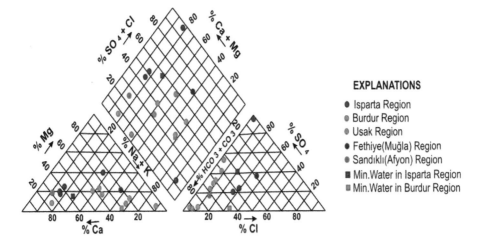

EXPLANATIONS

- Isparta Region
- Burdur Region
- Usak Region
- Fethiye(Muğla) Region
- Sandıklı(Afyon) Region
- Min.Water in Isparta Region
- Min.Water in Burdur Region

Figure 5.9. Piper diagram (in meq%).

of the spring are CO$_3$ > Cl > HCO$_3$ > SO$_4$, HCO$_3$ > Cl > SO$_4$. The major-ion contents of the Gebeler spring are Na > Ca > Mg > K and Cl > SO$_4$ > HCO$_3$. In Usak region, the major cations of the spring are Na > Ca > Mg > K and Ca > Na > Mg > K, and the major anions of the spring are HCO$_3$ > SO$_4$ > Cl (Table 5.1).

The geochemical evolution of groundwater can be understood by plotting the concentrations of major cations and anions in the Piper trilinear diagram (Piper, 1944; Fig. 5.9). The thermal water compositions are controlled by the rock composition in the basin and hydrologic characteristics such as flow paths and residence times. Therefore, hydrogeochemical facies of thermal and mineral waters in the eastern Mediterranean region are variable. The plot shows that the Arslandogmus and

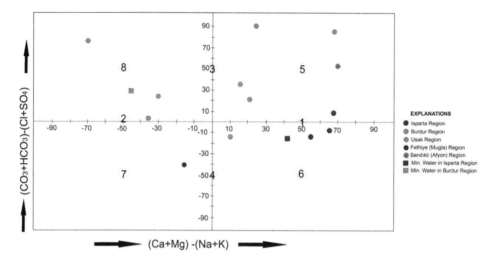

Figure 5.10.   Chadha diagram (unit in meq%).

Ilidere thermal waters have Ca–Mg–HCO$_3$ facies and the İcmeler mineral water has Ca–Mg–Na–HCO$_3$–SO$_4$ facies (Davraz *et al.*, 2006). The Degirmendere mineral spring water is of Ca–Mg–SO$_4$ type (Davraz, 2003). Hydrogeochemical facies of Barutlusu and Pınarbası springs are Na–CO$_3$–Cl and Ca–Mg–HCO$_3$, respectively (Varol and Davraz, 2010). The Gebeler thermal water is of Na–Ca–Cl–SO$_4$ type (Davraz and Sener, 2006). The hydrogeochemical facies of thermal waters in the Usak region are variable comprising Na–HCO$_3$–SO$_4$, Na–HCO$_3$, and Ca–Na–HCO$_3$–SO$_4$ types. Danadamları mineral water is Ca–Na–HCO$_3$ facies (Davraz, 2008). The Hüdai thermal and mineral water springs have Na–HCO$_3$ facies (Afsin, 1991; 1997; Afşin and Canik, 1996; Afsin *et al.*, 2012; Aksever, 2011). The geochemical facies in Piper diagram supports the dominance of alkaline earth elements over alkali metal in these thermal waters. The same results are also observed in Chadha and Durov diagrams (Figs. 5.10 and 5.11).

   In the Chadha diagram (Fig. 5.10), the rectangular field describes the overall character of the water. To define the primary character of water, the rectangular field is divided into eight subfields, each of which represents a distinct water type:

1. Alkaline earths exceed alkali metals.
2. Alkali metals exceed alkaline earths.
3. Weak acidic anions exceed strong acidic anions.
4. Strong acidic anions exceed weak acidic anions.
5. Alkaline earths and weak acidic anions exceed both alkali metals and strong acidic anions, respectively. Such water has temporary hardness. The positions of data points in this domain represent Ca^{2+}–Mg^{2+}–HCO$_3^-$ water type.
6. Alkaline earths exceed alkali metals and strong acidic anions exceed weak acidic anions. Such water has a permanent hardness and does not deposit residual sodium carbonate in irrigation use. The positions of data points in this domain represent Ca^{2+}–Mg^{2+}–Cl$^-$ type, Ca^{2+}–Mg^{2+} dominant Cl$^-$ type, or Cl$^-$ dominant Ca^{2+}–Mg^{2+} type waters.
7. Alkali metals exceed alkaline earths and strong acidic anions exceed weak acidic anions. The positions of data points in this domain represent Na$^+$–Cl$^-$ type and Na$^+$–SO$_4^{2-}$ type Na$^+$ dominant Cl$^-$ type, or Cl$^-$ dominant Na$^+$ type waters.
8. Alkali metals exceed alkaline earths and weak acidic anions exceed strong acidic anions (Chadha, 1999).

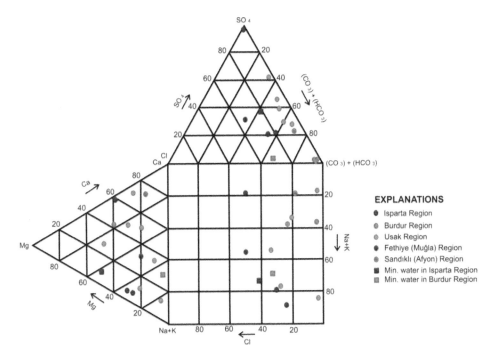

Figure 5.11.   Durov diagram (unit in meq%).

5.3   PROCESSES AFFECTING GEOTHERMAL FLUID COMPOSITION

The composition of the geothermal fluid in the reservoir is influenced by the chemical and physical processes. The mineral-fluid equilibria are the most important chemical processes. The dominant physical process is also boiling, conductive cooling, and mixing.

Mineral-fluid equilibria: Mineral-fluid equilibria play a fundamental role in determining the chemistry of discharging geothermal fluids. Hydrothermal alteration in geothermal systems occurs as a result of the interaction between rocks become unstable under the new physic-chemical conditions caused by thermal fluids. The rock-fluid system tends to attain a new chemical equilibrium between minerals and fluids by dissolving primary minerals and precipitating new phases, known as secondary or hydrothermal minerals. The fluid composition also changes during these processes. The type and amount of secondary minerals are controlled by several geochemical and physical factors, such as (i) temperature, (ii) chemical compositions of primary minerals and fluids (especially its pH), (iii) lithostatic and fluid pressure (mainly because it controls the depths at which boiling occurs), (iv) rock texture and permeability, (v) the duration of water-rock interaction, and (vi) kinetics of alteration processes (Browne, 1984).

Mineral solubilities played a major role in determining geothermal fluid composition which include quartz, calcite, and anhydrite. In most geothermal systems, the solubility of minerals follows one of three behavior patterns:

1. Mineral solubility increases with increasing temperature (e.g., alkali metal chlorides).
2. Mineral solubility decreases with increasing temperature, known as retrograde solubility (e.g., gypsum, anhydrite, calcite).
3. Mineral solubility increases with increasing temperature but only to a maximum value and then decreases with further temperature rises (e.g., silica; Nicholson, 1993).

Ion-exchange reactions: These reactions involve the transfer of ions between two or more aluminosilicate minerals and control the ratios of cations in solution, including H^+. This means
that solution pH can be buffered by a silicate mineral assemblage. Observations on geothermal
alteration assemblages show that the majority of secondary minerals are formed by reactions such
as (Nicholson, 1993):

Albite-K-feldspar

$$NaAlSi_3O_{8(s)} + K^+_{(aq)} = KAlSi_3O_{8(s)} + Na^+_{(aq)}$$

$$K = (a_{Na}/a_K)$$

K-feldspar-K-mica + quartz

$$3KAlSi_3O_{8(s)} + 2H^+_{(aq)} = KAlSi_3O_{10}(OH)_{2(s)} + 6SiO_{2(s)} + 2K^+_{(aq)}$$

$$K = (a_K/a_H)$$

The above equations show the reactions between minerals and the liquid phase. Reactions can
however also occur between minerals and gases, and the reaction of iron sulfide with hydrogen,
and hydrogen sulfide illustrates mineral-gas buffering reaction (Nicholson, 1993):

Pyrite-pyrrhotite

$$FeS_{2(s)} + H_{2(g)} = FeS_{(s)} + H_2S_{(g)}$$

$$K = P_{H_2S}/P_{H_2}$$

Activity diagrams: These diagrams depicting chemical equilibrium between minerals and aqueous species are important references for geochemists. Activity of the given species is shown in
brackets and the activity of the solid phase is assumed to be unity. The approach to chemical
equilibrium for natural water-rock systems can be tested by comparing the water composition
to theoretical composition of alteration minerals using thermodynamic data and activity phase
diagrams (Bowers *et al.*, 1984). Activity diagrams are also used to estimate the fluid-mineral equilibrium and water-rock interactions of thermal waters. These diagrams are based on the estimation
of the alteration minerals, which are formed because of water-rock interactions. Such diagrams
facilitate the prediction and interpretation of the chemical environment in which minerals are
formed by various geochemical processes. In order to construct an activity diagram, first, the
minerals to be included in the system are selected. Then, using the reaction between two mineral phases (e.g., muscovite and K-feldspar), equilibrium constant is calculated. The equilibrium
constant (K_{eq}) of any reaction at a given temperature and pressure is calculated.

$$KAl_3Si_3O_{10}(OH)_2 + 2K^+ + 6H_4SiO_4 \rightleftarrows 3KAlSi_3O_8 + 2H^+ + 12H_2O$$

muscovite   K-feldspar

$$\log K_{eq} = -2\log(a_K/a_H) - \log a_{H_4SiO_4}$$

In the above reaction, it is assumed that aluminum is conserved in the solid phase, but it is does
not mean that aluminum does not go into the solution at all. It is clear that the only solution
species are K^+ and H^+ and so it is possible to write the equilibrium constant as (a_K/a_H) assigning
a suitable value for $\log a_{H_4SiO_4}$. For Na-bearing minerals, the constants will be a_{Na^+}/a_{H^+} and for
the reactions between K- and Na-bearing minerals, the constants will be a combination of both
ratios. These reactions represent the phase boundaries under specified temperature and pressure
conditions and silica activity ($a_{H_4SiO_4}$).

Precipitation-dissolution reactions: The most important chemical reactions involving groundwater are those of dissolution and precipitation. Dissolution is the process whereby solids dissolve
into groundwater, and precipitation is the formation of solid phases from the water. Most dissolved
components in groundwater are ultimately derived from the dissolution of rock-forming minerals
in the aquifers. Generally, the longer the time of contact between the groundwater and the enclos-
ing rocks, the higher the content of dissolved materials in the water. Dissolved components in

groundwater may also be removed from the water through the process of precipitation (Runnels, 1993). Thermodynamically, the potential for dissolution or precipitation of solid phases can be expressed as the computed saturation index (*SI*):

$$SI = \log (IAP/K_{sp})$$

where *IAP* is the empirical ion activity product for a given mineral in the water of interest, and K_{sp} is the equilibrium solubility product constant for the same mineral at the temperature and pressure of the water. The value of the saturation index is only an indicator of the thermodynamic tendency for dissolution, precipitation, or equilibrium; it does not give us any information about the rates of reaction (Runnels, 1993). These calculations are useful in predicting the presence of reactive minerals and estimating mineral reactivity in a groundwater system (Deutsch, 1997; Gemici *et al.*, 2004). An *SI* of zero indicates that the ion activity product and the solubility product are equal, and that thermodynamic equilibrium exists with the solid phase. A negative (−) or positive (+) index indicates undersaturation and oversaturation, respectively. On the other hand, equilibrium constants for mineral dissolution often vary strongly with temperature (Kharaka *et al.*, 1988; Simsons and Browne, 2000). If the *SI*s with respect to several minerals converge to zero at a particular temperature, this temperature corresponds to the most likely mineral solution equilibrium temperature or at least the equilibrium temperature of that particular water (D'Amore and Mejia, 1998; Tole *et al.*, 1993). This method has the advantage of discriminating between equilibrated and non-equilibrated solutions. However, this method is sensitive to the choices of secondary minerals considered, the quality of thermodynamic data for the minerals, and to the quality of the analysis of elements such as Mg, Al, and Fe that occur in the geothermal solutions in very low concentrations.

Reed and Spycher (1984) have proposed that the best estimate of reservoir temperature can be attained by simultaneously considering the state equilibrium between specific water and many hydrothermal minerals as a function of temperature. Therefore, if a group of minerals converges to equilibrium at a specific temperature, the temperature corresponds to the most likely reservoir temperature. It should be noted that the range of equilibrium temperatures, the cluster equilibrium temperature, and the best equilibrium temperature were arbitrarily defined as the temperature interval between minimum and maximum equilibrium temperature for selected minerals. The temperature range over which most minerals appear to attain equilibrium, and the temperature at which the greatest number of minerals is in equilibrium with the aqueous solutions, respectively (Tole *et al.*, 1993).

Boiling (adiabatic) cooling: When a geothermal fluid rises to the surface, it may boil due to reduced hydrostatic pressure; where the flow rate is high, cooling can be considered to be approximately adiabatic. On boiling, gases and other volatile species partition into the vapor (steam) phase and move independently of the residual liquid water phase. This loss of steam, and therefore mass, produces a proportional increase in the concentration of the dissolved aqueous constituents. This change in concentration can be determined from mass-balance calculations. For example, upon boiling, carbon dioxide is lost from the liquid phase and this creates a rise in liquid pH, an increase in silica solubility, and an initial fall in calcite solubility (Nicholson, 1993).

Conductive cooling: If a fluid ascends to the surface only slowly, it may have sufficient time to lose heat by conduction to the cooler host rocks. As no steam will have been lost by boiling, then the concentration of chloride will be about the same as in the reservoir feeding the spring. The concentration of other solutes would be the same as that in the reservoir too if the water did not react with the rocks as it ascended. Unfortunately, rock-water reactions, both with the primary minerals and secondary alteration products, are likely to occur as the water moves to the surface at a slow rate. This gives plenty of time for reactions to take place and change the original composition of the water. It is therefore important to be able to recognize springs discharging conductively cooled waters, as these are unsuitable for geothermometry (Nicholson, 1993).

Mixing (dilution) with other waters: The deep geothermal fluid may mix with cold groundwater before being discharged in a spring or well. Since such meteoric waters are usually less saline than the geothermal fluid, this process is also known as dilution. In the interpretation of spring and well water chemistry, it is important to recognize mixed fluids and criteria which can indicate that dilution by meteoric water has occurred have been developed by Fournier (1979; 1991) and Arnosson (1985). Although dilution of the deep fluid with cold meteoric waters is the situation most commonly considered, mixing with steam-heated and other shallow-reservoir thermal waters can also occur. In such cases, mixing may be indicated by plots of Cl *vs.* SiO_2 and Cl *vs.* SO_4 (Nicholson, 1993).

5.3.1   *The saturation indices of geothermal waters in the eastern Mediterranean region*

Mineral saturation indices of hydrothermal minerals within the thermal water in the eastern Mediterranean region, which are possibly present in the reservoirs of the geothermal systems, are calculated at outlet temperature and pH by the SOLMINEQ 88 (Kharaka *et al.*, 1988) software in different researches (Tables 5.2 and 5.3). The Arslandogmus and Ilidere thermal waters are oversaturated with respect to adularia, aragonite, calcite, analcime, anhydrite, and gypsum minerals, and are undersaturated with respect to quartz and chalcedony (Table 5.2). The İcmeler mineral water is oversaturated regarding aragonite, calcite, dolomite, and quartz minerals (Davraz *et al.*, 2006). The water of Barutlusu spring is oversaturated regarding adularia, albite, analcime, aragonite, calcite, chalcedony, dolomite, quartz, kaolinite, microcline, muscovite, and illite. Due to interaction with limestone aquifer and ophiolite units; calcite, dolomite, and aragonite minerals are saturated in the Barutlusu spring water. Magnesite, anhydrite, and halite minerals are undersaturated. Adularia, albite, analcime, quartz, muscovite, illite, kaolinite, and microcline minerals are oversaturated for the Pınarbası spring. For this spring, anhydrite, aragonite, calcite, chalcedony, magnesite, and dolomite minerals are undersaturated (Table 5.2; Varol and Davraz, 2010).

Thermal springs in the Usak region are related to the Menderes metamorphic rocks. All thermal waters of this region are undersaturated in respect to gypsum and anhydrite indicating that dissolution of SO_4 is still taking place in the reservoir. Calcite, dolomite, aragonite, quartz, and chalcedony minerals are oversaturated or nearly in equilibrium at the discharge temperatures for

Table 5.2.   Saturation indices (*SI*s) of thermal waters in Isparta, Burdur, and Fethiye regions.

	Barutlusu (Burdur)	Pınarbası (Burdur)	Arslandog (Isparta)	Ilidere (Isparta)	İçmeler (Isparta)	Gebeler (Fethiye)
Adularia	1.279	1.200	−0.31	−0.15	2.82	1.551
Albite	2.251	2.431				2.213
Analcime	2.057	5.953	−0.52	−0.37	2.37	2.014
Anhydrite	−4.447	−4.331	−1.68	−1.71	−0.39	−0.876
Aragonite	0.846	−8.353	−0.461	−0.488	0.65	0.106
Calcite	0.992	−8.491	−0.31	−0.34	0.80	0.237
Chalcedony	0.413	−3.698	0.169	0.18	0.43	0.376
Dolomite	2.145	−18.100	0.035	−0.01	2.206	1.383
Halite	−7.328	1.594				−5.095
Muscovite	6.755	9.360				9.024
Quartz	0.578	0.691	0.355	0.378	0.592	0.606
Illite	3.318	5.218				
Kaolinite	4.118	6.861				6.271
Magnesite	−0.425	−0.728				−0.354
Microcline	3.200	3.208				3.244

the thermal waters. Muscovite, albite, kaolinite, adularia, analcime, and microcline minerals are supersaturated (Table 5.3). The increase in *SI* values of these minerals was observed in Usak area where carbonate rocks, gneiss, and schist as reservoir rocks indicate that the reservoir rocks have been influenced by the chemical composition of groundwaters and the data suggest a high residence time of the water in the reservoir rocks. Since waters of Usak area have high carbonate contents, the temperatures obtained from the equilibration of carbonate minerals such as calcite, dolomite, and aragonite have been ignored in determining the equilibrium temperatures. The equilibrium state between water and mineral is a function of temperature and the chemical composition of the solution. The saturation indices were recomputed to evaluate the equilibrium states of some hydrothermal minerals at different temperatures (Fig. 5.12). If the equilibrium lines ($SI = 0$) of a group of predicted minerals converge, it indicates a temperature corresponding to the most likely reservoir temperature (Tole *et al.*, 1993). The assessment of saturation indices shows that the reservoir temperatures of the Usak area can reach up to 120°C (Davraz, 2008).

Table 5.3. Saturation indices (*SI*s) of thermal waters in the Usak region (Davraz, 2008).

SI	Aksaz	Hasköy	Hamamb	Danadam	Örencik	Emirfakı
Calcite	0.229	0.150	0.461	0.228	0.284	−0.028
Dolomite	1.339	1.106	2.330	1.342	1.593	0.804
Aragonite	0.100	0.018	0.341	0.087	0.156	−0.157
Magnesite	−0.389	−0.548	0.280	−0.421	−0.190	−0.666
Gypsum	−0.469	−0.630	−0.512	−0.597	−1.348	−0.560
Anhydrite	−0.659	−0.853	−0.512	−0.894	−1.530	−0.742
Quartz	0.873	0.649	0.739	0.983	1.053	0.578
Chalcedony	0.638	0.424	0.484	0.787	0.817	0.342
Muscovite	8.874	9.030	7.171	8.275	7.231	6.656
Albite	2.109	1.409	1.686	1.708	2.230	0.612
Kaolinite	5.969	6.369	4.656	5.775	4.718	4.789
Adularia	2.249	1.534	1.973	1.991	2.225	0.629
Analcime	1.650	1.159	1.405	1.104	1.595	0.452
Microcline	3.920	3.248	3.228	3.814	3.886	2.290

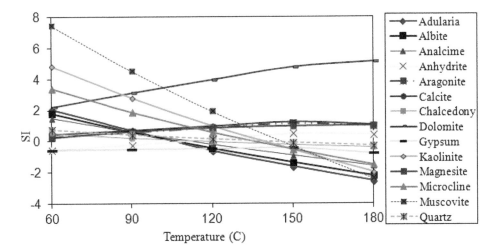

Figure 5.12. Mineral equilibrium diagrams of Hamambogazı spring in the Usak region (Davraz, 2008).

5.4   GEOTHERMOMETRY

Geothermometry is used as geochemical tool for the exploration and development of geothermal resources. Geothermometers can be applied to both natural spring discharges and well fluids. Chemical geothermometers are "path indicators" in geothermal exploration. During predrilling stage, it is important to assess a geothermal fields potential for exploratory drilling. Since drilling is expensive, chemical and geophysical methods are adopted to understand the field with low cost input. When the "indicators" show strong positive signals, the exploratory drill hole is drilled to gain more confidence and check the interpretation based on surface and subsurface tools (chemical and geophysical). If the exploratory drill holes data supports the interpretation based on chemical signatures, then this bore well becomes a production well some times and more wells are drilled depending on the power generation capacity of the field and power requirement of the region.

The geothermometers depend primarily on one or more dissolved constituents in the geothermal fluid whose concentrations vary depending on the temperature of the fluids. These constituents may be solutes, gases, or isotopes. Therefore, geothermometers have been classified into three groups such as solute (water), gas (steam), and isotope geothermometers. The solute and gas geothermometers are called chemical geothermometers (Yock, 2009).

5.4.1   *Chemical geothermometers*

Chemical geothermometry has become an important tool for estimating reservoir temperatures of hydrothermal systems and has proved to be very useful in determining the geothermal resource potential of a specific region. Chemical geothermometers depend on the existence of temperature-dependent mineral-fluid equilibrium at depth, which must be preserved during the passage of fluid to the surface. This statement must always be kept in mind, and it comprises the following assumptions (Ellis, 1979; Fournier, 1977; Nicholson, 1993; Truesdell, 1976; Verma, 2000; White, 1970; Yock, 2009):

- Fluid-mineral equilibrium at depth.
- A temperature-dependent reaction at depth.
- An adequate supply of solid phases to permit the fluid to become saturated with respect to the constituents used for geothermometry.
- Negligible re-equilibrium as the water flows to the surface. No dilution or mixing of hot and cold waters.

The first three assumptions are probably good for a few reactions that occur in many places. The last two are probably not valid for many geothermal fluids since the information obtained is only from the upper part of those systems (Yock, 2009).

The two main types of chemical geothermometers that are commonly used in geothermal exploration are based on mineral solubility (silica-SiO_2) and exchange reactions (Na–K, Na–K–Ca, etc.). These chemical geothermometers generally refer to silica and cation geothermometers.

5.4.1.1   *Silica geothermometry*
The silica geothermometry is based on experimentally determined variations in the solubility of different silica species (quartz, chalcedony, α-cristobalite, β-cristobalite, amorphous silica) in water, as a function of temperature (Fig. 5.13, Table 5.4). Quartz is the most stable and least soluble form of solid silica, and in general, it controls silica solubility in geothermal waters at temperatures exceeding 150°C. However, other solid silica compounds (chalcedony, amorphous silica) have higher solubilities than quartz and when they are in contact with the solution they control silica solubility preferentially over quartz below this temperature (Fournier, 1973). Quartz solubility increases with increasing of pH (alkaline solutions); however, this is not a significant problem for many geothermal fluids, despite it tends to be alkaline in surface discharges. This is because in most cases, alkalinity in chloride waters is due to boiling and CO_2 loss. Under this

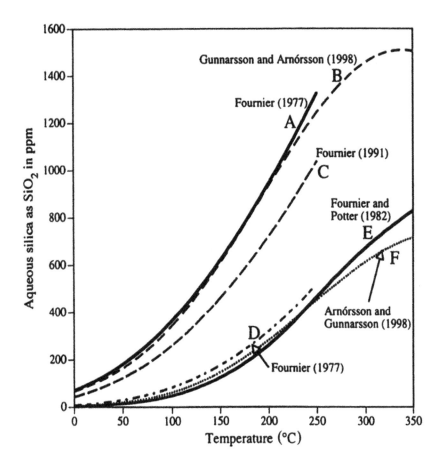

Figure 5.13. Temperature curves for the silica geothermometers.

condition, dissolved silica is gained in the reservoir, where pH tends to be neutral to slightly acidic, before significant boiling has occurred (Yock, 2009).

The silica geothermometry has been applied extensively to estimate geothermal reservoir temperature using the silica concentration of the fluid obtained from natural manifestations and drilled wells. White *et al.* (1956) proposed for the first time that the silica content of well discharges could be used as a geochemical indicator to estimate geothermal reservoir temperature (Verma, 2000). Since then enormous contributions have been made to gather more evidence and to create a systematic approach for the identification of geothermal reservoir characteristics from the silica fluid geochemistry (Arnorsson, 1970; 1975; Bödvarsson, 1960; Bödvarsson and Palmason, 1961; Fournier and Rowe, 1966; Levitte and Eckstein, 1979; Mahon, 1966; Morey *et al.*, 1962; Sigvaldason, 1966).

Fournier (1977) derived the first geothermometer equation from the experimental quartz solubility data (Table 5.4). Henley *et al.* (1984) compiled all the existing silica geothermometers for many silica phases, including the effects of adiabatic and conductive cooling processes. A limiting temperature for silica geothermometry tends to be about 250°C, since above this temperature silica dissolves and precipitates very rapidly. Fournier and Potter (1982) deduced the most acceptable geothermometer regression equation valid up to 330°C (Fournier and Rowe, 1977). The effects of added salts are significant only for concentrations greater than about 2–3wt% (Marshall, 1980; Fleming and Crerar, 1982; Fournier, 1985). However, above 300°C, small changes in pressure and salinity become important. Ragnarsdóttir and Walther (1983)

Table 5.4. Silica geothermometry equations.

Geothermometer	Equations (concentration of silica mg/L)	
SiO_2 (amorphous silica)	$T = 731 / (4.52 - \log SiO_2) - 273.15$	Fournier (1977)
SiO_2 (-Cristobalite)	$T = 1000 / (4.78 - \log SiO_2) - 273.15$	Fournier (1977)
SiO_2 (-Cristobalite)	$T = 781 / (4.51 - \log SiO_2) - 273.15$	Fournier (1977)
SiO_2 (Chalcedony)	$T = 1032 / (4.69 - \log SiO_2) - 273.15$	Fournier (1977)
SiO_2 (Quartz)	$T = 1309 / (5.19 - \log SiO_2) - 273.15$	Fournier (1977)
SiO_2 (Quartz steam loss)	$T = 1522 / (5.75 - \log SiO_2) - 273.15$	Fournier (1977)
SiO_2 (Chalcedony)	$T = 1112 / (4.91 - \log SiO_2) - 273.15$	Arnorsson *et al.* (1983)
SiO_2 (Quartz steam loss)	$T = 1264 / (5.31 - \log SiO_2) - 273.15$	Arnorsson *et al.* (1983)
SiO_2 (Quartz steam loss)	$T = 1021 / (4.69 - \log SiO_2) - 273.15$	Arnorsson *et al.* (1983)
SiO_2 (Quartz steam loss)	$T = 1164 / (4.9 - \log SiO_2) - 273.15$	Arnorsson *et al.* (1983)
SiO_2 (Quartz steam loss)	$T = 1498 / (5.7 - \log SiO_2) - 273.15$	Arnorsson *et al.* (1983)
SiO_2 (Quartz)	$T = 1175.7 / (4.88 - \log SiO_2) - 273.15$	Verma (2001)

determined the pressure dependence of quartz solubility at 250°C and concluded that pressure had significant effect on calculated quartz equilibrium temperatures from silica contents in waters from deep geothermal reservoirs. Recently, Verma and Santoyo (1997) applied a statistical data treatment method and theory of error propagation to improve the silica geothermometer equation presented initially by Fournier and Potter (1982). They also pointed out reasons for the spread in geothermometer temperatures, including errors in the regression coefficients of the geothermometric equations, accuracy, and precision of the analytical determinations of chemical species in a given sample, sampling errors, calibration errors of the geothermometers for high-temperature hydrothermal systems, and errors related to the geologic and thermodynamic processes of the chemical equilibria involved in the reactions (Verma, 2000).

The silica solubility determination has been performed using one of three methods: (i) weight loss of quartz in a known amount of water, (ii) chemical analysis of dissolved silica remaining in solution after rapid quenching and opening of the reaction vessel, and (iii) chemical analysis of dissolved silica in a small amount of solution extracted from the reaction solution while the vessel is maintained at the specified temperature and pressure (Verma, 2001). A regression relations established on a plot (log SiO_2 *vs.* $1/T_{(K)}$). In these graphs, silica solubility data which is determined from the above-mentioned methods are used.

For example, the regression expression for quartz is:

$$\log SiO_{2(ppm)} = [1175.7/T_{(K)}] + 4.88 \text{ (Verma, 2001)}$$

According to this regression relation, quartz geothermometer equation is:

$$T(°C) = 1175.7/(4.88 - \log SiO_2) - 273.15$$

Silica geothermometry equations that are derived from different researches are presented in Table 5.5.

At pH values above 7.8–9.3 (depending on temperature), the solubility of silica is also affected by pH. Quartz solubility increases with increasing pH (alkaline solutions); however, this is not much of a problem for many geothermal fluids, despite tending to be alkaline in surface discharges. This is because for most cases, alkalinity in chloride waters is due to boiling and CO_2 loss. Under this condition, dissolved silica is gained in the reservoir, where pH tends to be neutral to slightly acidic, before much boiling has occurred. In some very rare circumstances, a correction for pH may be required (Chandrasekharam and Bundschuh, 2008). However, since pH of geothermal reservoir liquids is generally constrained at value of 5–7 by water-rock reactions corrections for pH, effects are rarely needed in geothermometric calculations. For these reasons, dissolved silica in solutions of near-neutral pH from geothermal wells is a reliable geothermometer (Huenges,

2010). The interpretation of dissolved silica from hot springs is somewhat ambiguous because of uncertainties about the mineral controlling dissolved silica and the amount of steam possibly separated (Fournier, 1991).

Dilution is another factor affecting silica concentration of thermal water. Subsurface dilution of the geothermal fluid decreases the silica content. If equilibrium between fluid and rock is achieved, the silica geothermometer will give a temperature reflecting these cooler conditions; alternatively, if equilibrium is not attained then the silica geothermometer will give erroneous temperatures (too low estimates) (Yock, 2009).

As the geothermometer is dependent on an absolute concentration, rather than a ratio of concentrations, it is affected by physical processes such as boiling and dilution. These processes need to be recognized and, where possible, corrections should be made using alternative geothermometry equations or the silica mixing model (Nicholson, 1993). Boiling invariably causes changes in the composition of rising geothermal waters. These include degassing and an increase in the solute content of the water due to steam loss. The boiling mechanism affects the gas content of the steam that forms. The principle application of chemical and isotopic geothermometers during geothermal exploration involves estimation of reservoir temperatures below the zone of cooling. When applying these geothermometers, it is invariably assumed that no changes in water composition occur in conjunction with conductive cooling; boiling is taken to be adiabatic (Chandrasekharam and Bundschuh, 2008).

Subsurface dilution of the geothermal fluid decreases the silica content. If equilibrium between fluid and rock is achieved, the silica geothermometer will give a temperature reflecting these cooler conditions; alternatively, if equilibrium is not attained then the silica geothermometer will give erroneous temperatures that are too cool. Fournier and Truesdell (1974) described two mixing models that may be applied to springs with high flow rates and temperatures below boiling. These models are based on the relationship between the enthalpy and silica content of the ascending thermal water, the cold groundwater, and the resulting mixed thermal spring water. In the first model, the enthalpy of the hot water plus steam that mixes with and heats the cold water is the same as the initial enthalpy of the deep hot water. In the second model, the enthalpy of the hot water in the zone of mixing is lower than the enthalpy of the hot water at depth because of the escape of steam during ascent.

5.4.1.2 *Cation geothermometry*

The cation geothermometry is based on ion-exchange reactions that have temperature-dependent equilibrium constants. These are Na–K, Na–K–Ca, Na–Li, Li–Mg, K–Mg, and Na–K–Mg geothermometers (Table 5.5).

Na–K geothermometer: The Na/K geothermometer is generally thought to take longer to reach equilibrium at a given temperature than other commonly used geothermometers. Therefore, the Na/K ratio is commonly used to estimate possible highest temperatures in deeper parts of a system where waters reside for relatively long time periods, and other geothermometers are used to estimate lower temperatures that occur in shallower reservoirs where waters reside for relatively short periods of time (Fournier, 1989). This geothermometer is related to the variation of sodium and potassium in thermal waters due to ion exchange of these elements between coexisting alkali feldspars (Nicholson, 1993). The main advantage of this thermometer is that it is less affected by dilution and steam separation than other commonly used geothermometers provided there is little Na^+ and K^+ in the diluting water compared to the reservoir water. Na–K equations are adapted for reservoir temperatures in the range of 180–350°C (Ellis, 1979), but are limited at lower temperatures, notably below 120°C (Nicholson, 1993).

Na–K–Ca geothermometer: The Na–K–Ca geothermometer, developed by Fournier and Truesdell (1973), does not give high and misleading results for cold and slightly thermal, non-equilibrated waters (D'Amore and Arnorsson, 2000), compared to the Na–K geothermometer which generally gives very high calculated temperatures for low enthalpy systems when

Table 5.5.   Cation geothermometry equations.

Geothermometer	Equations (concentration of Na, K mg/L)	
Na–K (100–275°C)	$T = 856 / (\log Na/K + 0.857) - 273.15$	Truesdell (1976)
Na–K	$T = 883 / (\log Na/K + 0.780) - 273.15$	Tonani (1980)
Na–K (25–250°C)	$T = 933 / (\log Na/K + 0.993) - 273.15$	Arnorsson et al. (1983)
Na–K (250–350°C)	$T = 1319 / (\log Na/K + 1.699) - 273.15$	Arnorsson et al. (1983)
Na–K	$T = 1217 / (\log Na/K + 1.483) - 273.15$	Fournier (1979)
Na–K	$T = 1178 / (\log Na/K + 1.470) - 273.15$	Nieva and Nieva (1987)
Na–K	$T = 1390 / (\log Na/K + 1.750) - 273.15$	Giggenbach (1988)
K–Mg	$T = 2330 / (\log K^2/Mg + 7.35) - 273.15$	Fournier (1991)
K–Mg	$T = 1077 / (\log K^2/Mg + 4.033) - 273.15$	Fournier (1991)
K–Mg	$T = 4410 / (\log K/Mg + 14.0) - 273.15$	Giggenbach (1988)
K–Li	$T = 2200 / (\log Li/Mg + 5.470) - 273.15$	Kharaka and Mariner (1989)
Na–Li	$T = 1590 / (\log Na/Li + 0.779) - 273.15$	Kharaka et al. (1982)
Na–Li (Cl > 0.3 M)	$T = 1195 / (\log Na/Li + 0.130) - 273.15$	Fouillac and Michard (1981)
Na–Li (Cl < 0.3 M)	$T = 1000 / (\log Na/Li + 0.389) - 273.15$	Fouillac and Michard (1981)
Na–Li	$T = 1267 / (\log Na/Li + 0.07) - 273.15$	Verma and Santoya (1997)
Li–Mg	$T = 2200 / (\log Li/(Mg)^{1/2} + 5.47) - 273.15$	Kharaka and Mariner (1989)
Na–Ca	$T = 1096.7 / (\log Na/\sqrt{Ca} + 2.37) - 273.15$	Tonani (1980)
K–Ca	$T = 1930 / (\log K/\sqrt{Ca} + 2.920) - 273.15$	Tonani (1980)
Na–K–Ca	$T = 1647 / (\log(Na/K) + \beta[\log(\sqrt{Ca}/Na) + 2.06 + 2.47) - 273.15$ $\beta = 4/3$ for $T < 100°C$; $=1/3$ for $> 100°C$	Fournier and Truesdell (1973)

geothermal fluids have high calcium contents (Nicholson, 1993). The Na–K–Ca geothermometer is also based on ion-exchange reactions between feldspars.

This geothermometer is more complex as it is determined empirically from analyses of a large number of different fluids including geothermal and oil well waters. It assumes that fluid mineral equilibrium is established between Na and K feldspars, calcic minerals (calcium feldspar, epidote, calcite), and clay minerals. The following rules apply: first calculate the temperature using $\beta = 4/3$, and cation concentrations expressed either as mg/L or ppm (mg/kg). If that calculated temperature is less than 100°C and $[\log(c_{Ca} \cdot 1/2/c_{Na}) + 2.06]$ is positive, then this calculated temperature is appropriate. However, if the $\beta = 4/3$, calculated temperature is greater than 100°C or $[\log(c_{Ca} \cdot 1/2/c_{Na}) + 2.06]$ is negative, use $\beta = 1/3$ to calculate the temperature. Obviously, the Na–K–Ca geothermometer is applicable to a larger range of thermal fluids than the Na/K geothermometer (Yock, 2009).

Fournier and Potter (1979) noted that when the Na–K–Ca geothermometer is applied to Mg-rich waters, it commonly gives too high a temperature. For the solution of this problem, they devised the Mg correction equations (Table 5.6). The Mg concentrations in geothermal fluids decrease rapidly as temperature increases and all Mg-rich fluids found in nature have undergone water-rock reaction at a relatively low temperature. Furthermore, as geothermal fluid flows from high-temperature to lower temperatures environments, it appears to pick up significant amounts of Mg from surrounding rock relatively easily and quickly (Fournier, 1989). Therefore, low-temperature water may require Mg correction to give a correct reservoir temperature.

Na–Li geothermometer: Earlier observations were made by Ellis and Wilson (1960) and Koga (1970). In these observations, they showed that the lowest Na/Li values corresponded to the hottest portions of the geothermal field. The Na–Li geothermometer was developed by Fouillac and Michard (1981) from a statistic study about groundwaters in granitic and volcanic domains. They demonstrated an empirical relationship between log Na/Li and temperature for dilute

Table 5.6. Equations for calculating the Mg correction for the Na–K–Ca geothermometer (Fournier, 1989).

$$R = c_{Mg}/(c_{Mg} + 0.61c_{Ca} + 0.31c_K) \times 100$$

Do not apply a Mg correction if Δt_{Mg} is negative or less than 1.5

For R from 1.5 to 5	$\Delta t_{Mg}\ ^\circ C = 1.03 + 59.971 \log R + 145.05(\log R)^2 - 36711(\log R)^2/T - 1.67 \times 10^7 \log R/T^2$
For R from 5 to 50	$\Delta t_{Mg}\ ^\circ C = 10.66 - 4.7415 \log R + 325.87(\log R)^2 - 1.032 \times 10^5(\log R)^2/ T - 1.968 \times 10^7 \log R/T^2 + 1.605 \times 10^7(\log R)^3/T^2$

Cl < 0.2 M and more saline Cl > 0.3 M waters. Two other Na–Li geothermometers are documented in Kharaka *et al.* (1982) and Verma and Santoyo (1997). However, the geothermometer appears to be particularly sensitive to total dissolved solids and rock type. Because Li is generally a relatively minor constituent in a geothermal fluid and Na is generally a major constituent, slight changes in Li that result from exchange involving ions other than Na can significantly affect the Na/Li ratio.

K–Mg and Li–Mg geothermometers: As exchange reactions with Mg appear to be rapid at low temperatures, the K/Mg and Li/Mg ratios are taken to be representative of the conditions of the last rock-water reaction prior to discharge (Fournier, 1989; Nicholson, 1993). The K–Mg geothermometer was developed in the 1980s, and it was applied by Giggenbach (1988). The basis for this geothermometer assumes that fluids have equilibrated with K-feldspar (adularia), K-mica (illite, muscovite), chlorite (clinochlore), and chalcedony (a silica phase, which forms at slightly cooler temperatures than quartz). Another K–Mg geothermometer was presented by Fournier (1991). D'Amore and Arnorsson (2000) explain that progressive interaction between water and rock toward equilibrium changes Na/K ratios toward equilibrium with feldspars, and similarly, magnesium concentrations decrease because magnesium is incorporated into precipitating minerals such as smectite and chlorite. These processes cause K/Mg ratios to increase strongly. The K/Mg geothermometer was first applied to waters from a low enthalpy reservoir, which had not attained equilibrium with alkali feldspars (Nicholson, 1993). By considering only geothermal fluids in formations of the sedimentary cover, which consist of carbonate, evaporite, and detrital deposits, the K-Mg geothermometers cannot be applied. Indeed, these formations are poor in feldspars implying another origin for these elements such as dolomite dissolution and leaching of seawater brines (Sonney and Vuataz, 2010).

The Li–Mg geothermometer was developed in the same way as the K–Mg geothermometer, resting on exchange reactions with magnesium, for waters from a low enthalpy reservoir, which had not attained equilibrium with alkali feldspars (Nicholson, 1993). Thereby, these processes cause Mg/Li ratios to decrease strongly and conversely for K/Mg. The two equations of Kharaka *et al.* (1985) and Kharaka and Mariner (1989) were tested on the studied geothermal fluids, but they strongly overestimate temperature, especially for waters in the crystalline basement. K–Mg and Li–Mg geothermometers are valid from 50 to 300°C, and are of greatest use in the study of low to intermediate enthalpy systems when equilibrium has not been attained between the fluid and the complete mineralogical assemblage of the host rock (Nicholson, 1993).

Na–K–Mg geothermometer: A ternary plot of Na/1000–K/100–Mg$^{1/2}$ has been proposed by Giggenbach (1988) as a method to determine reservoir temperature and to recognize waters, which have attained equilibrium with the host lithologies. The Na/1000–K/100–Mg$^{1/2}$ ternary diagram was modified by Fournier (1990) and is most useful in determining which water is most suitable for geothermometry, eliminating those which is only partially equilibrated owing to dilution/mixing or near-surface water-rock reactions (Nicholson, 1993). Giggenbach (1988) divided thermal fluids into three main groups depending on the equilibrium of Na, K, and Mg ions as (i) fully equilibrated waters, (ii) partially equilibrated waters, and (iii) immature waters. The ternary diagram of Na/400–K/10–Mg$^{1/2}$ (Giggenbach and Corrales, 1992) was modified from

Giggenbach (1988) for lower temperature geothermal systems (in the range of 20–220°C) and used to estimate reservoir temperatures of waters that are most suitable for geothermometry. If studied water samples fall into the immature fields in these diagram, this indicates to continue water-rock interaction and do not attain fluid-mineral equilibrium. Therefore, the results obtained from the cation geothermometers should be taken into account as doubtful. In this instance, reservoir temperatures obtained from silica geothermometers are generally used as reliable results.

In addition, Giggenbach (1998) proposed an equation showing the degree of water-rock equilibrium attained in the reservoir. The equation is the maturity index (MI) of thermal waters:

$$MI = 0.315 \times \log(K^2/Mg) \log (K/Na) \quad \text{(ion concentrations in mg/L)}$$

MI values less than 2.0 are found in immature waters that do not attain equilibrium with their associated rocks, and cation geothermometers are not considered. If MI values are between 2.0 and 2.66, thermal water is partially equilibrated with reservoir rocks. Fully equilibrated conditions are accepted for $MI > 2.66$ values. Cation geothermometers (generally Na–K geothermometers) are reliable reservoir temperature results for this type of geothermal water.

Other cation geothermometers: Various cation geothermometers have been proposed that they may be of use in some situations. These are Na–Ca and K–Ca geothermometers developed by Tonani (1980) and Ca–Mg geothermometer of Marini *et al.* (1986). Na/Ca and K/Ca will be affected by variations in CO_2 partial pressure to an even greater extent than is the Na–K–Ca geothermometer, and Giggenbach (1988) uses K/Ca mainly as an indicator of CO_2 partial pressure. Note that precipitation of calcium carbonate from an ascending fluid will cause the Na–Ca and K–Ca geothermometer temperatures to be too high and also will cause the estimated CO_2 partial pressure to be too high (Fournier, 1989).

5.4.2   *Isotope geothermometers*

Isotope-exchange reactions, which achieve equilibrium in the natural system, are temperature dependent. Isotopes of elements are fractionated in the chemical processes operating in natural water rock systems (Ellis and Mahon, 1977). The fractionation is largest for lighter elements found in geothermal systems, such as helium, hydrogen, carbon, oxygen, and sulfur. Isotope-exchange reactions may be between gases and steam phase, a mineral and gas phase, water and a solute, or a solute and a solute. For using isotope-exchange reactions as geothermometer, certain conditions stated below have to be met (Nicholson, 1993):

- For exchange to occur, isotopes of the element in question must coexist in two different components in the fluid and/or solid phases of the geothermal reservoir.
- Isotopic equilibrium must be achieved between these components.
- The rate of isotopic exchange must be sufficiently rapid to permit equilibrium to be attained in the residence time of the fluid, but should also be slow enough to prevent re-equilibration as the fluid ascends to the surface.
- Temperature must be the dominant control on the fractionation equilibrium.
- The degree of fractionation must show a regular relationship with temperature and be detectable.

Although there are many isotope-exchange processes, few have been used because of the simplicity of sample collection and preparation, ease of isotopic measurement, a suitable rate of achieving isotopic equilibrium, and knowledge of the equilibrium constants (Karingithi, 2009). In general, isotope geothermometry equations have the form:

$$\Delta = A/T + B/T^2 + C$$

with T in Kelvin. The constants A, B, and C for these equations are specific to the phases (e.g., minerals) involved. The values of the constants are derived from experimentally determined values of the temperature-dependent fractionation factors (Nicholson, 1993).

Oxygen, hydrogen, carbon, and sulfur isotope geothermometry equations were proposed by different researches (Arnason, 1977; Hulston, 1983; Lyon and Hulston, 1984; Truesdell and Hulston, 1980; Truesdell *et al.*, 1977). However, oxygen isotope geothermometry is the most commonly used method.

5.4.2.1 Oxygen isotope geothermometry

Chemical elements with the same atomic number (protons) and different atomic mass are defined as isotopes. Isotopes have identical chemical behavior but different physical properties. Hydrogen has three isotopes, ^1H, ^2H, and ^3H, respectively, and oxygen also has three isotopes, ^{16}O, ^{17}O, and ^{18}O. Isotope geothermometers are generally involving oxygen or hydrogen exchange reactions with water assuming equilibrium. An isotopic fractionation occurs when steam separates from hot water. The isotopic compositions of both the steam and water in a well sample may be determined from the total discharge, whose steam and water fraction is known (Karingithi, 2009).

Oxygen isotope geothermometer is based on experimentally derived oxygen fractionation data for the sulfate-water system (Lloyd, 1968; Mizutani, 1972; Mizutani and Rafter, 1969). The kinetics of the oxygen exchange was shown by Lloyd (1968) to be dependent upon pH as well as temperature. However, McKenzie and Truesdell (1977) calculated that for a near-neutral pH chloride reservoir fluid, pH did not significantly influence the fractionation equilibrium position provided residence times which are in excess of the equilibration periods of 18 years for a 200°C fluid or 2 year for a 300°C fluid. The longer residence times of the geothermal fluid would therefore invariably ensure that isotopic equilibrium is established. Moreover, these slow re-equilibration periods also ensure that no isotopic exchange will occur as the fluids rise to the surface (Nicholson, 1993).

The two equations converge around 200°C and are in reasonable agreement (within 10°C) over the temperature range 0–300°C (Panichi and Gonfiantini, 1978). The version of Mizutani and Rafter (1969) was adopted by Fournier (1981) and Henley *et al.* (1984), and appears to be the equation most widely used. This equation is valid for subsurface temperatures in the range ∼140–350°C and may possibly be used down to 95°C (McKenzie and Truesdell, 1977; Truesdell, 1976). The low reservoir temperatures will require long residence times to establish isotopic equilibrium (∼500 years at 100°C), making the geothermometer unsuitable for low enthalpy systems unless residence times are extraordinarily lengthy (Nicholson, 1993).

Two of the commonly used empirical relationships are as follows:

$$10^3 \ln \alpha^{18}O_{SO_4^- - H_2O} = 3.251(10^6/T^2) - 5.1 \qquad \text{(Mizutani and Rafter, 1969)}$$

$$10^3 \ln \alpha^{18}O_{SO_4^- - H_2O} = 2.88(10^6/T^2) - 4.1 \qquad \text{(McKenzie and Truesdell, 1977)}$$

5.4.3 Gas geothermometers

Surface manifestations in most geothermal fields consist mainly of fumaroles, springs, and hot grounds (Arnorsson, 2000). The majority of gas geothermometers require that the gas/steam and, for a hot-water reservoir, the steam/water ratios are known (D'Amore and Panichi, 1987). Since steam and the corresponding water phase rarely discharge at the surface together, these ratios cannot be determined for hot springs or fumaroles. This has therefore limited the application of most gas geothermometers to well discharges (Nicholson, 1993). The gas geothermometers depend on gas–gas equilibria, gas–mineral equilibria, and mineral–gas equilibria.

D'Amore and Panichi (1980) proposed an empirical geothermometer to overcome the problem of unknown gas/water ratios, enabling estimates of the reservoir temperature to be made from knowledge of CO_2, H_2S, H_2, and CH_4 concentrations in natural discharges. Another CO-based geothermometer was developed by D'Amore *et al.* (1987) and Saracco and D'Amore (1989). This utilizes a series of equations, derived from gas equilibria, to calculate the reservoir temperature and only requires that the concentrations of CO_2, H_2S, H_2, CH_4, and CO in dry gas of the discharge are known. In addition, CO_2 and H–Ar geothermometers were developed in different researches. The gas geothermometers can be applied to both natural and well steam discharges.

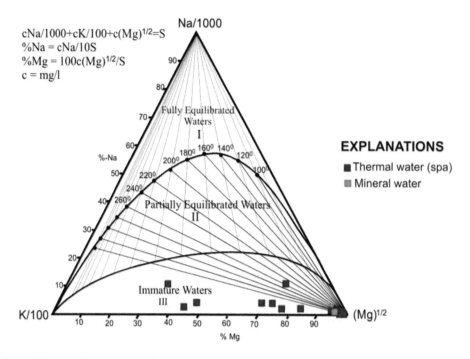

Figure 5.14.   Na–K–Mg ternary diagram.

5.4.4   *Geothermometer applications in the eastern Mediterranean region*

Discharge temperature of geothermal water is lower than reservoir temperature. Decreasing of discharge temperature is related to mixing cold water, heat transfer with rocks, thickness of cap rock, etc. According to Na–K–Mg ternary diagram which is proposed by Giggenbach and Corrales (1992), all thermal waters in the eastern Mediterranean region fall into the immature water field (Fig. 5.14). This indicates that no fluid–mineral equilibrium has been attained and that water-rock interactions continue. Therefore, the results obtained from the cation geothermometers should be taken into account as doubtful, and silica geothermometers are expected to give more consistent results.

The silica geothermometer is improved by connecting to mineral dissolution. Generally, dissolution of silica in the natural water is not affected from factors such as other ions, formation of complex molecule, and separation of volatile substances (Sahinci, 1991). Some of silica geothermometers (i.e., amorphous silica, α-cristobalite, β-cristobalite, chalcedony) give unreasonable temperatures for Arslandogmus, Ilidere, and Icmeler springs. The quartz geothermometers yield reservoir temperatures ranging from 30 to 54°C for Arslandogmus thermal water and 44 to 63°C for Icmeler mineral water (Davraz et al., 2006). Silica geothermometers give reliable results for the Degirmendere spring. The reservoir temperature was as 49°C which is the result of amorphous silica geothermometry (Davraz, 2003).

The reservoir temperature of Pınarbası thermal water is also determined as 30–83°C with silica geothermometers. The amorphous silica and cristobalite geothermometers give unreasonable temperatures for this water (Varol and Davraz, 2010). For Gebeler thermal water, amorphous silica and cristobalite geothermometers are not taken into account since they give temperatures that are lowered than the discharge temperatures. According to chalcedony and quartz geothermometers, reservoir temperature of Gebeler thermal water is ranging from 46 to 84°C (Davraz and Sener, 2006). The reservoir temperature of Sandıklı thermal waters is also determined as 107°C with quartz geothermometers (Afşin et al., 2010).

Some silica geothermometers (i.e., amorphous silica, α-cristobalite, β-cristobalite) give unreasonable temperatures for Usak thermal waters. Compared with quartz geothermometers, the chalcedony geothermometers display moderately lower temperatures between 50 and 105°C. According to quartz geothermometers, reservoir temperatures of thermal waters being lacustrine limestone and marble reservoir rocks range from 73 to 105°C. In addition, reservoir temperatures of other thermal waters being gneiss and schist reservoir rocks also range from 106 to 130°C according to quartz geothermometers (Davraz, 2008).

5.5   STABLE ISOTOPE APPLICATIONS

Isotopic investigation of thermal water is applied to determine water origin, water-rock interaction, and the age of thermal waters. Isotopes are two forms of the same element, which differ only in the number of neutrons (uncharged atomic particles) in the nucleus of the atom. This means that different isotopes of the same element will differ only in their relative mass. This mass difference governs their kinetic behavior and allows isotopes to fractionate during the course of certain chemical and physical processes occurring in nature.

The four stable isotopes that have proven most useful in water resource evaluation are hydrogen (^{1}H or H), deuterium (^{2}H or D), oxygen-16 (^{16}O), and oxygen-18 (^{18}O). These isotopes make up 99.9% of all water molecules. Isotopic compositions are reported in "δ" notation in parts per thousand (per mil: ‰) relative to Standard Mean Ocean Water (SMOW) as defined by Craig (1961). Based on a large number of analyses of meteoric waters collected at different latitudes, Craig (1961) showed that the δ^{18}O and δD values relative to SMOW are linearly related and can be represented by the equation. When δ^2H is plotted as a function of δ^{18}O, the relationship is linear (at the global scale). This relationship is known as the Meteoric Water Line (MWL). The equation is:

$$\delta D = 8\delta^{18}O + 10$$

However, the MWL may vary in a local or regional situation. For example, in the eastern Mediterranean, the deuterium excess parameter (defined as $d = \delta D - 8\delta^{18}O$) is higher than 10 because evaporation processes at the sea-surface occur into low-humidity air masses of continental origin. Here, the oxygen–hydrogen isotope relationship is better expressed by the Mediterranean MWL (MMWL):

$$\delta D = 8\delta^{18}O + 22 \quad \text{(Gat, 1980; Gat and Carmi, 1987; 1970)}$$

In a different study, which investigated the Lake District (LD), containing several lakes around Isparta province, the MWL of Lake District (LDMWL) was determined as follows:

$$\delta D = 8\delta^{18}O + 14.6 \quad \text{(SHW, 1989)}$$

It has been demonstrated in numerous studies that isotopic composition of local precipitation is controlled by evaporation of surface ocean water, and the progressive raining out of the vapor masses as they move toward regions with lower temperatures, i.e., higher latitudes and altitudes (Craig, 1961; Fisher, 1990; Merlivat and Jouzel, 1979; Rozanski *et al.*, 1982). The value of deuterium excess "$d = \delta^2H - 8\delta^{18}O$" is believed to relate to the relative humidity over the evaporating surface and wind speed under which evaporation takes place at the source area for atmospheric moisture (Merlivat and Jouzel, 1979). The subsequent rainout from air mass, which is the source of water vapor, is invariant with regard to d-value. However, re-evaporation of falling rain droplets under dry conditions may decrease the value of the deuterium excess in the residual rainwater. The causes of spatial and temporal variations of δ^2H or $\delta^{18}O$ in precipitation and water vapor are isotopic fractionation associated with evaporation and condensation process during the global water vapor circulation. In this processes, water molecule species (H^2HO and $H_2^{18}O$) remain in the liquid phase or are transferred into the liquid phase. During the process of water vapor adiabatic cooling, precipitating air masses are progressively being depleted in ^{2}H and ^{18}O.

This shows the correlation between δ^2H and δ^{18}O content of precipitation and local temperature variations. To understand the isotopic variation processes in precipitation, it is necessary to determine the isotopic composition of recent precipitation related to the climatic condition of today (IAEA, 2005).

Precipitation is the ultimate source of groundwater in virtually all systems. Hence, knowledge of the factors that control the isotopic compositions of precipitation before and after recharge allows the use of oxygen and hydrogen isotopes as tracers of water sources and processes. On a regional scale, the distribution of isotopic compositions is controlled by several factors:

- *Altitude effect*: On the windward side of a mountain, the δ^{18}O and δD values of precipitation decrease with increasing altitude. Typical gradients are $-0.15‰$ to $-0.5‰$ per 100 m for ^{18}O, and $-1.5‰$ to $-4‰$ per 100 m for D. This pattern is often not *observed in* interior mountains, for snow, or on the leeward side of mountains.
- *Latitude effect*: The δ^{18}O and δD values decrease with increasing latitude because of *the increasing* degree of "rainout."
- *Continental effect*: The ratios decrease inland from the coast.
- *A mount effect*: The greater the amount of rainfall, the lower the δ^{18}O and δD values of the rainfall; this effect is not seen in snow.

At a given location, the seasonal variations in δ^{18}O and δD values of precipitation and the weighted average annual δ^{18}O and δD values of precipitation remain fairly constant from year to year. This happens because the annual range and sequence of climatic conditions (temperatures, vapor source, direction of air mass movement, etc.) remain fairly constant from year to year. In general, rain in the summer is isotopically heavier than rain in the winter. This change in average isotopic composition is principally caused by seasonal temperature differences but is also affected by seasonal changes in moisture sources and storm tracks. Shallow groundwater δ^{18}O and δD values reflect the local average precipitation values but are modified to some extent by selective recharge and fractionation processes that may alter the δ^{18}O and δD values of the precipitation before the water reaches the saturated zone (Gat and Tzur, 1967). Some of these processes include evaporation of rain during infiltration, selective recharge, interception of precipitation by the tree canopy, and exchange of infiltrating water with atmospheric vapor. In the case of snow, various post-depositional processes, such as melting and subsequent infiltration of surface layers and evaporation, may alter the isotopic content of the snowpack, often leading to meltwater δ values that become progressively enriched (IAEA, 1981).

Generally, a geothermal reservoir is recharged by cold waters from precipitation falling close to the thermal area as well as by precipitation falling at higher altitudes at more distant areas. Although both of waters are meteoric origin, isotopic compositions are different due to altitude effect. Gat (1971) reported that incongruous results in isotope hydrology studies have generally been interpreted to mean (i) geographic displacement of groundwaters by flow, (ii) recharge from partially evaporated surface waters, (iii) recharge under different climatic conditions, (iv) mixing with non-meteoric water bodies-brines, seawater, connate, metamorphic, or juvenile waters, (v) differential water movements through soils or aquifers which result in fractionation processes (membrane effects), and (vi) isotopic exchange or fractionation between water and aquifer materials.

Water isotopes (δ^2H and δ^{18}O) are generally used in geothermal studies to identify the origin of geothermal fluids and to define the physical and chemical parameters of the geothermal reservoir for optimal exploitation. They can also be applied to monitor geothermal systems, study responses to exploitation, and sometimes as geothermometers. Geothermal waters cool as they rise to the surface and this can occur by conduction, mixing, and boiling (Henley *et al.*, 1984). Each of these processes can be traced isotopically. Mixing with cold water before and after boiling is very common in geothermal systems and this can be detected by a change in isotopic composition. Deviations of the isotopic composition of geothermal waters from that of the local meteoric water usually results from mixing of waters of different origin. Admixture of highly saline formation water or of connate water is frequently detected in lower temperature mineral springs, but may also play an important role in high-temperature systems situated in sedimentary rocks.

Water-rock interaction is the most important and extensive of the natural processes. An active hydrothermal system is an open or partially open geochemical system where various kinds of water-rock interaction are taking place. Isotope exchange and chemical alteration of geothermal fluids and rocks result largely from water-rock interaction. Therefore, experimental study of interaction between geothermal water and rock will help to (i) determine the origin of geothermal fluids, (ii) explore the mechanism of many complicated geochemical reactions in geothermal systems, and (iii) provide useful information to the exploitation and utilization of the geothermal resources.

Oxygen is a dominant element in the earth's crust. Oxygen-18 isotope content of thermal water is enriched depending on the relationship between reservoir rock, deep circulation, and meteoric waters due to water-rock interaction. The increase of oxygen-18 isotope is generally directly proportional with water temperature. Meteoric water that infiltrated to depth and formed a geothermal reservoir is in isotopic disequilibrium with the rock: oxygen-isotope exchange with the surrounding rocks results in an enrichment of the heavy isotopes with respect to the initial water, controlled by the temperature of the geothermal reservoir. Craig (1963) showed that high-temperature geothermal systems have waters with the same δ^2H value as local precipitation but have variable amounts of enrichment in ^{18}O.

This "oxygen shift" is attributed to exchange of oxygen in the water molecule with oxygen in silicate and carbonate minerals in the confining rocks ($\sim$50% of the rock mass is oxygen atoms). Oxygen mass balance equations, together with isotope fractionation factors between water and minerals, can be used to calculate the "oxygen shift" in both water and rocks (Kharaka and Thordsen, 1992). Results show that a high "oxygen shift" occurs in water of high-temperature systems with fresh rock and relatively little water throughput. Small or negligible oxygen shifts are observed in old systems where the isotopic composition of the rock has been shifted to equilibrium with the water and in low-temperature systems where the rate of exchange between the water and rock is too slow to cause a change in the isotopic composition of water as it passes through the system. Volcanic rocks such as basalts and andesites have a relatively narrow range of $\delta^{18}O$ values that reflects solidification at high temperatures. Granites have slightly higher values reflecting solidification at lower temperatures and contribution of various source materials, whereas sedimentary rocks have a wide range of variations and generally characterized by very high $\delta^{18}O$ values resulting from a variety of source materials and low formation temperatures (Chandrasekharam and Bundschuh, 2008). The water-rock interaction usually results in the oxygen shift, because (i) oxygen isotopic composition of rocks is generally higher than that of meteoric water, (ii) the amount of oxygen in rocks, in contact with water, is generally greater than that of water, (iii) the amount of hydrogen in rocks is, in general, much smaller than that of contacting water, and (iv) oxygen isotopic fractionation between silicates and water becomes smaller as temperature increases as shown later. The change in isotopic compositions of water and rocks, in a closed system, can be calculated from the equation describing mass balance, isotopic balance, and isotopic fractionation (Sakai and Matsuhisa, 1996; Yoshida, 2001).

The rate of oxygen-isotopic exchange between water and rock decreases rapidly with declining temperature. The presence of an ^{18}O shift in water is, therefore, usually taken as evidence for a present or former high temperature ($>250°C$) within the geothermal system. However, the magnitude of any oxygen-isotope shift depends, beside on temperature, on the degree of isotopic exchange and the proportions and $\delta^{18}O$ values of the exchanging water and rock (Geyh *et al.*, 2000).

Oxygen-isotope shift is the significant property of any geothermal water in high-temperature systems. There are two physical processes which might have produced such isotopic pattern (Geyh *et al.*, 2000):

1. *Liquid-vapor isotopic exchange* at 220°C. At this temperature, the two phases at equilibrium have the same hydrogen isotopic composition while there is still appreciable oxygen isotope fractionation. The heavy oxygen isotope becomes enriched in the liquid phase.

This phenomenon is only observed if the temperature of any water did exceed 220°C in the area (Geyh *et al.*, 2000).

2. Oxygen-isotope shift due to *isotopic exchange* with minerals forming the rock matrix through which the water has moved. Meteoric water that infiltrated to depth and formed a geothermal reservoir is in isotopic disequilibrium with the rock: oxygen-isotope exchange with the surrounding rocks results in an enrichment of the heavy isotopes with respect to the initial water, controlled by the temperature of the geothermal reservoir. The fast continuous removal of heavy isotopes leads to a steady depletion in ^{18}O of the rock (Clayton and Steiner, 1975; Geyh *et al.*, 2000).

Deuterium values of water do not usually change significantly by water-rock isotope exchange because most rock has very little hydrogen (<1% by mass). However, geothermal water in sedimentary or other rocks high in clay minerals can undergo significant deuterium shifts (Kharaka and Mariner, 2005; Kharaka and Thordsen, 1992; Kharaka *et al.*, 2002).

Radioactive isotope chemistry has been used to attempt to determine the age of the water in geothermal systems, using techniques similar to those for dating of rocks. Numerous methods exist for dating groundwater, including carbon-14, krypton-85, chlorine-36, and chlorofluorocarbon analyses. Many of these methods require large quantities of sampled water, have complex chemical analysis, or require instrumentation found in only a few laboratories. However, the simplest, most frequently used, and currently most popular method involves analyzing water for isotopes of hydrogen and ^{14}C. The environmental radioisotopes of ^{3}H and ^{14}C have transient concentrations in the hydrological system due to both their radioactive decay properties (which is a function of time) and variable input concentrations. This facilitates the study of water movement dynamics in the "time" domain. In general, the basic information to be obtained from these isotopes refers to "travel time" of water within a given system and/or to its distribution.

The most successful applications have used tritium, which has a half-life of 12.26 years. Minor amounts of tritium are naturally produced continually in the stratosphere by the action of cosmic radiation on hydrogen in the air. However, major amounts of tritium have been put into the atmosphere by thermonuclear weapons testing. Tritium concentration is expressed in terms of the tritium unit (TU), which is equivalent to a ratio of tritium to hydrogen-1 of 10–18. In the continental climates, in the temperature zone, cosmic radiation produced about 10 TU. As many as 10,000 TU were measured in 1963 following extensive atmospheric nuclear weapons testing. Ambient tritium levels thereafter decreased until about 1968, and since then have remained fairly constant. The following generalizations can be made concerning the age of geothermal water in the absence of mixing. A tritium content of less than 3 TU indicates that no water younger than 25 years is present. Values of 3–20 TU suggest that some amount of thermonuclear tritium is present, which indicates that the fluids entered the groundwater environment in the 1954–1961 period. If more than 20 TUs are found, the water entered the system after 1963. Many geothermal reservoir waters are older, some much older, than 25–50 years of useful dating range available with tritium. Typically, convecting hydrothermal fluids move at speeds measured in meters. However, tritium dating of water can indicate rapid movement in a system (Wright, 1991).

Tritium decays to ^{3}He by beta particle emission, and knowing the decay rate allows for a more accurate shallow groundwater recharge age determination. $T/^{3}He$ ratios are useful for groundwater dating if ages are several months to about 30 years (but not exceeding 50 years). $T/^{3}He$ ratios have an accuracy of 1–3 years. Groundwater ages can be estimated using the following equation:

$$\text{Groundwater age (in years)} = -17.8 \ln (1 + {}^{3}\text{He}_{\text{trit}}/{}^{3}\text{H})$$

where:

$^{3}He_{\text{trit}}$ = component of ^{3}He from the decay of tritium corrected for other ^{3}He sources such as the earth's atmosphere, small contributions from spontaneous fission of lithium-6, and from uranium and thorium decay.

^{3}H = tritium concentration in TU.

Because ^{3}He is also present within the mantle, in the ratio of 200–300 parts of ^{3}He to a million parts of ^{4}He, ratios of ^{3}He/^{4}He in excess of atmospheric concentrations are indicative of a contribution of ^{3}He from the mantle. This commonly occurs in geothermal areas and crystalline crustal sources dominated by ^{4}He, which is produced by the decay of radioactive elements in the crust and mantle. Therefore, in other than alluvial terrain, terragenic-produced helium may give anomalous results.

The radioactive isotope ^{14}C is also naturally produced in the atmosphere by cosmic radiation. It is readily oxidized to carbon dioxide and enters into the carbon cycle. Its natural production is rather constant and its input to hydrological systems can be assumed to be steady state for practical purposes. The concentration of ^{14}Cr is expressed as "percent of the ^{14}C of Modern Carbon" (pmc). It has a half-life of 5730 years.

Unlike tritium, ^{14}C is not a conservative tracer providing direct indication of the travel time, due to complex chemical reactions involved during the transport process. The concentration of this isotope in water is often measured as the ^{14}C activity in the DIC, which is altered due to interactions of water with the aquifer matrix. Various chemical and isotopic models have been suggested to account for these isotope (Fontes, 1983). The main sources of carbon element in groundwater are (i) atmospheric CO_2, (ii) biologic CO_2 which is produced as a result of organic activities in the percolation zone, (iii) geogenic CO_2 originated from metamorphism of carbonate rocks and escape of CO_2 from the earth's crust and mantle, and (iv) carbonate (CO_3) ions which are present in water due to dissolved of carbonate minerals.

5.5.1 *Results from the stable isotope analysis in the eastern Mediterranean region*

In the eastern Mediterranean region, the isotopic ratios of hydrogen and oxygen of hot spring waters show characteristics of meteoric water and with different degrees of shift conventional δ notation as the per mil (‰) deviation of the isotope relative to the known standards ($\delta = (R_{sample}/R_{standard} - 1)*1000$, where R_{sample} and $R_{standard}$ are the D/H and $^{18}O/^{16}O$ ratios in the sample and in the standard, respectively). The standard water for hydrogen and oxygen isotopes is Vienna Standard Mean Ocean Water (VSMOW).

The oxygen (^{18}O) and hydrogen isotope (deuterium, ^{2}H or D) compositions of the thermal waters from the eastern Mediterranean region show narrow ranges of variation. The relationship between δ^{18}O and δD (Fig. 5.6) for the thermal waters of the study area shows that these waters fall on and just below the MWL, $\delta^2H = 8\delta^{18}O + 10$, given by Craig (1961) with a d-excess value of $\sim$10‰. Figure 5.6 also shows that the thermal waters plot away from the Eastern Mediterranean Meteoric Water Line (MMWL), $\delta^2H = 8\delta^{18}O + 22$, given by Gat *et al.* (1969) and IAEA (1981), which represents modern Mediterranean precipitation. This line indicates deuterium enrichment of thermal groundwater from the modern Mediterranean precipitation with a d-excess value of $\sim$22‰. It has been suggested that low d-excess values of around 10‰ indicate paleo-recharge ($>$30 ka BP), while higher d-excess values ($>$20‰) indicate modern-day recharge (Vengosh *et al.*, 2007).

The isotopic composition of the thermal waters in the eastern Mediterranean region shows that they are of meteoric origin. The isotopic values of the mineral waters and are low-temperature thermal water very close to the Region of Lakes Meteoric Water Line equation ($\delta^2H = 8\delta^{18}O + 14.6$; Dilsiz, 2006). The oxygen-18 isotope contents of the all thermal waters are increased in the eastern Mediterranean area (Fig. 5.15). The increase of δ^{18}O is related to the δ^{18}O exchange between the deeply circulating meteoric waters and reservoir rocks of the geothermal systems. Oxygen shifts may also be observed in geothermal waters with temperatures as low as 40°C (Qin *et al.*, 2005). The degree of the shift depends on the reservoir temperature, residence time and water, and rock-water interactions (Truesdell and Hultson, 1980). The oxygen-18 shift must result from one of the following processes: (i) exchanges of oxygen isotopes of geothermal waters with reservoir carbonates over long residence times, and low water-rock ratios, or (ii) exchanges of oxygen isotopes of geothermal waters with silicate minerals at high ($>$250°C) reservoir temperatures (Qin *et al.*, 2005).

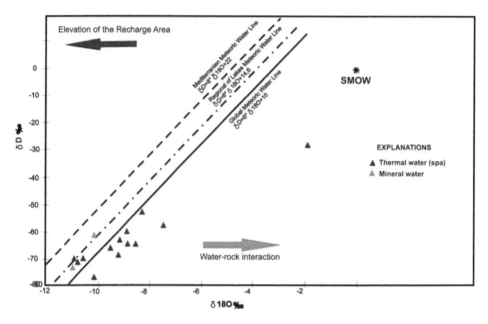

Figure 5.15.   Plot of δ^2H and δ^{18}O compositions of thermal and mineral waters.

Tritium is commonly used to estimate the residence time and mixing of waters in geothermal systems. The tritium values of Pınarbaşı (0.4 TU), Barutlusu (1.05 TU), Hüdai (0.5 TU), İçmeler (0.9), and Ilidere (5.6 TU) springs are low (Örmeci, 2006; Varol and Davraz, 2010; Aksever, 2011). This shows that, this thermal and mineral waters are deep circulation and have long residence time. However, there is mixture with young water in the Ilidere thermal water (5.6 TU).

5.6   CONCLUSIONS

Hydrogeochemical techniques provide important input for geothermal systems investigations. The major goals of hydrogeochemical techniques are to obtain the subsurface composition of the fluids in a geothermal system and use this to obtain information on temperature, origin, and flow direction. Furthermore, chemistry of geothermal fluid is an important factor during exploration and development of geothermal resources. In addition, during development of geothermal resources, it can be used to predict scaling of different components by comparing the water chemistry to theoretical solubility of minerals. All of the thermal and mineral water springs in eastern Anatolia are clearly related to young volcanic activity and block faulting. The hydrogeochemical facies of thermal and mineral waters in the eastern Mediterranean region are variable due to the differences in rock composition of the region and hydrologic characteristics such as flow paths and residence times. High-temperature geothermal waters (maximum 68°C) are observed in Uşak and Sandıklı regions related to Menderes Metamorphic rocks.

ACKNOWLEDGMENTS

The author thanks Prof. Dr. Jochen Bundschuh (University of Southern Queensland, Toowooba) and Prof. Dr. Alper Baba (Izmir Institute of Technology, Izmir, Turkey) for their thoughtful and constructive reviews of this chapter, and he wishes to thank Dr. Simge Varol and Dr. Fatma Aksever for their technical assistance.

REFERENCES

Afşin, M.: *Hydrogeological investigations of Kuruçay plain (Afyon-Sandıklı) and Hüdai spa water.* PhD Thesis, Ankara University, Ankara, Turkey, 1991.

Afşin, M.: Hydrochemical evolution and water quality along the groundwater flow path in the Sandıklı plain, Afyon, Turkey. *Environ. Geol.* 31:3–4 (1997), pp. 221–230.

Afşin, M. & Canik, B.: Hydrogeology of Hüdai (Sandıklı/Afyon) thermal and mineral water and genetic interpretation. *Yerbilimleri J.* (Geosound) 28 (1996), pp. 69–86.

Afşin, M., Davraz, A., Hınıs, M.A. & Karakaş, Z.S.: Hydrogeological, hydrogeochemical and isotopical investigations of Hüdai (Sandıklı/Afyon) geothermal area and surroundings. The Scientific and Technological Council of Turkey (TÜBİTAK) – 111Y034 (continuing), 2010.

Afsin, M., Dağ, T., Davraz, A., Aksever, F., Karakaş, Z. & Hınıs, M.A.: The origin and sustainability of Hudai geothermal waters, Sandikli, Afyonkarahisar, Turkey. *39th IAH (International Association of Hydrogeologists) Congress*, 16–21 September 2012, Canada.

Aksever, F.: *Hydrogeological investigations of Sandıklı (Afyonkarahisar) basin.* PhD Thesis, Suleyman Demirel University, Isparta, Turkey, 2011.

Arnason, B.: The hydrogen-water isotope applied to geothermal areas in Iceland. *Geothermics* 5 (1977), pp. 75–80.

Arnorsson, S.: Underground temperatures in hydrothermal areas in Iceland as deduced from the silica content of the thermal water. *Geothermics* 2 (1970), pp. 536–541.

Arnorsson, S.: Application of the silica geothermometer in low temperature area in Iceland. *Am. J. Sci.* 275 (1975), pp. 763–784.

Arnorsson, S.: The use of mixing models and chemical geothermometers for estimating underground temperatures in geothermal systems. *J. Volcanol. Geoth. Res.* 23 (1985), pp. 299–335.

Arnorsson, S.: Isotopic and chemical techniques in geothermal exploration, development and use: sampling methods, data handling, interpretation. International Atomic Energy Agency, Vienna, Austria, 2000.

Arnorsson, S., Gunnlaugsson, E. & Svavarsson, H.: The chemistry of geothermal waters in Iceland. III. Chemical geothermometry in geothermal investigations. *Geochim. Cosmoschim. Acta* 47 (1983), pp. 567–577.

Barbier, E.: Geothermal energy technology and current status: an overview. *Renew. Sust. Energy Rev.* 6 (2002), pp. 3–65.

Bödvarsson, G.: Exploration and exploitation of natural heat in Iceland. *Bull. Volcanol.* 2:23 (1960), pp. 241–250.

Bödvarsson, G. & Palmason, G.: Exploration of subsurface temperatures in Iceland. *Proceedings UN Conference New Sources Energy*, G/24, Rome, Italy, 1961, pp. 82–90.

Bowers, T S., Jackson, K.J. & Helgeson, H.C.: Equilibrium activity diagrams for coexisting minerals and aqueous solutions at pressures and temperatures to 5 kb and 600°C. Springer-Verlag New York, 1984.

Browne, P.R.L.: Subsurface stratigraphy and hydrothermal alteration of eastern section of the Olkaria geothermal field, Kenya. *Proceedings of the 6th New Zealand Geothermal Workshop*, Geothermal Institute, Auckland, New Zealand, 1984, pp. 33–41.

Chadha, D.K.: A proposed new diagram for geochemical classification of natural waters and interpretation of chemical data. *Hydrogeol. J.* 7 (1999), pp. 431–439.

Chandrasekharam, D. & Bundschuh, J.: *Low-enthalpy geothermal resources for power generation.* Taylor & Francis Group, London, UK, 2008.

Clayton, R.N. & Steiner, A.: Oxygen isotope studies of the geothermal system at Wairakei, New Zeland, *Geochem. Cosmochim. Acta* 39 (1975), pp. 1179–1186.

Craig, H.: Isotopic variations in meteoric water. *Science* 133 (1961), pp. 1702–1703.

Craig, H.: The isotope geochemistry of water and carbon in geothermal areas. In: E. Tongiorgi (ed): *Nuclear geology on geothermal areas.* Proceedings International Symposium, Spoleto, Italy, Consiglio Nazionale delle Ricerche, Laboratorio di Geologic Nucleare, Pisa, Italy, 1963, pp. 17–53.

D'Amore, F. & Arnórsson, S.: Geothermometry. In: S. Arnórsson (ed): *Isotopic and chemical techniques in geothermal exploration, development and use. Sampling methods, data handling interpretation.* International Atomic Energy Agency, Vienna, Austria, 2000, pp. 152–199.

D'Amore, F. & Mejia, J.T.: Chemical and physical reservoir parameters at initial conditions in Berlin geothermal field, El Salvador: a first assessment. *Geothermics* 28 (1998), pp. 45–73.

D'Amore, F. & Panichi, C.: Evaluation of deep temperatures in hydrothermal systems by new gas geothermometer. *Geochim. Cosmochim. Acta* 44 (1980), pp. 549–556.

D'Amore, F. & Panichi, C.: Geochemistry in geothermal exploration. In: M. Economides & P. Ungemach (eds): *Applied geothermics*. Wiley & Sons, New York, 1987, pp. 69–89.

D'Amore, F., Fancelli, R. & Caboi, R.: Observation on the application of chemical geothermometers to some hydrothermal systems in Sardinia. *Geothermics* 16 (1987), pp. 271–282.

Davraz, A.: A study on water-rock interaction: Keçiborlu-Degirmendere spring (Turkey). *Suleyman Demirel University Journal of Natural and Applied Sciences* 7:2 (2003), pp. 327–335.

Davraz, A.: Hydrogeochemical and hydrogeological investigations of thermal waters in the Usak area (Turkey). *Environ. Geol.* 54:3 (2008), pp. 615–628.

Davraz, A. & Sener, E.: Hydrogeology and hydrogeochemistry of Gebeler spa water (Fethiye-Mugla). *Proceedings Symposium on Recent Applications Engineering Geology*, Denizli, Turkey, 2006, pp. 347–356.

Davraz, A., Varol, S. & Ismailov, T.: Hydrogeochemical investigations of Sarkikaraagac (Isparta/Turkey) thermal and mineral water spring. *The III International Scientific and Practical Conference 'Use of the Water Resources and its Integrational Management in Globalization Processes'*, 6–7 July 2006, Bakü, Azerbaijan, 2006, pp. 148–152.

Deutsch, W.J.: *Groundwater geochemistry: fundamentals and applications to contamination*. Lewis Publisher, 1997.

Dilsiz, C.: Conceptual hydrodynamic model of the Pamukkale hydrothermal field, southwestern Turkey, based on hydrochemical and isotopic data. *Hydrogeol. J.* 14 (2006), pp. 562–572.

Durov, S.A.: Classification of natural waters and graphic representation of their composition. *Dokl. Akad. Nauk SSSR*, v. 59, 1948, pp. 187–190.

Ellis, A.J.: Chemical geothermometry in geothermal systems. *Geothermics* 25 (1979), pp. 219–226.

Ellis, A.J. & Mahon, W.A.J.: *Chemistry and geothermal systems*. Academic Press, 1977.

Ellis, A.J. & Wilson, S.H.: The geochemistry of alkali metal ions in the Wairakei hydrothermal system. *N.Z.J. Geol. Geophys.* 3 (1960), pp. 593–617.

Ersoy, Y.E., Helvacı, C. & Palmer, M.R.: Stratigraphic, structural and geochemical features of the NE-SW trending Neogene volcano-sedimentary basins in western Anatolia: Implications for associations of supra-detachment and transtensional strike-slip basin formation in extensional tectonic setting. *J. Asian Earth Sci.* 41 (2011), pp. 159–183.

Fisher, D.A.: A zonally-averaged stable-isotope model coupled to a regional variable elevation stable isotope model. *Ann. Glaciol.* 14 (1990), pp. 65–71.

Fleming, B.A. & Crerar, D.A.: Silicic acid ionization and calculation of silica solubility at elevated temperature and pH. Application to geothermal fluid processing and reinjection. *Geothermics* 11 (1982), pp. 15–29.

Fontes, J.C.: Dating of groundwater. In: *Guidebook on nuclear techniques in hydrology*. Technical Report Series, 91, IAEA, Vienna, Austria, 1983, pp. 285–318.

Fouillac, R. & Michard, S.: Sodium/lithium ratio in water applied to geothermometry of geothermal reservoirs. *Geothermics* 10 (1981), pp. 55–70.

Fournier, R.O.: Silica in thermal waters: laboratory and field investigations. *International Symposium on Hydrogeochemistry and Biogeochemistry*, Tokyo, 1970, vol.1, *Hydrochemistry*, Washington, DC, 1973, pp. 122–139.

Fournier, R.O.: A review of chemical and isotopic geothermometers for geothermal systems. In: *Proceedings of the Symposium on Geothermal Energy*, Cento Scientific Programme, Ankara, Turkey, 1977, pp. 133–143.

Fournier, R.O.: A revised equation for the Na/K geothermometer. *Geoth. Resour. Council Trans.* 3 (1979), pp. 221–224.

Fournier, R.O.: Application of water geochemistry to geothermal exploration and reservoir engineering. In: L. Rybach & L.J.P. Muffler (eds): *Geothermal systems: principals and case histories*. Wiley, Chichester, 1981, pp. 109–143.

Fournier, R.O.: The behavior of silica in hydrothermal solutions. In: B.R. Berger & P.M. Bethke (eds): Geology and geochemistry of epithermal systems. *Rev. Econ. Geol.* 2 (1985), pp. 45–61.

Fournier, R.O.: Lectures on geochemical iinterpretation of hydrothermal waters. UNU Geothermal Training Programme Reykjavik, Iceland Report 10, 1989.

Fournier, R.O.: The interpretation of Na–K–Mg relations in geothermal waters. *Geoth. Resour. Council Trans.* 14 (1990), pp. 1421–1425.

Fournier, R.O.: Water geothermometers applied to geothermal energy. In: F. D'Amore (ed): *Application of geochemistry in geothermal reservoir development*. United Nations Institute for Training and Research, Rome, Italy, 1991, pp. 37–69.

Fournier, R.O. & Potter, R.W. II: Magnesium correction to Na–K–Ca geothermometer. *Geochim. Cosmochim. Acta* 43 (1979), pp. 1543–1550.

Fournier, R.O. & Potter, R.W. II: A revised and expanded silica (quartz) geothermometer. *Geoth. Resour. Council Bull.* 11:10 (1982), pp. 3–12.

Fournier, R.O. & Rowe, J.J.: Estimation of underground temperatures from the silica content of water from hot springs and wet steam wells. *Am. J. Sci.* 2:64 (1966), pp. 685–691.

Fournier, R.O. & Rowe, J.J.: The solubility of amorphous silica in water at high temperatures and high pressures. *Am. Mineral.* 62 (1977), pp. 1052–1056.

Fournier, R.O. & Truesdell, A.H.: An empirical Na–K–Ca geothermometer for natural waters. *Geochim. Cosmochim. Acta* 37 (1973), pp. 1255–1275.

Fournier, R.O. & Truesdell, A.H.: Geochemical indicators of subsurface temperatures, 2. Estimation of temperature and fraction of hot water mixed with cold water. US Geological Survey, *J. Res.* 2:3 (1974), pp. 263–269.

Gat, J.R.: Comments on the stable isotope method in regional groundwater investigations. *Water Resour. Res.* 7 (1971), pp. 980–993.

Gat, J.R.: The isotopes of hydrogen and oxygen in precipitation. In: P. Fritz & J.Ch. Fontes (eds): *Handbook of environmental isotope geochemistry*, 1. 1980, pp. 22–48.

Gat, J.R. & Carmi, I.: Evolution of the isotopic composition of atmospheric waters in the Mediterranean Sea area. *J. Geophys. Res.* 75 (1970), pp. 3039–3048.

Gat, J.R. & Carmi, I.: Effect of climate changes on the precipitation patterns and isotope composition of water in a climate transition zone – Case of the eastern Mediterranean Sea area. In: The influence of climate change and climate variability on the hydrologic regime and water resources. *IAHS* 168, 1987, pp. 513–523.

Gat, J.R., Mazor, E. & Tzur, Y.: The stable isotope composition of mineral waters in the Jordan Rift Valley, Israel. *J. Hydrology* 76 (1969), pp. 334–352.

Gat, J.R. & Tzur, Y.: Modification of the isotopic composition of rainwater by processes which occur before groundwater recharge. *Proceedings Isotope Hydrology Symposium*, 1966, IAEA, Vienna, Austria, 1967, pp. 49–60.

Gemici, U., Tarcan, G., Çolak, M. & Helvacı, C.: Hydrogeochemical and hydrogeological investigations of thermal waters in the Emet area (Kutahya, Turkey). *Appl. Geochem.* 19 (2004), pp. 105–117.

Geyh M., D'Amore, F., Paces, T., Pang, Z. & Šilar, J.: Groundwater saturated and unsaturated zone vol. 4. In: Environmental isotopes in the hydrological cycle principles and applications. IAEA, Vienna, Austria, 2000, pp. 311–424.

Giggenbach, W.F.: Geothermal solute equilibria derivation of Na–K–Mg–Ca geoindicator. *Geochim. Cosmoschim. Acta* 52 (1988), pp. 2749–2765.

Giggenbach, W.F. & Corrales, R.S.: The isotopic and chemical composition of waters and steam discharges from volcanic magmatic-hydrothermal systems of the Guanacaste geothermal province, Costa Rica. *Appl. Geochem.* 7 (1992), pp. 309–332.

Giggenbach W.F. & Goguel R.L.: Collection and analysis of geothermal and volcanic water and gas discharges. Report No. CD 2401. Department of Scientific and Industrial Research, Chemistry Division, Petone, New Zealand, 1989.

Güneş, N.: *Hydrothermal alteration in the surrounding Keçiborlu sulphur mine.* Master Thesis, Akdeniz University, Isparta, Turkey, 1993.

Heasler, H.P., Jaworowski, C. & Foley, D.: Geothermal systems and monitoring hydrothermal features. In: R. Young, & L. Norby (eds): *Geological monitoring.* Geological Society of America, Boulder, CO, 2009, pp. 105–140.

Hem, J.D.: Study and interpretation of the chemical characteristics of natural water. US Geological Survey Water-Supply Paper, 2254, 1985.

Henley, R.W. & Ellis, A.J.: Geothermal systems, ancient and modern: a geochemical review. *Earth Sci. Rev.* 19 (1983), pp. 1–50.

Henley, R.W., Truesdell, A. & Barton, P.B. Jr. H.: Fluid mineral equilibrium in hydrothermal systems. Society of Economic Geologists. *Rev. Econ. Geol.* 1, 1984.

Huenges, E.: *Geothermal energy systems, exploration, development and utilization.* Wiley VCH, Weinheim, Germany, 2010.

Hulston, J.R.: Environmental isotope investigations of New Zeland geothermal systems, a review. *Geothermics* 12 (1983), pp. 223–232.

IAEA (International Atomic Energy Agency): Stable isotope hydrology, deuterium and oxygen-18 in the water cycle. Edited by J.R. Gat & Gonfiantini. *Technical Reports series* No. 210, Vienna, Austria, 1981.

IAEA (International Atomic Energy Agency): Isotopic composition of precipitation in the Mediterranean basin in relation to air circulation patterns and climate. IAEA-TECDOC-1453, ISBN 92-0-105305-3, ISSN 1011-4289, Vienna, Austria, 2005.

Karingithi, C.Y.: Chemical geothermometers for geothermal exploration. Presented at Short Course IV on Exploration for Geothermal Resources, organized by UNU-GTP, KenGen and GDC, at Lake Naivasha, Kenya, 2009.

Kharaka, Y.K. & Mariner, R.H.: Chemical geothermometers and their application to formation waters from sedimentary basins. In: N.D. Naser &T.H. McCollin (eds): *Thermal history of sedimentary basins.* Springer-Verlag, New York, 1989, pp. 99–117.

Kharaka, Y.Y. & Mariner, R.H.: Geothermal systems. In: P.K. Aggarwal, J.R. Gat & K.F.O. Froehlich (eds): *Isotopes in the water cycle: past, present and future of a developing science.* Springer 2005, pp. 243–270.

Kharaka, Y.K. & Thordsen, J.J.: Stable isotope geochemistry and origin of water in sedimentary basins. In: N. Clauer & S. Chaudhuri (eds): *Isotope signatures and sedimentary records.* Springer-Verlag, Berlin, Germany, 1992, pp. 411–466.

Kharaka, Y.K., Lico, M.S. & Law, L.M.: Chemical geothermometers applied to formation waters, Gulf of Mexico and California basins. *Bulletin American Association of Petroleum Geologists* 66 (1982).

Kharaka, Y.K., Specht, D.J. & Carothers, W.W.: Low to intermediate subsurface temperatures calculated by chemical geothermometers. *American Association of Petroleum Geologists, Annual Meeting* 69, New Orleans, 1985.

Kharaka, Y.K., Gunter, W.D., Affarwall, P.K., Perkins, E.H. & De Braall, J.D.: SOLMINEQ (a computer program code for geochemical modelling of water-rock interactions). US Geological Survey Water Investigations, Report 88-4227, 1988.

Kharaka, Y.K., Marnier, R.H., Bullen, T.D., Kennedy, B.M. & Sturchio, N.C.: Geochemical investigations of hydraulic connections between Corwin springs known geothermal resources area and adjacent parts of Yellowstone National Park. In: M.L. Sorey (ed): *Effects of potential geothermal development in the Corwin springs known geothermal resources area, Montana on the thermal features of Yellowstone National Park.* US Geological Survey Water Resources Investigations Report, 91-4052, 1991, pp. F1–F38.

Kharaka, Y.K., Thordsen, J.J. & White, L.D.: Isotope and chemical compositions of meteoric and thermal waters and snow from the greater Yellowstone National Park region. US Geological Survey Open-File Report 02-194, 57, 2002.

Koçak, A.: Water-rock interaction, alteration and hydrothermal alteration, Lecture notes of geological practice in geothermal, 2002.

Koga, A.: Geochemistry of the waters discharged from drillholes in Qtake and Hatchobaru areas (Japan). *Geothermics,* Special Issue 2, 2:2 (1970), pp. 1422–1425.

Langelier, W. & Ludwig, H.: Graphical methods for indicating the mineral character of natural waters. *J. Am. Water Ass.* 34 (1942), pp. 335–352.

Levitte, D. & Eckstein, Y.: Correlation between the silica concentration and the orifice temperature in the warm springs along the Jordan-Dead-Sea rift valley. *Geothermics* 7 (1979), pp. 1–8.

Lloyd, R.M.: Oxygen isotope behavior in the sulphate water system. *J. Geophys. Res.* 73:18 (1968), pp. 6099–6110.

Lyon, G.L. & Hulston, J.R.: Carbon and hydrogen isotopic compositions of New Zealand geothermal gases. *Geochim. Cosmochim. Acta* 48 (1984), pp. 1161–1171.

Mahon, W.A.J.: Silica in hot water discharged from drillholes at Wairakei, New Zealand. *New Zeal. J. Sci.* 9 (1966), pp. 135–144.

Marini, L.: Geochemical techniques for the exploration and exploitation of geothermal energy. Lecture notes. http://www.dipteris.unige.it/geochimica/Pesto/lectures/chile.pdf (2000) (accessed May 2013).

Marini, L., Chiodini, G. & Cioni, R.: New geothermometers for carbonate-evaporite geothermal reservoirs. *Geothermics* 15:1 (1986), pp. 77–86.

Marshall, W.L.: Amorphous silica solubilities – III. Activity coefficient relations and predictions of solubility behavior in salt solutions, 0–350°C. *Geochim. Cosmochim. Acta* 44 (1980), pp. 925–931.

McKenzie, W.F. & Truesdell, A.H.: Geothermal reservoir temperatures estimated from the oxygen isotope compositions of dissolved sulfate and water from hot springs and shallow drillholes. *Geothermics* 5 (1977), pp. 51–61.

Merlivat, L. & Jouzel, J.: Global climatic interpretation of the deuterium-oxygen-18 relationship for precipitation. *J. Geophys. Res.* 84 (1979), pp. 5029–5033.

Mizutani, Y.: Isotopic composition and underground tempwerature of the Otake geothermal water, Kyushu, Japan. *Geochem. J.* 6 (1972), pp. 67–73.

Mizutani, Y. & Rafter, T.A.: Oxygen isotope composition of sulphates. Oxygen isotope fractionation in the bisulphate ion-water system. *New Zeal. J. Sci.* 12:1 (1969), pp. 54–59.

Morey, G.W., Fournier, R.O. & Rowe, J.J.: The solubility of quartz in water in the temperature interval from 29 to 300°C. *Geochim. Cosmochim. Acta* 26 (1962), pp. 1029–1043.

Nicholson, K.: *Geothermal fluids, chemistry and exploration techniques.* Springer, Berlin, Germany, 1993.

Nieva, D. & Nieva, R.: Developments in geothermal energy in Mexico, Part 12 – Acationic composition geothermometer for prospecting geothermal resources. *Heat Recovery Syst.* 7 (1987), pp. 243–258.

Örmeci, S.: *Hydrogeological investigations of Sarkikaraağaç basin* (Isparta). Master Thesis, Süleyman Demirel University, Isparta, Turkey, 2006.

Panichi, C. & Gonfiantini, R.: Environmental isotopes in geothermal studies. *Geothermics* 6 (1978), pp. 143–161.

Piper, A.M.: A Graphic procedure in the geochemical interpretation of water analyses. *Trans. Am. Geophys. Union* 25 (1944), pp. 914–923.

Poisson, A., Yagmurlu, F., Bozcu, M. & Senturk, M.: New insights on the tectonic setting and evolution around the apex of the Isparta angle (SW Turkey). *Geol. J.* 38 (2003), pp. 257–282.

Qin, D., Turner, J.V. & Pang, Z.: Hydrogeochemistry and groundwater circulation in the Xi'an geothermal field, China. *Geothermics* 34 (2005), pp. 471–494.

Ragnarsdosttir, K.V. & Walther, J.W.: Pressure sensitive "silica geothermometer" determined from quartz solubility experiments at 250°C. *Geochim. Cosmochim. Acta* 47 (1983), pp. 941–946.

Reed, M.H. & Spycher, N.F.: Calculation of pH and mineral equilibria in hydrothermal waters with application to geothermometry and studies of boiling and dilution. *Geochim. Cosmochim. Acta* 48 (1984), pp. 1479–1492.

Rodgers, K.A., Browne, P.R.L., Buddle, T.F., Cook, K.L, Greatrex, R.A., Hampton, W.A., Herdianita, N.R., Holland, G.R., Lynne, B.Y., Martin, R., Newton, Z., Pastars, D., Sannazarro, K.L. & Teece, C.I.A.: Silica phases in sinters and residues from geothermal fields of New Zealand. *Earth-Sci. Rev.* 66:1–2 (2004), pp. 1–61.

Rozanski, K., Sonntag, C.H. & Munnich, K.O.: Factors controlling stable isotope composition of modern European precipitation. *Tellus* 34 (1982), pp. 142–150.

Runnels, D.D.: Inorganic chemical processes and reactions, Chapter 6. In: W.M. Alley (ed): *Regional Groundwater quality.* John Wiley & Sons, 1993.

Sahinci, A.: Geochemistry of natural waters. Reform Published, 33, İzmir, Turkey, 1991 (in Turkish).

Sakai, H. & Matsuhisa, Y.: Stable Isotope Geochemistry, pp. 403, University of Tokyo Press, Tokyo, 1996 (in Japanese).

Saracco, L. & D'Amore, F.: CO$_2$ B: A computer program for applying a gas geothermometer to geothermal systems. *Comput. Geosci.* 15 (1989), pp. 1053–1065.

Sarıız, K.: *Occurence of Keçiborlu sulphur mine and geology of the region.* Publication of Anadolu Unv., 91, PhD Thesis, Eskişehir, Turkey, 1985.

Sigvaldason, G.E.: Chemistry of thermal waters and gases in Iceland. *Bull. Volcanol.* 29 (1966), pp. 589–604.

Simmons, S.F. & Browne, P.R.L.: Hydrothermal minerals and precious metals in the Broadlands- Ohaaki geothermal system: implications for understanding low-sulfidation epithermal environments. *Econ. Geol.* 95:5 (2000), pp. 971–999.

Sonney, R. & Vuataz, F.D.: Validation of chemical and isotopic geothermometers from low temperature deep fluids of northern Switzerland. *Proceedings World Geothermal Congress 2010,* Bali, Indonesia, 2010.

State Hydraulic Works (SHW): Research on the origin of the karst water in upper Curuksu plain using isotope techniques. DSI-IAEA Project report, No: 4476/RB, Ankara, Turkey, 1989.

Tole, M.P., Armannsson, H., Zhong-He, P. & Arnorsson, S.: Fluid/mineral equilibrium calculations for geothermal fuids and chemical geothermometry. *Geothermics* 22:1 (1993), pp. 17–37.

Tonani, F.: Some remarks on the application of geochemical techniques in geothermal exploration. *Proceedings. Adv. Eur. Geoth. Res., Second Symposium,* Strasbourg, France, 1980, pp. 428–443.

Truesdell, A.H.: Summary of section III geochemical techniques in explanation. *Proceedings, Second United Nations Symposium on the Development and Use of Geothermal Resources,* San Francisco, V.1, US Government Printing Office, Washington DC, 1976.

Truesdell, A.H., Nathenson, M. & Rye, R.O.: The effects of subsurface boiling and dilution on the isotopic compositions of Yellowstone thermal waters. *J. Geophys. Res.* 82 (1977), pp. 3964–3704.

Truesdell, A.H. & Hulston, J.R.: Isotopic evidence of geothermal systems. In: P. Fritz & J.Ch. Fontes (eds): *Handbook of environmental geochemistry, the terrestrial environment.* Elsevier, Amsterdam, The Netherlands, 1980, pp. 179–226.

Varol, S. & Davraz, A.: Hydrogeochemical investigations of Barutlusu ve Pınarbaşı spring water (Tefenni/Burdur). Süleyman Demirel University, *Journal of Natural and Applied Sciences* 14:2 (2010), pp. 156–167.

Vengosh, A., Hening, S., Ganor, J., Mayer, B., Weyhenmeyer, C.E., Bullen, T.D. & Paytan, A.: New isotopic evidence for the origin of groundwater from the Nubian sandstone aquifer in the Negev, Israel. *Appl. Geochem.* 22 (2007), pp. 1052–1073.

Verma, M.P.: Limitations in applying silica geothermometers for geothermal reservoir evaluation. *Twenty-Fifth Workshop on Geothermal Reservoir Engineering*, Stanford University, Stanford, CA, SGP-TR-165, 2000.

Verma, M.P.: Silica solubility geothermometers for hydrothermal systems. *Water-Rock Interaction*. Swets & Zeitlinger, Lisse, The Netherlands, 2001, pp. 349–352.

Verma, S.P. & Santoyo, E.: New improved equations for Na/K, Na/Li and SiO_2 geothermometers by outlier detection and rejection. *J. Volcanol. Geoth. Res.* 79:1–2 (1997), pp. 9–23.

White, D.E.: Geochemistry applied to the discovery, evalution and exploitation of geothermal energy resources. *Geothermics*, Special Issue 2:1 (1970), pp. 58–80.

White, D.E., Brannock, W.W. & Murata, K.J.: Silica in hot-spring waters. *Geochim. Cosmochim. Acta* 10 (1956), pp. 27–59.

Wright, P.M.: Geochemistry, engineering and design guidebook, Chapter 4. *GHC Bulletin*, 1991.

Yagmurlu, F. & Sentürk, M.: The actual tectonic structure of southwest of Anatolia. *Turkey Quaternary Symposium, TURQUA 5*, 2005, pp. 51–61.

Yağmurlu, F., Savaşcın, Y. & Ergun, M.: Relation of alkaline volcanism and active tectonism within the evolution of Isparta Angle, SW-Turkey. *J. Geology* 105 (1997), pp. 717–728.

Yock, A.: Geothermometry. Presented at "Short Course on Surface Exploration for Geothermal Resources", organized by UNU-GTP and LaGeo, in Ahuachapan and Santa Tecla, El Salvador, 2009.

Yoshida, N.: Hydrogen and oxygen isotopes in hydrology. *The Textbook for the Eleventh IHP Training Course in 2001 – International Hydrological Programme*, 2001.

CHAPTER 6

Hydrochemical investigations of thermal and mineral waters in the Turgutlu-Salihli-Alaşehir plain (Gediz graben), western Turkey

Tuğbanur Özen & Gültekin Tarcan

6.1 INTRODUCTION

Turkey is the seventh richest country in the world in terms of geothermal potential (2705 MW$_t$ for direct use and 114.2 MW$_e$ for power production annually (Baba, 2012; 2013). The Aegean and Marmara regions show substantial potential for geothermal energy utilization (Faulds *et al.*, 2009). The first geothermal research and investigations in Turkey were started by the General Directorate of Mineral Research and Exploration (MTA) in the 1960s. Since then 222 geothermal areas, 95% of which are low- and medium-enthalpy fields, widely used for direct heat applications, have been discovered by MTA (Mertoğlu, 2010; Dağıstan, 2012). Turkey has around 1000 natural thermal and mineralized springs and 527 wells. Their installed capacity is 4764 MW$_t$ (Dağıstan, 2012; MTA, 2011). Geothermal areas in the western Turkey are located mainly along the major grabens such as Büyük Menderes, Gediz, and Küçük Menderes (Fig. 6.1). Geothermal fields in the study area are located within the Gediz graben (Fig. 6.2). Hydrogeological and hydrochemical characteristics of the thermo-mineral waters in the study area are described in this chapter. The study area is geographically divided into three main groups: Turgutlu, Salihli, and Alaşehir geothermal areas. The thermo-mineral waters are used for spa facilities, therapeutical purposes, and greenhouse and district heating applications. Additionally, several new deep wells have been drilled for power production.

6.2 GEOLOGICAL AND HYDROGEOLOGICAL SETTINGS

Western Anatolia is one of the most seismically active and rapidly expanding regions in the world (Baba and Sözbilir, 2012). This region has been experiencing N–S directed extension since at least the Late Oligocene–Early Miocene and is currently under the influence of forces exerted by subduction of the African Plate beneath the southern margin of Anatolian Plate along the Aegean-Cyprean subduction zone and the dextral slip on the North Anatolian Fault System (Fig. 6.1a). In response to pervasive crustal extension, geology of the southwest Turkey is dominated by numerous graben basins that are filled with Miocene to recent continental clastic rocks with volcanic rocks and minor carbonates. Based on the trend, these basins can be grouped into two as E–W-oriented and N–S-oriented, usually referred as cross-grabens (Fig 6.1b; Çiftçi and Bozkurt, 2009). The evolution of the N–S extension along the Gediz graben was characterized by two episodes: (i) rapid exhumation of metamorphic core complexes under presently low-angle normal faults from latest Oligocene to early Miocene and (ii) late stretching of crust producing E–W grabens along high-angle normal faults (rift mode) during Pliocene–Quaternary times (Bozkurt and Sözbilir, 2004).

The Menderes metamorphic rocks constitute the pre-Neogene basement and are exposed extensively over the horst blocks rising up to ∼2000 m elevation. They are composed mainly of schist, quartzite, phyllite, and marble. The Menderes metamorphic rocks are located in the southern part of the study area. Within the graben basins, pre-Neogene basement lies unconformably underneath the graben-fill. This fill varies in age from Miocene to recent and forms the cover units

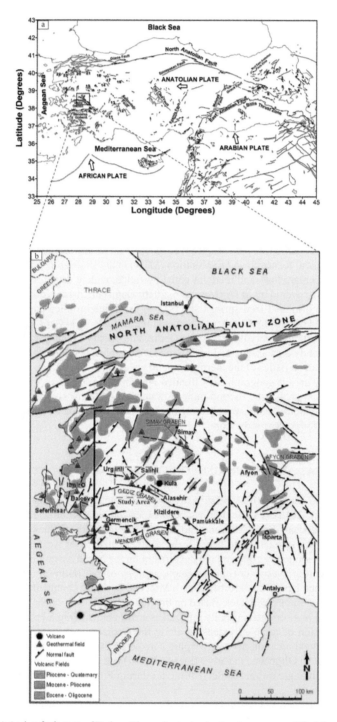

Figure 6.1.   (a) Active fault map of Turkey. The major tectonic structures are modified from Bayrak *et al.*,
2009. (b) Generalized geological map of western Turkey and graben structures (modified from
Faulds *et al.*, 2009). Black box surrounds the area in Faulds *et al.* (2009). Green box shows the
study area in this chapter.

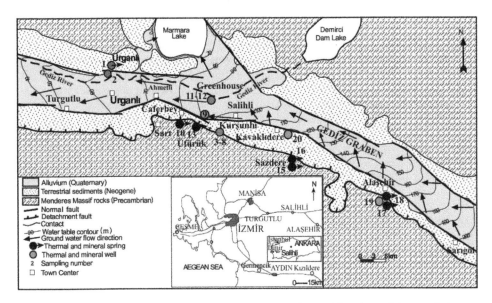

Figure 6.2. Simplified hydrogeological map of the Gediz graben and locations of the sampled waters (Tarcan *et al.*, 2005b).

(Bozkurt and Sözbilir, 2004; Çiftçi and Bozkurt, 2009; Cohen *et al.*, 1995; Koçyiğit *et al.*, 1999; Purvis and Robertson, 2005; Seyitoğlu *et al.*, 2002; Yılmaz *et al.*, 2000). Laterally, fluvial and alluvial conglomerates, sandstones, and mudstones constitute the predominant lithologies of the basin fill. Sedimentary rocks have been studied by many authors (Cohen *et al.*, 1995; Emre, 1996; Seyitoğlu and Scott, 1996; Yılmaz *et al.*, 2000) and they have created new formation names for sediments from the Early Miocene to Late Pliocene.

In the present study, the hydrogeological mapping of the Gediz graben was carried out to study the geological structure of the area (Fig. 6.1a,b). Geological units were divided into three main groups in accordance with the lithological and hydrogeological properties of the area. The basement of the study area consists of Menderes massif rocks that are made up of high- to low-grade metamorphics (gneiss, mica schists, phyllites, quartz schists, and marbles) and locally intruded granodiorites. Neogene terrestrial sediments, which are mainly composed by alluvium fan deposits (pebbles, pebbly sandstones, claystone–mudstones, interbedded conglomerate, claystones, siltstones, conglomerates comprising sandstone intercalations, and limestone), unconformably cover the basement rocks in different facies in the northern and southern parts of the Gediz graben. The topography of this unit shows a hard base-relief construction in the southern section. The northern section of Gediz graben shows a flat topography. In the southern section of the Gediz graben, terrestrial Neogene sediments overlie the Menderes massif rocks with a low-angle normal fault. This normal fault was identified as the "detachment fault" (Emre, 1996). High-angle normal faults are found in the middle section of the graben. The Quaternary alluvium, which is made up of unconsolidated granular sediments, is the youngest unit.

The carbonates (marbles and dolomitic marbles) of the Menderes massif rocks are highly fractured and karstified and act as an aquifer for both cold groundwater and thermal waters depending on the location. Fractured parts of granodiorite, gneiss, and quartz-schist units of the Menderes massif act as aquifers for low-salinity cold groundwater and for thermal waters. The terrestrial Neogene sediments, which are made up of alluvial fan deposits including poorly cemented clayey levels, have very low permeability and locally may act as cap rocks for the geothermal systems. Clayey levels of the Neogene sediments occur as impermeable barrier rocks. Sandy to gravelly and limestone sections of this Neogene unit form small aquifers. It is possible to withdraw groundwater with low-flow rates through shallow wells (5–15 m depth). The shallow regional aquifer consists of Holocene alluvial deposits in the middle of the study area (Fig. 6.2). Alluvium is the most important and favorable unit for cold groundwater production. It is possible to supply groundwater with 5–30 L/s discharges from shallow (20–150 m) wells. The circulation of thermal waters may probably be related to the high thermal gradient (3–5°C/10 m; Erentöz and Ternek, 1969; Uzunlar, 2009) owing to the graben structure (Çiftçi and Bozkurt, 2009).

Meteoric waters descending through the fractured zones are heated in the aquifer rocks, and rise to the surface along the faults. The stable oxygen isotope ($\delta^{18}O$) composition of the thermo-mineral waters ranges from -1.00% to -10.00% $\delta^{18}O$-VSMOW and deuterium isotope (δD) composition ranges from -40.00% to -63.00% δD-VSMOW. Tritium values of the waters vary from 0.20 to 6.15 TU. Isotopic data (^{18}O and D) suggest that the waters are of meteoric origin but shifted in ^{18}O, indicating deep circulation and water-rock interactions in the Salihli geothermal area (Bülbül, 2009; Filiz *et al.*, 1993; Özen, 2009; Özen *et al.*, 2008; 2010; 2012).

6.3   GENERAL INFORMATION OF THE GEOTHERMAL AREAS

Turgutlu, Salihli, and Alaşehir have become the largest industrial and agricultural towns in the Gediz graben, with a total population exceeding 300,000. Geothermal areas have been physically subdivided into eight main groups from west to east as follows: Urganlı, Sart, Caferbey, Üfürük, Greenhouse, Kurşunlu, Kavaklıdere, and Alaşehir thermal and mineral waters (Fig. 6.2). All the geothermal areas with the exception of Urganlı are located along the southern rim of the graben, which is more active than the northern rim. Outlet temperatures have been measured in the thermal waters from 22 to 98°C with electrical conductivity values of about 906–6020 $\mu S/cm$ and the pH ranging from 4.9 to 8.3.

Summary of geothermal wells is given in Table 6.1. Thermal springs in the Turgutlu-Urganlı area are well distributed and emerge with outlet temperatures between 20 and 83°C, and with 50–100 L/s total discharge (MTA, 2005; Tarcan and Gemici, 2005). U-1 well (Urganlı) was drilled by MTA in 2002 with a depth of 460 m. The water of the artesian well has a temperature of 62°C and discharge is 20 L/s. This well was used for supplying the greenhouse plants with CO_2, which was the first application of its kind in Turkey (Tarcan *et al.*, 2005a). At the present, thermal waters produced from six additional wells are used for this purpose. Sart, Caferbey, Üfürük, and Kurşunlu geothermal areas are located in the southern part of the Gediz graben (Fig. 6.2). Sart thermal springs are situated in westernmost and within the ancient city of Sardes. Thermal waters having 52°C and 5 L/s yields have been used for spa facilities since ancient times. Caferbey geothermal area is one of the most important areas in terms of temperature. The first well was drilled in 1990 by MTA and the maximum bottom-hole temperature was found to be 155°C with 2 L/s discharge at 1189.1 m depth (Karamanderesi, 1997). Because of the low discharge, fluid production has not been considered economical; however, new exploration studies continued for several years in the northern field. The Üfürük (meaning "puffs out" in Turkish) geothermal area has several mineral springs and natural gas (e.g., CO_2, H_2S) outlets. The mineral springs have an outlet temperature between 20 and 30°C with an average discharge of 0.4 L/s.

The most important geothermal area in the study region is located within Kurşunlu valley. Since 1969,20 wells have been drilled in Kurşunlu geothermal area. Most of the wells are shallow in depth (between 40 and 400 m). The K-1 well at 42.5 m, was firstly drilled in 1976 by MTA, and yielded a discharge of 20 L/s at 90°C. In the years between 1992 and 1998, four shallow wells were drilled with varying depths (70–114 m). These wells have temperatures between 83 and 94°C and discharges from 40 to 80 L/s. The geothermal district heating project was initiated at Kurşunlu geothermal area in 2000. At present, 5900 residences, hotels, and greenhouses are heated with geothermal energy in the surroundings of Salihli town. Total discharge of production obtained from the wells in a day measured 160 L/s during the winter season. New wells (950–1500 m) have been drilled by the Salihli Municipality since 2009. These wells have varying temperatures between 90 and 92°C and discharges from 30 to 35 L/s. The two reinjection wells (K-1 and K-17)

Table 6.1. Geothermal drilling information of the Gediz graben (Yılmazer *et al.*, 2010, well information is before 2010).

Geothermal area	Number of wells	Depth (m)	Temp. (°C)	Total discharge (L/s)	Reservoir rocks	Cap rocks
Turgutlu	10	280-605	62–83°C (well head temp.)	2–40	Paleozoic to Mesozoic Menderes massif rocks (marble, schist, and quartzite)	Neogene intercalated siltstone, claystone, sandstone, conglomerate
Salihli	25	40–1189	57–155°C (bottom-hole temp.)	5–40	Paleozoic to Mesozoic Menderes massif rocks (marble, schist, and quartzite)	Neogene intercalated siltstone, claystone, sandstone, conglomerate
Alaşehir	7	450–1507	56–215°C (bottom-hole temp.)	8–12	Paleozoic to Mesozoic Menderes massif rocks and quartzite)	Neogene intercalated siltstone, claystone, sandstone, conglomerate

are located in this area. Thermal waters have been widely used for swimming pools, bathing, and medicinal purposes in addition to heating the district and its greenhouses as mentioned above in Kurşunlu. The greenhouse geothermal area is located in the north of the study region. At present, thermal water of about 85°C and 60 L/s discharge is used for cultivation.

The Alaşehir geothermal area is located in the easternmost part of the southern rim of the Gediz graben (Fig. 6.2). There are two springs with low temperatures (27–31°C) with a discharge of 10 L/s. KG-1 and AK-2 (215°C) were drilled by MTA in 2002 and 2004, respectively. The bottom-hole temperature in KG-1 was found to be 183°C with 12 L/s discharge at 1447 m depth (Bülbül, 2009). The AK-2 well has a temperature of 213°C with discharge rate of 6.74 L/s artesian flow at 1501 m depth. The bottom-hole temperature of KG-1 is 215°C, the third highest-temperature well in Turkey (Karahan *et al.*, 2002). The other important deep wells reported in Alaşehir have a depth ranging from 1750 to 2750 m with a discharge of 5–90 L/s and a temperature of 159–287.5°C (well names MAK2010/14–15 and MAK2011/3 [MTA, 2011]). Sarıkız mineral spring (sample 17) has for many years produced bottled mineral water for consumption. Alaşehir thermal spring and wells have been used for greenhouse heating and could also be used for electricity production in the future. At present, many deep geothermal wells are being drilled and several power plant installations are tendered by private companies for electricity production.

6.4 HYDROCHEMICAL SETTINGS

For the investigation and comparison of the major chemical characteristics, water samples were collected from natural springs and wells. The results obtained from the chemical analyses for this study and those of some analyses of previous investigations are shown in Table 6.2. Table 6.2 also includes the classification of waters according to the principles published by IAH (1979). The total equivalents for cations and anions are recorded as 100%, respectively, and ions contributing more than 20 meq% are used for the classification. Five different thermal water types corresponding to $Na–HCO_3$, $Ca–Na–HCO_3$, $Ca–Mg–SO_4$, $Na–Mg–HCO_3$, and $Mg–Ca–HCO_3$ waters, were identified in Urganlı-Kurşunlu-Caferbey- Greenhouse-Alaşehir, Sart, Üfürük, Sazdere, and Sarıkız

Table 6.2. Hydrochemical compositions of the waters in the Gediz graben*.

No.	Location	Sampling date	T (°C)	pH	EC (μS/cm)	Depth (m)	K (mg/L)	Na (mg/L)	Ca (mg/L)	Mg (mg/L)	As (μg/L)	B (mg/L)	Li (mg/L)	Cl (mg/L)	SO4 (mg/L)	HCO3 (mg/L)	SiO2 (mg/L)	H2S (mg/L)	Water type
1	[1]Urganlı thermal spring	8.08.2002	55	6.53	2620	–	50	525	45	16	<0.03	6.70	1.09	85	55	1576	42.15	–	Na–HCO3
2	[1]Urganlı thermal well	8.08.2002	60	6.93	2870	460	49	511	91	23	<0.03	7.10	1.17	84	69	1679	30.81	–	Na–HCO3
3	[2]K–1 reinjection well	29.09.2006	39	7.25	2230	40	49.71	407.33	83.19	16.48	106.2	43.77	2.45	77	80.9	1403.4	231.1	28.70	Na–HCO3
4	[2]K-2 well	29.09.2006	44.5	7.91	2630	73	84.42	660.22	46.82	13.92	367.0	60.41	5.14	125	68.9	1339.9	322.1	24.45	Na–HCO3
5	[2]K-5 well	30.08.2007	54.9	6.59	2270	115	51.07	388.89	96.46	18.73	571.8	40.22	2.62	65	86.2	1460.7	187.2	31.89	Na–HCO3
6	[3]K-8 well	12.09.2008	57	6.67	1967	112	31.20	241.87	191.40	37.50	49.9	22.98	1.58	53	140.7	1249.6	104.9	52.09	Na–Ca–HCO3
7	[2]K-15 well	29.09.2006	38.6	6.52	2610	250	82.14	639.90	31.34	7.86	45.3	61.40	4.49	130	26.9	1684.0	396.1	9.57	Na–HCO3
8	[3]K-19 well	13.02.2008	35.1	7.34	2010	160	47.68	340.24	97.92	17.42	151.0	33.55	2.94	65	143.8	1161.9	82.7	143.81	Na–HCO3
9	[4]Caferbey-1 well	1990	90	7.8	2700	1189	70	680	42	6.1	–	67		115	34	1983	214	–	Na–HCO3
10	[3]Sart thermal spring	18.09.2008	43	6.33	1506	–	24.99	155.30	197.70	24.63	21.2	10.59	1.36	27	71.9	976.3	98.6	25.51	Ca–Na–HCO3
11	[3]Greenhouse deep well-1	30.08.2007	54	7.54	1991	958	7.34	494.18	8.69	2.40	198.8	18.35	0.22	81	209.7	906.7	113.9	77.60	Na–HCO3
12	[3]Greenhouse deep well-2	30.08.2007	36.3	7.75	1963	1000	7.09	461.73	6.18	1.35	69.2	18.00	0.22	79	209.66	984.8	94.00	77.60	Na–HCO3
13	[2]Üfürük mineral water	18.09.2008	22	4.90	3560	–	11.95	56.72	709.84	249.84	14.9	<0.05	0.06	54	3591.9	20.7	67.5	1274.30	Ca–Mg–SO4
14	[2]Kurşunlu mineral water	12.09.2008	26	6.71	1906	–	5.186	20.45	388.24	63.19	173.9	.14	<0.0001	20	284.3	1037.3	29.8	105.24	Ca–Mg–HCO3–SO4
15	[1]Sazdere mineral spring	8.08.2002	29	6.35	2450	–	33	353	119	134	<0.03	36.4	2.41	95	78	1991	63.97	–	Na–Mg–HCO3
16	[1]Sazdere thermal spring	8.08.2002	34	6.35	2450	–	66	796	130	68	<0.03	85.8	0.96	179	91	1471	24.60	–	Na–HCO3
17	[1]Sarıkız mineral spring	8.08.2002	27	6.01	906	–	5.7	26	90	57	<0.03	2	0.15	16	104	544	12.84	–	Mg–Ca–HCO3
18	[1]Alaşehir thermal spring	8.08.2002	31	6.83	874	–	12	219	122	135	<0.03	13.30	0.96	67	64	1471	24.60	–	Mg–Na–Ca–HCO3
19	[1]Alaşehir thermal well	17.10.1997	37	6.38	4270	–	49	506	311	162	–	1.26	–	101	4	3115	71.24	–	Na–Ca–Mg–HCO3
20	[5]KG-1 thermal well	20.09.2002	95	8.31	6020	1447	78.9	2027	12.8	3.1	0.009	86.6	5.28	165	1843	2950	261	–	Na–HCO3–SO4
21	[6]AK-2 thermal well	2005	98	8.20	3180	1507	99.8	810	12.3	<1		131	–	162	119	2127	242	–	Na–HCO3

*Sample numbers are the same as in Figure 6.2. Blanks refer no records and concentration limits.

[1]Tarcan et al. (2005b)

[2]Özen (2009)

[3]Özen et al. (2012)

[4]Karamanderesi (1997)

[5]Tarcan (2005)

[6]Yılmazer et al. (2010)

geothermal areas, respectively. Figure 6.3a shows the relative concentrations of Na+K, Ca, and Mg of waters. Most of the waters plot in the Na+K area except samples 9, 10, 13, 14, and 17, which are located in the Ca area. The relative Cl, HCO_3, and SO_4 concentrations of the waters are shown on Figure 6.3b. The main components of the waters are plotted in the trilinear Cl–SO_4–HCO_3 diagram (Giggenbach, 1988, 1991; Fig. 6.3b). All of the samples are plotted in the peripheral waters (HCO_3-rich part) of the diagram. Na–HCO_3-type thermal waters are produced by calcite and dolomite and silicate dissolution and ion exchange reactions in deep aquifers and have high temperature (Gemici and Tarcan, 2002). Aquifer rocks in the study area are composed by metamorphics (marble, schist, and quartzite) and granodiorite. The waters emerging from such aquifers are rich in calcium, sodium, and bicarbonate ions. The solution of calcareous materials in geothermal systems leads to an increase in Ca^{2+} which is then exchanged for Na^+ from clay minerals which are associated with the cap rocks (Neogene sediments) of the geothermal systems in the Gediz graben. The geochemical processes in thermal waters are the result of the combined effects from the dissolution of silicates, carbonates, mixing phenomena, and ion exchange reactions. The expected type of thermal waters in the deep geothermal reservoir is Na–HCO_3 at high temperatures (Özen *et al.*, 2012). Some of the thermal waters, such as Üfürük mineral spring (sample 11) and KG-1 well (sample 20), contain relatively high sulfate concentrations (3591.9 and 1843 mg/L in Üfürük and KG-1 wells, respectively), the source of which may be dissolution of minerals like gypsum (Gökgöz, 1998; Tarcan *et al.*, 2005b). Hydrogeochemical processes for the cold waters seem to be carbonate solution and silicate weathering reactions. As shown in Table 6.1, As and B concentrations of the thermal waters of the Salihli geothermal fields are very high reaching 86 mg/L for B and 571.8 µg/L for As. The source of the boron and arsenic in thermal waters can be explained by the combined effects of geological formations, water-rock interaction, and degassing of magma intrusions (Gemici and Tarcan, 2002; Mutlu *et al.*, 2008; Özen *et al.*, 2012; Özgür *et al.*, 1998; Pasvanoğlu and Chandrasekharam, 2011; Tokçaer, 2007).

6.4.1 *Geothermometry applications*

6.4.1.1 *Chemical geothermometers*
Aquifer temperatures of all the thermal waters in the study area were estimated using chemical geothermometers, which are based on temperature-dependent water-rock equilibria and give the last temperature of water-rock equilibrium attained in the aquifer (Nicholson, 1993). Table 6.3 shows the results of some of the chemical geothermometer applications.

As can be seen in Table 6.3, the temperatures of silica geothermometers vary from 40 to 232°C. The Na–Li geothermometer (Kharaka *et al.*, 1982) results for all samples gave temperatures higher than measured bottom-hole temperatures. The K–Mg geothermometer (Giggenbach *et al.*, 1983) results range between 44 and 107°C. This geothermometer gives the best match with measured temperature of the greenhouse-2 well (sample 12). The Na–K geothermometers gives mostly temperatures higher than measured down-hole for Urganlı (sample 1 and 2), Kurşunlu (samples 3–8), Caferbey (sample 9), Sart (sample 10), Sazdere (sample 18), and Alaşehir (sample 19). The Mg–Li (Kharaka and Mariner, 1989) and the Na–K geothermometer (Arnorsson *et al.*, 1983; Truesdell, 1976) results are very close to the measured temperature for KG-1 and AK-2 wells (182 and 215°C, respectively; samples 20 and 21). To summarize, the chemical geothermometry results indicate that reservoir temperatures of the thermo-mineral waters vary between 80 and 250°C.

6.4.1.2 *The ternary (Na–K–Mg) diagram*
The Na–K–Mg diagram (Fig. 6.4) proposed by Giggenbach (1988) is often used to evaluate aquifer temperatures and to recognize waters, which have attained equilibrium with the host rocks and waters affected by mixing and/or reequilibration at low temperatures along their circulation path. Immature waters plot rather close to the Mg corner. Most of the waters fall into the immature waters field, because these waters are not in equilibrium with reservoir rocks, and are probably dominated by rock dissolution, mixing with cold groundwater, and ion exchange reactions. Moreover, the application of cation geothermometers must be questioned. The thermal waters in the study area

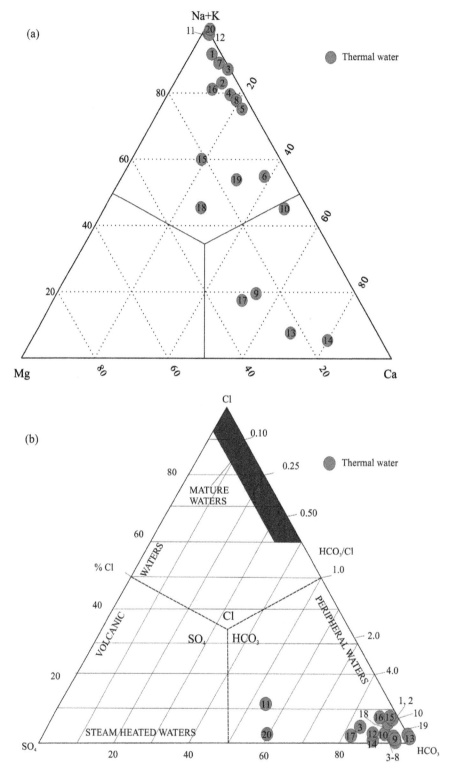

Figure 6.3. (a) The Na + K–Ca–Mg diagram (modified from Giggenbach, 1991) for the waters, (b) the Cl–HCO₃–SO₄ diagram (modified from Giggenbach, 1991) for the waters (sample numbers are the same as in Table 6.2 corresponding to the locality numbers as shown in Fig. 6.2).

Table 6.3. Reservoir temperatures (°C) of the thermal waters in the Gediz graben estimated using chemical geothermometers[1].

No.	Measured T (°C)	SiO₂ Chalcedony[a]	SiO₂ Chalce. steam loss[b]	SiO₂ Quartz[a]	SiO₂ Quar. steam loss[a]	Na/Li[c]	K–Mg[d]	Mg/Li[e]	Na/K[b]	Na/K[f]	Na/K[g]	Na/K[h]
1	80	*	*	94	96	182	102	90	197	183	212	213
2	77	*	*	80	84	192	96	87	198	183	212	213
3	90	170	100	190	176	292	105	131	235	228	261	258
4	94	200	98	215	196	287	107	131	232	225	259	256
5	83	154	96	174	165	285	99	120	234	228	261	258
6	57	118	109	140	135	264	79	89	227	217	253	250
7	115	220	110	232	219	290	*	138	235	229	262	259
8	52	99	95	127	124	288	100	122	234	227	261	258
9	155	164	160	185	172	*	*	*	205	191	235	229
10	52	115	106	142	136	290	78	91	249	246	273	272
11	80	119	115	145	139	108	*	*	86	*	115	99
12	83	107	103	134	130	111	83	*	88	*	117	101
13	22	88	84	103	115	150	38	*	282	285	300	302
14	26	48	44	79	83	*	35	*	302	317	320	326
15	29	85	81	85	113	267	74	93	210	181	227	220
16	34	101	*	129	125	274	91	122	292	168	201	201
17	27	*	*	48	55	*	38	95	286	291	304	307
18	31	*	36	48	55	234	44	57	141	291	172	295
19	37	40	170	71	76	*	72	*	*	131	213	170
20	182	*	176	*	*	200	*	180 *	184	*	214	
21	215	*	*	*	*	*	*	211	217	250	*	245

[1] Sample numbers are as in Figure 6.2 and Table 6.2. Measured down-hole temperatures and spring discharge temperatures are shown in second column for comparison. * denotes "not applicable."
[a] Fournier (1977).
[b] Arnorsson *et al.* (1983).
[c] Kharaka *et al.* (1982).
[d] Giggenbach *et al.* (1983).
[e] Kharaka and Mariner (1989).
[f] Truesdell (1976).
[g] Giggenbach *et al.* (1988).
[h] Fournier (1979).

have a fast hydrological circulation and/or mix with the cold groundwater in different depths. Hence, thermo-mineral waters are not equilibrium with reservoir rocks. Additionally, greenhouse thermal waters (samples 11 and 12), KG-1 well (sample 20), and AK-2 well (sample 21) samples fall into the partially equilibrated waters field. This combination geothermometer diagram (Fig. 6.4) shows that the reservoir temperatures of the thermal systems range between 100°C and 180°C.

Saturation indices can be used as geothermometers by plotting temperature *versus* saturation index (*SI*) diagrams (Reed and Spycher, 1984). If the *SI* with respect to several minerals converges to zero at a particular temperature, this temperature corresponds to the most likely mineral solution equilibrium temperature or at least the equilibrium temperature of that particular water. Using the results of the aqueous speciation calculations, the *SI* of minerals in aqueous solutions at different temperatures was estimated:

$$SI = \log Q/K = \log Q - \log K$$

where Q is the calculated ion activity product, and K is the equilibrium constant. The *SI* value for each mineral gives an estimate of the equilibrium. Values of *SI* greater than, equal to, and

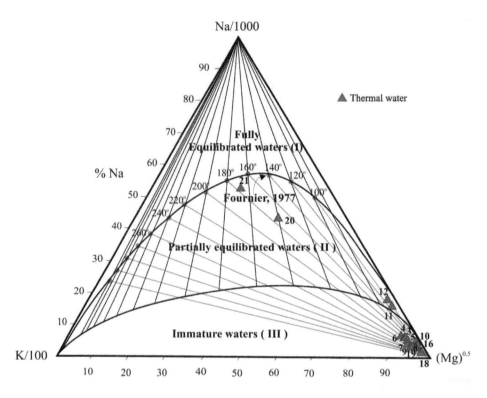

Figure 6.4.   Water composition of the thermal waters in a Na–K–Mg ternary diagram (Giggenbach, 1988; sample numbers are the same as in Table 6.2 and Fig. 6.1).

less than zero represent oversaturation, equilibrium, and undersaturation, respectively, of a fluid regarding a specific mineral (Xilai *et al.*, 2002).

Version 2.1A of the WATCH (Arnórsson *et al.*, 1982; Bjarnason, 1994) is a tool for interpreting the composition of thermal fluids. It is applied to calculate their chemical speciation for different geothermometer temperatures and is also applicable to the analysis of other geochemical problems. The ion activity products (log Q) and solubility products (log K) of selected geothermal minerals are also computed for the thermal waters. Log (Q/K) *versus* temperature diagrams for the thermal waters are presented in Figure 6.5. These diagrams show that solute-mineral equilibria for Urganlı, Kurşunlu, Greenhouse, Sazdere, and Alaşehir geothermal areas seem to be nearly attained, and the equilibrium temperatures are in the range of 70–180°C. In Figure 6.5, the curves for Caferbey-1 well, Sart, and Alaşehir AK-2 well do not converge on the $SI = 0$ line. These samples are not in equilibrium with the minerals probably owing to mixing with cold water. The methods for studying water-mineral equilibria indicate that there may be an exploitable reservoir with a probable temperature of 70–180°C for the Turgutlu and Salihli geothermal fields. In addition, Bülbül *et al.* (2011) performed same methods, and reservoir temperatures for the Alaşehir geothermal field were estimated to vary between 125 and 225°C.

6.4.2   Mineral saturation

Prediction of scaling tendencies of geothermal waters is important when evaluating the production characteristics of geothermal aquifers and for taking the necessary precautions to prevent or control scale formation. Assessment of scaling tendencies involves calculation of the saturation state of the scale-forming minerals.

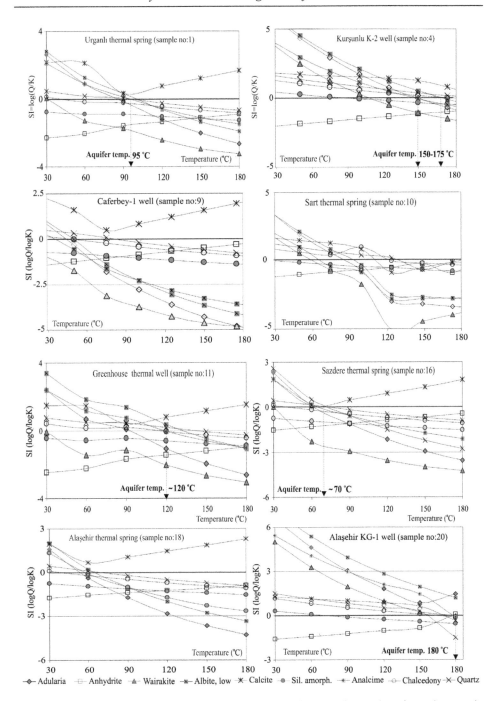

Figure 6.5. Changes in the saturation states of thermal waters in the study area (sample numbers are the same as in Table 6.2 and Fig. 6.2).

Saturation indices in the aquifer waters were calculated at the outlet temperature, and the measured using the PHREEQC computer code (Parkhurst and Appelo, 1999) with the LLNL database for the following minerals: anhydrite, aragonite, calcite, dolomite, amorphous silica, chalcedony, gypsum, and quartz. Saturation indices for each mineral were then plotted

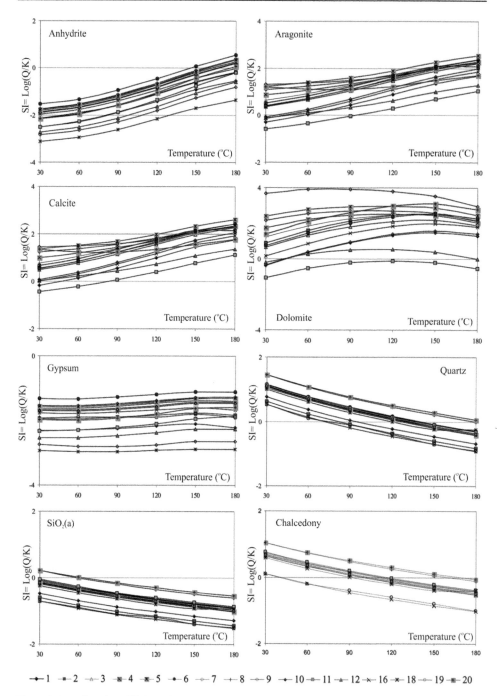

Figure 6.6.   Mineral equilibrium diagrams for the thermal waters from the study area (Sample numbers are the same as in Table 6.2 and Fig. 6.2).

versus temperature and trend curves depicted (Fig. 6.6). In Figure 6.6, saturation indices for anhydrite in samples 1, 2, 6, 9, 10, 11, 12, and 18 are positive at temperatures above ~120–170°C. The other samples are undersaturated with respect to these minerals at all temperatures.

All thermal waters are oversaturated with respect to calcite except for samples 1 and 18. Saturation indices for calcite for the samples are positive at temperatures above 50 and 85°C, respectively. Aragonite shows a similar trend as the calcite for most of the waters. Saturation indices for aragonite in samples 1, 10, 11, 12, and 16 are positive at temperatures above 60, 35, 92, 50, and 48°C, respectively. The other samples are oversaturated with respect to aragonite at all temperatures. The thermal waters are oversaturated with respect to dolomite except for samples 1, 10, 11, and 12. Saturation indices for this mineral in samples 1, 10, and 12 are positive at temperatures above ~40°C. The thermal water in sample 11 is undersaturated at any temperature. All of the thermal waters are undersaturated with respect to gypsum at all temperatures.

Saturation indices for quartz in samples 7, 20, and 10 are positive at all temperatures. The other samples are oversaturated with respect to this mineral at temperatures below ~70–140°C. All thermal waters are undersaturated with respect to amorphous silica except for samples 9, 20, and 21. The samples are oversaturated at below ~60°C. The saturation indices of chalcedony for 9 and 20 are oversaturated at above ~160 and 162°C, respectively. The other samples are oversaturated with respect to this mineral at temperatures below ~45–170°C.

Calcium is the main common ion present in the processes controlling the geochemical evolution of the Turgutlu, Salihli, and Alaşehir geothermal systems. In the shallow and deep geothermal wells, aquifer water compositions are always close to carbonate mineral saturation (Bülbül *et al.*, 2011; Özen *et al.*, 2010; 2012). Calcite, aragonite, and dolomite oversaturations are maintained to low temperatures, and the degree of oversaturation increases with rising temperature. At lower temperatures, the relative effect of retrograde solubility of these minerals becomes more important. The results show that carbonate scaling risk may start at temperatures above ~40–90°C.

6.5   CONCLUSIONS

Hydrogeological units in the Gediz graben were divided into three main groups: Precambrian to Paleocene Menderes massif basement rocks, overlying Neogene terrestrial sediments, and Quaternary alluvium, from bottom to top. The Menderes massif rocks, which are composed of metamorphics and granodiorite, host the aquifer of the thermal waters. The Neogene sediments act as cap rock of the geothermal reservoirs. Heat sources of the systems might be related to the high thermal gradient caused by the graben structure of the area.

Most of the thermo-mineral waters are of Na–HCO_3 type. The dominant anion and cation are HCO_3^- and Na^+, respectively. Hydrogeochemical processes for the cold waters seem to be carbonate solution and silicate weathering reactions. As and B concentrations of the thermal waters of the study area are very high reaching 86 mg/L for B and 571.8 μg/L for As. Water-rock interaction and degassing of magma can explain the high amount of boron and arsenic in the thermal waters. The high B and As concentrations of thermo-mineral waters cause environmental problems in groundwater aquifers, surface waters, and soils in agricultural areas (especially vinicultures) of the Gediz graben. To prevent boron and arsenic contamination of the water used for irrigation in the study area, reinjection of the produced thermal waters into the geothermal reservoir is necessary.

Geothermometer results show that the minimum and the maximum aquifer temperatures of the study area are likely to be 80 and 250°C, respectively. Aquifer temperatures of deep wells in the study area support geothermometer results.

During the extraction and using stages of thermal waters in the study area, precipitation of carbonate minerals (calcite, aragonite, and dolomite) is the most critical scaling problem. The results show that carbonate scaling risk may start above ~40–90°C. To avoid scaling problems in reinjection wells and surface equipment, effluent water should be reinjected at suitable temperatures.

ACKNOWLEDGMENTS

The authors acknowledge the partial financial supports of TÜBİTAK Research Fund (Project no. YDABAG 102Y039) and DEÜ BAP Research Fund (Project no. KB.FEN.016).

REFERENCES

Arnórsson, S., Gunnlaugsson, E. & Svavarsson, H.: The chemistry of geothermal waters in Iceland–I. Calculation of aqueous speciation from 0 to 370 C. *Geochim. Cosmochim. Acta* 46 (1982), pp. 1513–1532.

Arnórsson S., Gunnlaugsson E. & Svavarsson, H.: The chemistry of geothermal waters in Iceland. III. Chemical geothermometry in geothermal investigations. *Geochim. Cosmoschim. Acta* 47 (1983), pp. 567–577.

Baba, A.: Present energy status and geothermal utilization in Turkey, *39th IAH Congress*, 16–21 September 2012, Niagara Falls, Canada, (2012), pp.1–3.

Baba, A.: Environmental impact of the utilization of geothermal areas in Turkey. *Geopower TURKEY Conference*, Istanbul, Turkey, 2013, pp. 1–61.

Baba, A. & Sözbilir, H.: Source of arsenic based on geological and hydrogeochemical properties of geothermal systems in western Turkey, *Chem. Geol.* 334 (2012), pp. 364–377.

Bayrak, Y., Öztürk, S., Çınar, H, Kalafat, D., Tsapanos, T.M., Koravos, G. Ch., & Leventakis, G.A.: Estimating earthquake hazard parameters from instrumental data for different regions in and around Turkey. *Eng. Geol.* 105 (2009), pp. 200–210.

Bjarnason, J.O.: The speciation program WATCH (version 2.1). Orkustofnun, Iceland, 1994.

Bozkurt, E. & Sözbilir, H.: Geology of the Gediz graben: new field evidence and tectonic significance. *Geol. Mag.* 141 (2004), pp. 63–79.

Bülbül, A.: *Hydrogeological and hydrochemical studies of the Alaşehir (Manisa) thermal and cold waters systems.* PhD Thesis, Dokuz Eylül University, Institute of Natural and Applied Sciences, Geological Engineering, İzmir, Turkey, 2009 (in Turkish).

Bülbül, A., Özen, T. & Tarcan, G.: Hydrogeochemical and hydrogeological investigations of thermal waters in the Alasehir-Kavaklidere area (Manisa-Turkey). *Afr. J. Biotechnol.* 10:75 (2011), pp. 17,223–17,240.

Cohen, H.A., Dart, C.J., Akay, Z. H.S. & Barka, A.: Syn-rift sedimentation and structural development of the Gediz ve Büyük Menderes Graben, western Turkey. *J. Geol. Soc. London* 152 (1995), pp. 629–638.

Çiftçi, N.B. & Bozkurt, E.: Pattern of normal faulting in the Gediz graben, SW Turkey. *Tectonophysics* 473 (2009), pp. 234–260.

Dağıstan, H.: Geothermal sources potential in Turkey, applications, sectorel progress and projection of 2015. *Geothermal Sources in Turkey, Explorations and Applications Symposium,* 8–9 November 2012, Istanbul Technical University, Istanbul, Turkey, 2012, p. 1.

Emre, T.: Tectonism and geology of the Gediz graben. *Turkish J. Earth Sci.* 5 (1996), pp. 171–185 (in Turkish).

Erentöz, C. & Ternek, Z.: Thermomineral sources and geothermic energy etudes in Turkey. General Directorate of Mineral Research and Exploration (MTA), Ankara, Turkey, 1969 (in Turkish).

Faulds, J.E., Bouchot, V., Moeck, I., & Oguz, K.: Structural controls of geothermal systems in western Turkey: A preliminary report. *Geotherm. Resour. Coun. Trans.* 33 (2009), pp. 375–383.

Filiz, S., Gökgöz, A. & Tarcan, G.: Hydrogeologic comparisons of geothermal fields in the Gediz and Büyük Menderes grabens. *Congress of the World Hydrothermal Organisation,* 13–18 May 1992, İstanbul, Turkey, 1993, pp. 129–153.

Fournier, R.O.: Chemical geothermometers ve mixing models for geothermal systems. *Geothermics* 5 (1977), pp. 41–50.

Fournier, R.O.: A Revised equation for the Na–K geothermometer. *Geotherm. Resour. Counc. Trans.* 3 (1979), pp. 221–224.

Gemici, Ü. & Tarcan, G.: Hydrogeochemistry of the Simav geothermal field, western Anatolia, Turkey. J. *Volcanol. Geotherm. Res.* 116:3–4 (2002), pp. 215–233.

Giggenbach, W.F.: Geothermal solute equilibria: derivation of Na–K–Mg–Ca geoindicators. *Geochim. Cosmochim. Acta* 52 (1988), pp. 2749–2765.

Giggenbach, W.F.: Chemical techniques in geothermal exploration. In: F. D'Amore (ed): *Application of geochemistry in geothermal reservoir development.* UNITAR/UNDP, Rome, Italy, 1991, pp. 119–143.

Giggenbach, W.F., Gonfiantini, R., Jangi, B.L. & Truesdell, A.H.: Isotopic and chemical composition of Parbati valley geothermal discharges, NW Himalaya, India. *Geothermics* 5 (1983), pp. 51–62.

Gökgöz, A.: Geochemistry of the Kizildere-Tekkehamam-Buldan-Pamukkale geothermal fields, Turkey. Reports, UNU GTP, Orkustofnun, Reykjavik, Iceland, 1998, pp. 115–156.

IAH: Map of mineral ve thermal water of Europe. Scale 1:500.000. International Association of Hydrogeologists, United Kingdom, 1979.

Karahan, Ç., Barkraç, S. & Dünya, H.: Evaluation of the KG-1 exploratory drilling in the Alaşehir - Kavaklıdere Göbekli. In: A.H. Deliormanlı & C. Seçkin, C. (eds): *Drilling Symposium,* 10–11 April 2002, İzmir, Turkey, 2002, pp. 39–43 (in Turkish).

Karamanderesi, İ.H.: Potential and future of the Salihli-Caferbey (Manisa, Turkey) geothermal fields. *Seventh Energy Congress of Turkey,* World Energy Council Turkish National Committee, İzmir, Turkey, 1997, pp. 247–261 (in Turkish).

Kharaka, Y.K. & Mariner, R.H.: Chemical geothermometers and their application to formation waters from sedimentary basins. In: N.D. Naeser & T.H. McCulloh (eds): *Thermal history of sedimentary basins, methods and case histories.* Springer-Verlag, New York, 1989, pp. 9–117.

Kharaka, Y.K., Lico, M.S. & Law, L.M.: Chemical geothermometers applied to formation waters, Gulf of Mexico and California basins. *Am. Assoc. Petrol. Geol. Bull.* 66 (1982), p. 588.

Koçyiğit, A., Yusufoğlu, H. & Bozkurt, E.: Evidence from the Gediz graben for episodic two-stage extension in western Turkey. *J. Geol. Soc. London* 156 (1999), pp. 605–616.

Mertoğlu, O., Şimşek, Ş., Dağıstan, H., Bakır, N. & Doğdu, N.: Geothermal country update report of Turkey (2005–2010). *Proceedings World Geothermal Congress,* 25–30 April 2010, Bali, Indonesia, 2010, Abstract no 0119.

MTA: Geothermal inventory of Turkey. General Directorate of Mineral Research and Exploration, Ankara, Turkey, 2005, pp. 561–587 (in Turkish).

MTA: Activity report of 2011. General Directorate of Mineral Research and Exploration, Ankara, Turkey, 2011, pp.1–71 (in Turkish).

Mutlu, H., Güleç, N., Hilton, D.R.: Helium-carbon relationships in geothermal fluids of western Anatolia, Turkey. *Chem. Geol.* 247 (2008), pp. 305–321.

Nicholson, K.: *Geothermal fluids, chemistry and exploration techniques.* Springer-Verlag, Berlin Germany, 1993.

Özen, T.: *Hydrogeological and Hydrochemical studies of the Salihli geothermalo fields.* PhD Thesis, Dokuz Eylül University, Institute of Natural and Applied Sciences, Geological Engineering, İzmir, Turkey, 2009, (in Turkish).

Özen, T., Bülbül, A. & Tarcan, G.: Hydrogeochemical and isotopic studies of the Salihli geothermal areas. *33rd International Geological Congress-33th IGC,* 6–14 August 2008, Oslo, Norway, 2008, Abstract no: 1349279.

Özen, T., Bülbül, A. & Tarcan, G.: Reservoir and hydrogeochemical characterizations of the Salihli geothermal fields in Turkey. *World Geothermal Congress,* 25–30 April 2010, Bali, Indonesia, 2010. Paper no: 1517.

Özen, T., Bülbül, A. & Tarcan, G.: Reservoir and hydrogeochemical characterizations of geothermal fields in Salihli, Turkey. *J. Asian Earth Sci.* 60 (2012), pp. 1–17.

Özgür, N., Vogel, M., Pekdeğer, A., Halback, P. & Sakala, W.: Geochemical, hydrochemical, and isotopic geochemical signatures of thermal fields in Kızıldere in the continental rift zone of the Büyük Menderes, western Anatolia, Turkey. *Third International Turkish Geology Symp.*1998, Ankara, Turkey, Abstract 31, p. 144.

Parkhurst, D.L. & Appelo, C.A.J.: User's guide to PHREEQC (version 2) – A computer program for speciation, batch-reaction, one-dimensional transport, and inverse geochemical calculations: U.S. Geological Survey Water-Resources Investigations Report, 99–4259, 1999.

Pasvanoğlu S. & Chandrasekharam, D.: Hydrogeochemical and isotopic study of thermal and mineralized waters from the Nevşehir (Kozakli) area, Central Turkey. J. *Volcanol. Geotherm. Res.* 202 (2011), pp. 241–250.

Purvis, M. & Robertson, A.H.F.: Miocene sedimentary evolution of the NE–SW trending Selendi ve Gördes basins, western Turkey: implications for extensional processes. *Sediment. Geol.* 174 (2005), pp. 31–62.

Reed, M.H. & Spycher, N.F.: Calculation of pH and mineral equilibria in hydrothermal water with application to geothermometry and studies of boiling and dilution. *Geochim. Cosmochim. Acta* 48 (1984), pp. 1479–1490.

Seyitoğlu, G. & Scott, B.C.: The age of the Alaşehir graben (West Turkey) and its tectonic implications. *Geol. J.* 31 (1996), pp. 1–11.

Seyitoğlu, G., Tekeli., O., C. Emen, I., Şen, S. & Işık, V.: The role of the flexural rotation/rolling hingemodel in the tectonic evolution of the Alaşehir graben, western Turkey. *Geol. Mag.* 139 (2002), pp. 15–26.

Tarcan, G.: Mineral saturation and scaling tendencies of waters discharged from wells (>150°C) in geothermal areas of Turkey. *J. Volcanol. Geoth. Res.* 142:3–4 (2005), pp. 263–283.

Tarcan, G. & Gemici, Ü.: Effects of the contaminants from Turgutlu-Urganlı thermomineral waters on the cold ground and surface waters. *Bull. Environ. Contam. Toxicol.* 74 (2005a), pp. 485–492.

Tarcan, G., Gemici, Ü. & Aksoy, N.: Hydrogeological and geochemical assessment of the Gediz Graben geothermal areas, western Anatolia, Turkey. *Environ. Geol.* 47 (2005b), pp. 523–534. Tokçaer, M.: *Geochemical cycle of boron and isotope fractionation in geothermal fluids of western Anatolia.* PhD

Thesis, Dokuz Eylül University, Institute of Natural and Applied Sciences, Geological Engineering, İzmir, Turkey, 2007 (in Turkish).

Truesdell, A.H.: Summary of section III geochemical techniques in exploration. *Proceedings of the 2nd United Nations Symposium on the Development and use of Geothermal Resources*, San Francisco, Vol. 1, US Government Printing Office, Washington, DC, 1976.

Uzunlar, Z.: *Determining the geothermal gradient distribution in Turkey using variogram analysis on deep well temperature data.* MSc Thesis, İstanbul Technical University, Institute of Natural and Applied Sciences, Petroleum and Natural Gas Engineering, İstanbul, Turkey, 2006 (in Turkish).

Xilai, Z., Armannsson, H, Yongle, L. & Hanxue, Q.: Chemical equilibria of thermal waters for the application of geothermometers from the Guanzhong basin, China. *J. Volcanol. Geoth. Res.* 113 (2002), pp. 119–127.

Yılmaz, Y., Genç, S.C., Gürer, O.F., Bozcu, M., Yılmaz, K., Karacık, Z., Altunkaynak, S. & Elmas, A.: When did the western Anatolian grabens begin to develop? In: E. Bozkurt, J.A. Winchester & J.D.A. Piper (eds.): Tectonics and magmatism in Turkey and the surrounding area. *Geological Society London Special Publication* 173, 2000, pp. 353–384.

Yılmazer, S., Pasvanoğlu, S. & Vural, S.: The relation of geothermal resources with young tectonics in the Gediz graben (west Anatolia, Turkey) and their hydrochemical analysis. *World Geothermal Congress,* 25–30 April 2010, Bali, Indonesia, 2010, pp. 1151–1161.

CHAPTER 7

Electrically conductive structures and geothermal model in Sakarya-Göynük area in eastern Marmara region inferred from magnetotelluric data

İlyas Çağlar

7.1 INTRODUCTION

The use of geoelectrical and geo-electromagnetic (e.g., magnetotelluric) methods in geothermal exploration is based on the fact that the resistivity of hydrothermal groundwater in the rocks decreases significantly at high temperatures and that geothermal activity can produce conductive alteration minerals. The resistivity of these rocks observed in geothermal areas is lower than in surrounding rocks, indicating the presence of a considerable resistivity contrast that can be investigated by the magnetotelluric method (Brown, 1994).

Based on the deep electrical structure from a magnetotelluric survey along the profile LG around Göynük (Fig. 7.1a), an anomaly region with high temperature or an upwelling of the asthenosphere was also suggested from an electrically very conductive ($\sim$1–10 m) zone at depths about 20–30 km (Gürer, 1996). Although this finding can be interpreted in other ways, this high conductivity may be the signature of a geothermal event within the earth's crust in the Göynük area not far from such famous geothermal sites as Bursa, Akyazi, and Tarakli (Fig. 7.1b).

The regional structural features, controlled by the major fault zones, such as North Anatolian Fault Zone (NAFZ), Geyve Fault Zone (GFZ), and Adapazari Fault Zone (AFZ), generally trend NE-SW around Sakarya Zone as can be seen in the gravity map (Fig. 7.1b). The low gravity around Geyve basin and Sapanca lake is probably due to a deep depression of a dense basement filled with less dense rocks such as Miocene and younger sediments (Ates *et al.*, 1999; Eroskay, 1965; Yilmaz *et al.*, 1995).

The magnetic anomaly map (Aydin, 1986) is represented in Figure 7.2 to correlate geology with geophysical data of the Göynük area. The magnetic anomaly map in general shows broadly about E-W trending contours. Most of the Gölpazari basin and surrounding area is represented by low intensity ($\leq$50 γ) that correlates with the largely deposited sedimentary units. The relatively intense three anomaly groups (up to 150 γ) around Aliplar upthrust in the north correlate with the extensive basal Paleozoic metamorphic and magmatic rocks (basement). The intense anomaly in the southwest of the Koyunlu hot spring site may owe its origins from metamorphic basement rock found in this region.

The mineral spring in the area is Ahibaba, and the hot springs are Tarakli and Koyunlu, located farther south (Fig. 7.2). The maximum temperatures of the geothermal fluids discharged from the Tarakli and Koyunlu springs are 41 and 32°C, respectively. The site of Tarakli has been known as the Tarakli spa for public bath since historical times. Today, a modern building with baths stands near the ruins of the ancient spa (Didik, 2009). The spring waters of the Tarakli spa are of therapeutic value, and are being used by several people. Our field observations show that the Tarakli thermal spring represents the outflow of a deep groundwater that has been recharged and circulated in a possible fracture zone between flysch and sandstone of Eocene age (Saner, 1978). The site of the Koyunlu hot spring is less known by people. Two main hot springs and about four leakages have been found around the Koyunlu by our field surface observations. Surface flows around these springs are observed by our field studies. The total discharges in Tarakli and Koyunlu

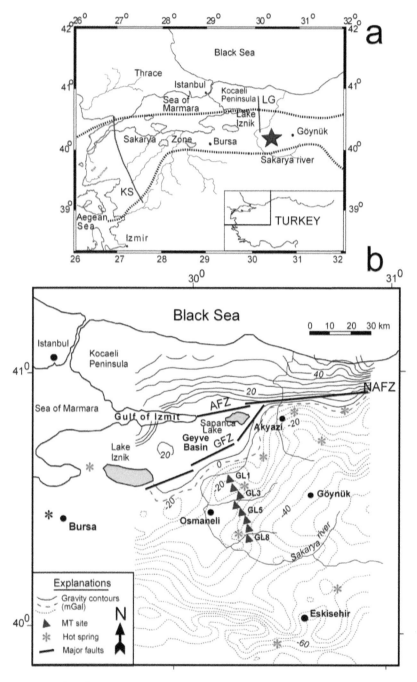

Figure 7.1. Map of northwestern Anatolia showing the magnetotelluric measuring sites. (a) The magnetotelluric lines LG and KS were measured by Gürer (1996) and Çaglar (2001), respectively. The
red big star is used to indicate Göynük area. (b) Locations of the magnetotelluric measuring
sites in the Göynük area and the Bouguer anomaly map sampled using grid $x = 4$ km from
gravity data (Ates *et al.*, 1999). Dashed lines represent negative values of the Bouguer anomaly
and solid lines positive values. Contour level is 5 mGal. Several well-known geothermal sites
and hot springs are represented in this map. Abbreviations: NAFZ, North Anatolian Fault Zone;
GFZ, Geyve Fault Zone; AFZ, Adapazari Fault Zone.

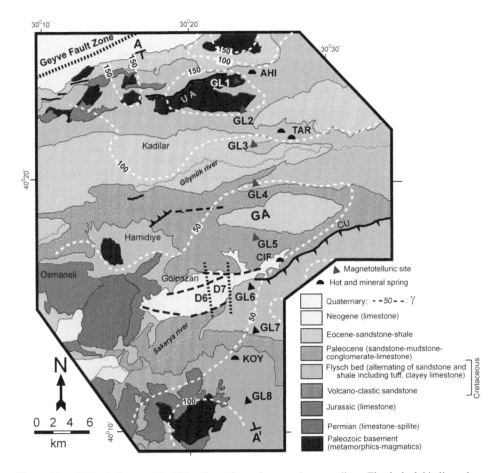

Figure 7.2. Geological map of the Göynük geothermal area and surroundings. The dashed thin lines show magnetic anomaly contours. Units are in milligamma (γ) and the contour interval is 50 γ. D6 and D7 denote the Schlumberger dc electrical profiles. The line A–A' line indicates gravity profile evaluated by the present study. Abbreviations: *Upthrusts*–UA, Aliplar; CU, Cemiski; GA, Güdümü anticline; *Mineral springs*–AHI, Ahibaba mineral spring; *Hot springs*–TAR, Tarakli; CIF, Ciftlik; KOY, Koyunlu.

are 15 and 26 L/s, respectively. On the other hand, the Ahibaba mineral spring located farther north is well known to many people, and its water is drunk. The spring waters in Tarakli and Ahibaba are alkaline (pH is 7.2 as average value) and total mineralization expressed as total dissolved solids (TDS) is about 8000 mg/kg (Yenal *et al.*, 1976). The spring waters of both sites probably have the same origin, since a similar chemical composition is observed in both geothermal systems. However, low temperature suggests that the spring water in Ahibaba is a mixture containing a component of low-mineralized groundwater and meteoric water.

7.2 NEAR-SURFACE AND DEEP ELECTRICAL STRUCTURE

The near-surface electrical structures obtained from direct current (dc) profiles D6 and D7 parallel to magnetotelluric line (Fig. 7.2) are shown in Figure 7.3. The Schlumberger electrode configuration was used for vertical electrical sounding data acquisition to image these structures.

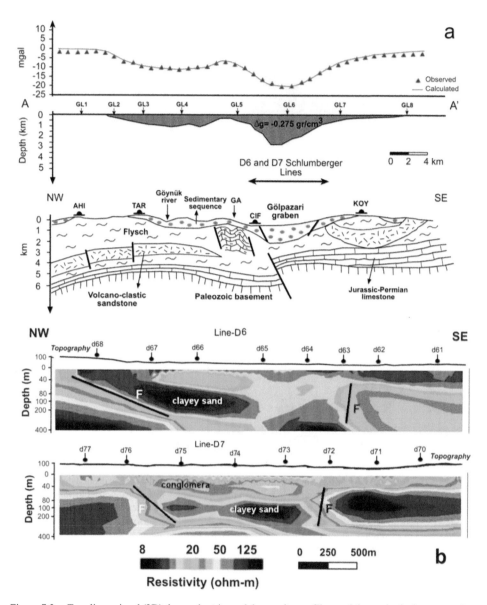

Figure 7.3.   Two dimensional (2D) dc geoelectric models, gravity profiles, and the geological cross-section. (a) Bouguer gravity anomaly across AA′ line with computed curve and retrieved model (lower panel)–the locations AHI, Ahiler spring; TAR, Tarakli spring. (b) The cross-section along AA′ line derived from geological data (Saner, 1978). (c) Electrical resistivity structures along lines D6 and D7. The thick solid lines (F) in the sections show interpreted faults.

Both electrical structures over D6 and D7 profiles are imaged to qualitatively explain small-scale features of the near-surface electrical resistivity structures in some parts of our magnetotelluric line. From the interpretation, it can be seen that there are many regions where subsurface geological formations are not horizontal and reveal the presence of graben structure in the Gölpazari basin.

The apparent resistivities within graben are in a low-to-moderate range (6–40 m) comparable to the resistivity of the basin deposits (clayey-sand and conglomerates) of Quaternary age in the area. The clayey sands are underlain by more resistive (40 m) material consisting of conglomerates. However, the conglomerates are more regularly developed in profile D6 compared to the profile D7. On the other hand, the depth to the Neogene (~150Ωm: limestone) within the Gölpazari graben was interpreted as deeper than $AB/2 = 400$ m for both sections. On the north-west side of the graben, the lower boundary between Quaternary deposits and Neogene smoothly dips to the south while on the southwest side of the graben it sharply dips to the north representing small faults. A gravity profile AA′ (Fig. 7.2) taken from the General Directorate of Mineral Research and Exploration of Turkey (MTA) was evaluated for basin modeling.

Figure 7.3a presents the Bouguer gravity anomaly with computed curve. According to dc Schlumberger data, an average value of the density contrast (-0.275 g/cm³) was used in the interpretation.

The combined interpretation of the magnetotelluric results with both dc geoelectric and gravity results will be presented in the following paragraphs to understand the complete resistivity structure and its geothermal implications.

The data were collected along one profile over the area that shows strong heterogeneity in the geology. 2D modeling seems more suitable for obtaining the electrical structure due to the distribution of the measuring sites. The magnetotelluric data were therefore modeled with 2D modeling code. Transverse Magnetic (TM) and Transverse Electric (TE) modes of magnetotelluric data were inverted using the 2D inversion code of Mackie (Mackie *et al.*, 1993).

Figure 7.4 presents the final models obtained from TE and TM data inversion. Here, the electrical resistivity model is calculated from the individual inversion of magnetotelluric data for TE and TM modes. It can be seen that the calculated 2D models fit very well with the experimental (observed) data.

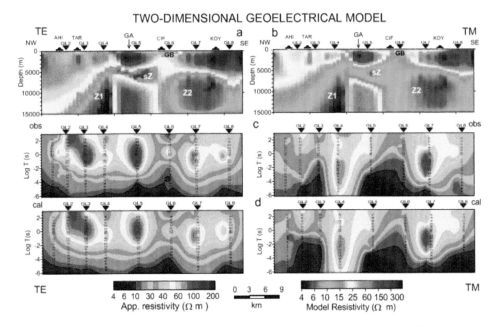

Figure 7.4.   The 2D electrical resistivity model beneath Göynük geothermal area obtained for independent 2D inversions on TM and TE data.

At greater depths within the upper crust, two main conductive zones labeled as Z1 and Z2 appear in both TE and TM models. The other one (zone sZ) is developed at a more shallow depth between Güdümü anticline and Ciftlik spring. The models for both modes show a general decrease in the resistivity values for depths greater than about 5 km beneath the site GL7 (Fig. 7.4; zone Z2).

Another conductive zone with resistivity range (4–10 Ωm) was similarly found for depths greater than about 4 km beneath GL3–GL4 (Fig. 7.4; zone Z1). The origin of these zones needs further explanation. One feature of the electrical resistivity model was the highly resistive dipped dyke-like region starting at about surface to the lower crust (except zone sZ) and located below GL5. The resistivity increased to relatively high values ($\sim$300 Ωm) at about 10 km depth in the TE model.

The irregular distribution of the low-to-moderate resistivity values (commonly 20–40 Ωm) within the depth level 0–6 km under the sites GL1–GL3 in both modes can hardly correspond to the hydrothermally altered flysch and volcano-clastic rocks. However, this zone seems to be less conductive in TE mode and becomes gradually thicker toward GL1 having moderate resistivity. However, a conductive cap centered at a depth of 3–5 km below, about GL5, appears in both models (Fig 7.4; zone sZ) but narrower in the TE probably due to electrical macro-anisotropy effect. This zone could have a characteristic geothermal origin. The circulation of hydrothermal fluids probably altered this zone. The petrophysical change within the rock that caused the above events has easily lowered the electrical resistivity.

7.3 GEOELECTRIC STRUCTURE AND GEOTHERMAL MODEL

Figure 7.5 shows electrical and geothermal model based on the interpretation of joint inversion results of magnetotelluric (MT) data. The near-surface regions under GL7–GL8 and GL1–GL3 are significant because of a considerable high resistivity observed in electrical model (Fig. 7.4). The compartments with relatively high resistivity value (50–250 Ωm) in both TE and TM models are diagnostic of these shallow flysch, limestone and underlying resistive metamorphic units. The flysch and other rocks were probably fractured, karstified, and watered over the area especially under GL1–GL4 (Fig. 7.5) as inferred by unexpected moderate- to low-resistivity values in both TE and TM models (Fig. 7.4).

Moreover, the rocks under GL1–GL3 where the Ahiler and Tarakli springs occur on the earth's surface also seem to be a geoelectrically characteristic place. The circulation of the hydrothermal fluids of both springs from the conducting zone Z1 to the surface probably comes from fractured and altered flysch and underlying basement rocks (Fig. 7.5). These alterations principally control the bulk resistivity (Hyndman and Hyndman, 1968; Telford et al., 1976) and specific resistivity values therefore significantly decreased to lower values (here below 80 Ωm) than in the case of massive or dry flysch and limestone rocks. Shale units within flysch also play a role in decreasing the specific resistivity.

The fault on the north-west side of the graben and the nearby conductive zone beneath points d68–d67 on line D6 and d77–d76 on line D7 probably allows the outflow of the hydrothermal fluids in the Ciftlik hot spring (Fig. 7.3b). More resistive structures developed in the northern and southern areas outside of the basin and several southward and northward-dipping faults or fractured zones could be also suggested coincidentally by the interpreted magnetic anomaly map. Indeed, the amplitude of the magnetic anomalies is also higher in the north and south than elsewhere suggesting a strong possibility for the existence of near-surface compact rocks around these regions (Fig. 7.2).

In the deeper levels under Gölpazari graben flysch units were thickly deposited on the basement. Here the Güdümü anticline consisted of limestone–sandstone and the flysch rocks were probably fractured, hydrothermally altered and watered. Although sedimentary units around Gölpazari basin could have lowered the magnetic field, there is a low-intensity ($\leq$50 γ), the magnetic anomaly can be also explained by a hydrothermal demagnetization of the geological rocks in the middle of the Göynük area. Such anomalies with low intensity caused by geothermal activity within the earth were found in Indonesia and New Zealand from aeromagnetic measurements and in California from ground measurements (Soengkono, 2001).

Finally, the original motivation for the geoelectric models presented by both TE and TM modes was to examine the electrically conducting zones Z1, Z2, and sZ (4–6 m) and to explain their possible geothermal origins. The zones Z1, Z2, and sZ (Fig. 7.5) were therefore significant for the geothermal origin since their high conductivity was associated with hydrothermal fluid of the geothermal system and with alteration of the rock minerals. The high values of Fe, Na, and Cl observed in the waters of the hot springs (the signs of Fe were easily seen on the surface around the springs) coincidentally explain the presence of limonite and kaolin produced only by this alteration.

These zones are highly permeable due to alteration. In fact, considering equation $\rho_b = \rho_f \phi^{-\eta}$ (adapted for the resistivity parameter; Archie, 1942), it is known that the low resistivity of pore fluid (ρ_f) in the rock with porosity (ϕ) has an active role in decreasing the bulk resistivity of the rock, ρ_b, in geothermal areas (η is an experimental constant with values between 1.8 and 2.2). The specific resistivities of hydrothermal waters (ρ_f) in the Ahiler and Tarakli springs, according to laboratory measurements at a temperature of 18°C, were 0.45 and 0.28 Ωm, respectively (0.52 Ωm for Koyunlu). The effective porosity within the rocks of the zones Z1 and Z2 could be estimated as $\phi = 30.3 \pm 3.5\%$ from Archie's equation using an average value for ρ_f and taking $\eta = 2$. The role of low-resistivity values of the hydrothermal waters in decreasing rock's bulk resistivity strengthens our hypothesis on the interpretation of the origins of all three conductive zones Z1, Z2, and sZ. In the area, the distribution of the hot springs occurs along magnetotelluric line trending north-south. It is probable that the origin of these hot springs is associated with the same geothermal system.

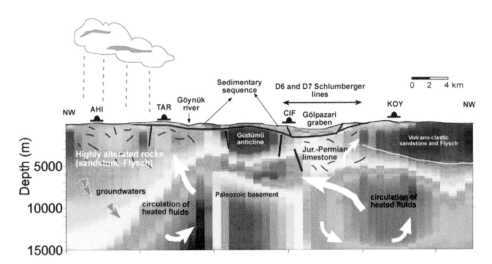

Figure 7.5. Electrical structure and geothermal model based on the interpretation of joint inversion of MT data. Typical zones are filled by blue color. Zone Z1 is below TAR hot sping; Zone Z2 is below KOY hot sping; and Zone sZ is below Güdümü anticline. Springs are AHI, Ahiler; TAR, Tarakli; CIF, Ciftlik; KOY, Koyunlu. D6 and D7 lines are taken for electrical sounding measurements (compare with Figs. 7.2 and 7.3).

7.4   CONCLUSION

The electrical resistivity of fractured rocks, depending on hydrothermal fluid circulation, decreases to a very low value. The resistivity of these rocks observed in geothermal areas is lower than in surrounding rocks. The presence of these considerable resistivity contrast can be investigated by magnetotelluric method. Magnetotelluric measurements were made in the Göynük site to find sufficient geophysical implications about geothermal activity around the area. The electric model computed from the magnetotelluric data was presented for three conducting zones denoted as Z1, Z2, and sZ beneath the area. These zones are interpreted as possible geothermal regions since the resistivity decreased to very low levels. The low-resistivity value ($\sim$4–6 Ωm) of these zones could be explained by the circulation of hydrothermal fluid of geothermal system of the studied area. The hydrothermal alteration that probably produced conductive minerals could have an effect on earth resistivity around the three zones mentioned above. Considering all of the data, further geological and geophysical explorations are necessary to obtain more detailed implications about geothermal activity for the Göynük area.

REFERENCES

Archie, G.E.: The electrical resistivity log as an aid in determining some reservoir characteristics. *AIEM Trans.* 146 (1942), pp. 54–62.

Ates, A., Kearey, P. & Tufan, S.: New gravity and magnetic anomaly maps of Turkey. *Geophys. J. Int.* 136 (1999), pp. 499–502.

Aydin, I.: Magnetic map of northwest Anatolia and its interpretation. Int. Rep. of General Directorate of Mineral research and Exploration (MTA), Report 8310, Ankara, Turkey, 1986.

Brown, C.: Tectonic interpretation of regional conductivity anomalies. *Surv. Geophys.* 15 (1994), pp. 123–157.

Çaglar, I.: Electrical resistivity structure of the northwestern Anatolia and its tectonic implications for the Sakarya and Bornova zones. *Phys. Earth Planet. Int.* 125 (2001), pp. 95–110.

Didik, S.: Revitalized millennial mineral springs: Taraklı thermal clay baths. Abstract book of the *Symposium on Geothermal potential and exploration methods*, 22–23 October 2009, Istanbul University, Engineering Faculty, Istanbul, Turkey, 2009.

Eroskay, S.O.: Geology of the Pasalar Gorge Gölpazarı area. *J. Istanbul Univ. Sci. Fac.* B 30 (1965), pp. 135–170.

Gürer, A.: Deep conductivity structure of the North Anatolian Fault Zone and the Istanbul and Sakarya zones along the Gölpazari-Akcaova profile, northwest Anatolia. *Int. Geol. Rev.* 38 (1996), pp. 727–736.

Hyndman, R.D. & Hyndman, D.W.: Water saturation and high electrical conductivity in the lower continental crust. *Earth Planet. Sci. Lett.* 4 (1968), pp. 427–432.

Mackie, R.T., Madden, A., Manzella, A. & Rvieven, S.: Two-dimensional modelling of magnetotelluric data from Larderello geothermal area, Italy. ENEL Internal Report, Italy, 1993.

Saner, S.: Geology and the environments of deposition of Geyve-Osmaneli-Gölpazarı-Taraklı area. *J. Istanbul Univ. Sci. Fac.* B 43 (1978), pp. 63–91.

Soengkono, S.: Interpretation of magnetic anomalies over the Waimangu geothermal area, Taupo volcanic zone, New Zealand. *Geothermics* 30 (2001), pp. 443–459.

Telford, W.M., Geldart, B.P. & Sheriff, R.E.: *Applied geophysics*. Cambridge University Press, London, UK, 1976.

Yenal,O., Osman, N., Bilecen, L., Kanan, E., Öz, G., Öz, U., Bonce, T., Göksel, A., Alkan, H. & Sezginer, N.: Mineral waters of Turkey. Istanbul Univ. Med. Hydrology Climant. Dept., Istanbul, Turkey, 1976.

Yilmaz, Y., Genç, Ş.C., Yiğitbaş, E., Bozcu, M. & Yılmaz, K.: Geological evolution of the late Mesozoic continental margin of northwestern Anatolia. *Tectonophysics* 243 (1995), pp. 155–171.

CHAPTER 8

Use of sulfur isotopes on low-enthalpy geothermal systems in Ayaş-Beypazarı (Ankara), central Anatolia, Turkey

Mehmet Çelik

8.1 INTRODUCTION

The study area is located around the Ayaş, Beypazarı, and Mihalıççık settlements and 90 km west of Ankara (Fig. 8.1). Sakarya river, Ankara creek, Kirmir creek, and Ilhan creek run to the Sarıyar dam lake. The Beypazarı granitoid is surrounded by Kirmir creek and Sakarya river in the study area. These granitoid units are most identical with the granite supersuite of central Anatolian crystalline complex in central Turkey (Kadıoğlu et al., 2006).

The Çoban bath spring (CBS-CBW-BT 3), Ayaş mineral spring (AMS-BT 2), Dutlu bath spring (DBS-BT 1), Kapullu bath spring (KBS-BT 5), and Karakoca mineral spring (KMS-BT 6) are the thermal and mineral waters in the area (Figs. 8.1 and 8.2). Karakaya bath and well waters (AK1) are thermal. The Ayaş waters have been used for balneological purposes since 1982, and the facility has recently started operation as a spa as well as a physical therapy and rehabilitation center. Similarly, the Dutlu waters are also used as a spa physical therapy and rehabilitation center. The Çoban bath was the site of a site in the 1970s but all the springs are dry at present. Five wells were drilled around the spring site, two of them artesian wells, but the other wells were found to be not productive and were closed. The Kapullu bath spring has been used for bathing. The water of Karakoca mineral spring has been sold in the market in the brand of Beypazarı Karakoca mineral water.

8.2 GEOLOGY AND HYDROGEOLOGY

The basement of the study area consists of metamorphic rocks of Paleozoic age. Known as Eskişehir metamorphics, these rocks were formed under amphibolite, green schist and horn-fels facies and outcrop around Mihalıççık and southern part of Ayaş settlement. The rocks consist of gneiss, garnet-schist, biotite-schist, calc-schist, mica-schist, graphite-schist, dolomitic-calcareous-crystalline limestone, phyllite, amphibolite, and metabasite (Diker, 2005; Helvacı and Bozkurt, 1994;) (Fig. 8.1).

Since magmatic rocks such as granite, granodiorite, monzonite, and quartz monzonite have the same origin and are found in a single intrusive body, they are called "Beypazarı Granitoid" by Kadıoğlu (2004). The Beypazarı granitoid intruded into the Eskişehir metamorphic rocks during the Late Cretaceous. The units with Early Miocene–Eocene age consist of proclastic volcanic, proclastic tuff, basalt, and agglomerates. They are collectively known as İnözü proclastic tuff, which is located at the bottom of Porsuk formation. Adaviran basalt is located at upper part of the proclastics (Helvacı and Bozkurt, 1994).

Clastic cover units: The Beypazarı granitoid is overlain by the Middle Eocene Kızılbayır formation of a transgressive character. The formation, consisting of claystone, sandstone, and conglomerate, is found at western and eastern parts of the study area (Fig. 8.1). With its typical red color, the Kızılbayır formation is easily distinguished from other units.

Evaporitic cover units: These cover units are formed by Mio–Pliocene age of Kirmir formation. Kirmir formation is formed as three parts. Conglomerate, sandstone, and marl are located at the

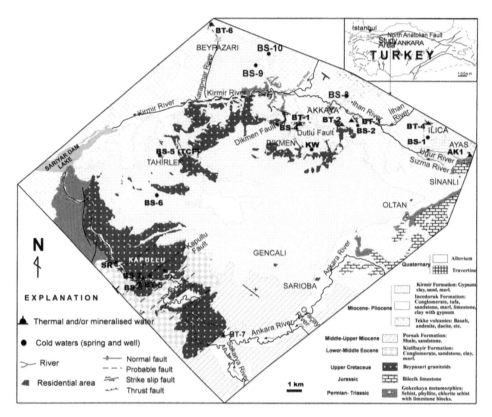

Figure 8.1.   Location and geological maps of the study area (modified after Çelmen and Çelik, 2009; Diker, 2005; MTA, 1963).

bottom of the formation; sandstone, marl, clayey limestone, clay, silty, and gypsum levels located at middle level of the formation. At the upper part of the formation it consists of gypsum beds and clay-marl levels intercalate (Canik, 1970). These units were precipitated in a lacustrine environment. Because of the lowering of the lake level, evaporitic units increase to the top of the lacustrine environment. Alluvium units are observed along the Sakarya river, Kirmir creek, and Ankara creek and their tributaries (Figs. 8.1 and 8.2).

Basement rocks of Paleozoic age are generally impermeable. However, there are permeable normal faults between granitoid and metamorphic units. Jurassic limestone is observed in the basement units around the Ayaş region. These units are the aquifers of the Karakaya bath spring. AK1 and AK2 (AK3 is located at the same place as AK1) wells have been penetrated into the limestone and cover units. The cumulative flow of these two wells is 40L/s (Özeke, 1987). Direction and dips of the joints of the granitoid reveal that the Beypazarı granitoid witnessed NE–SW compressional stress and NW–SE extensional stress in the Kapullu bath spring area (Diker *et al.*, 2006). The thermal and mineral springs discharge from fault zones cutting the Beypazarı granitoid. The granitoids are generally impermeable, but faulting and cracking upper zones have generated secondary porosity and permeability.

Conglomerates and sandstones of cover units are permeable. However, intercalated thick marls strata are impermeable. Therefore, these cover units are impermeable throughout the vertical direction to the stratum plane. Gypsum strata at the upper part of the cover units have limited reservoir for brine cold springs.

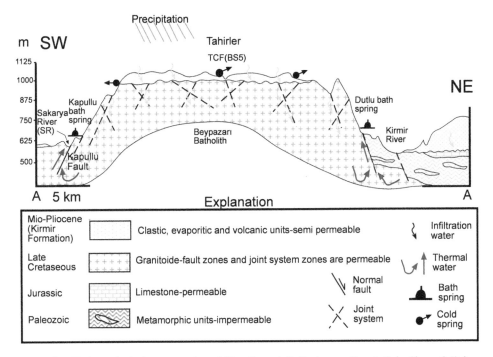

Figure 8.2.   Hydrogeological cross-section of Kapullu and Dutlu (across line A–B in Figure 8.1) low-enthalpy geothermal systems.

The geothermal reservoir is fed by precipitation, river and cold groundwater infiltrating into the generally low-permeable cover units , faults and jointed crystalline rocks. The base of the low-enthalpy reservoir system is the very low-permeable crystalline (massive) and metamorphic basement that contains negligible groundwater resources. Groundwater flow in this crystalline terrain is slow, and groundwater is estimated to be about 43 years old (1963–2006: 43). In addition, the tritium values of the waters were zero in 2006 (Çelik, 2007).

The fluids of low-enthalpy geothermal system waters are of meteoric origin and generally heated by natural geothermal gradient, volcanic intrusive, and partial radiation of radioactive minerals in granitoid.

8.3   HYDROCHEMICAL AND ISOTOPIC STUDIES

8.3.1   *Hydrochemical and isotopic evaluation*

Dutlu bath springs have dominant ions of Na^+ and $Cl^- \geq SO_4^{2-}$. Ayaş mineral and Kapullu bath springs have dominant ions of Na^+, SO_4^{2-}, and Cl^-. Karakoca mineral spring has dominant ions of Na^+ and HCO_3^-. Dutlu bath spring has a temperature of 46°C with discharge of 6 L/s (Özbek, 1984). Kapullu bath spring has a temperature of 40.5°C with discharge of 2.49 L/s (Diker *et al.*, 2006). Wells have been drilled in these thermal areas. Ayaş wells are artesian and have discharge temperatures of 54.5 and 51°C. Spring water temperatures of the Ayaş are 52 and 51°C (Canik, 1970). The temperature of Karakaya bath spring was measured as 30°C (Çetin, 2006).

Cold waters have two main types of water chemistry. The first water type is dominated by of Ca^{2+} and HCO_3^- ions, the second type has Ca^{2+} and SO_4^{2-} as dominant ions. A few cold waters have Mg^{2+} and Na^+ as dominant ions (Figs. 8.1–8.4).

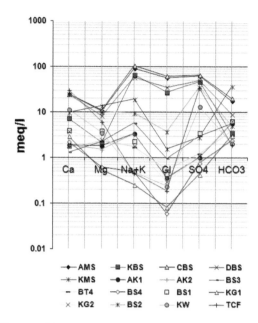

Figure 8.3.   Schoeller diagram of the waters.

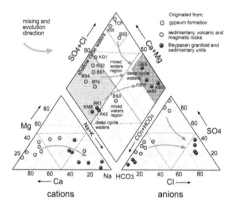

Figure 8.4.   Piper diagram of the waters.

If gypsum ($CaSO_4 \cdot 2H_2O$) is a major source of sulfate, then the Ca/SO_4 ratio (in mg/L) in the corresponding groundwater should be approximately 0.4 and the $\delta^{34}S$ of SO_4 should be very close to that of the gypsum (Mayer *et al.*, 1995). Only Tahirler cold spring (BS5-TCF) satisfies these requirements, thus the elevated sulfate in the gypsiferous Kirmir formation (Fig. 8.5) is unlikely due to gypsum dissolution. According to Sacks *et al.* (1995), the sources of sulfate in the Upper Floridan aquifer were able to identify gypsum dissolution, based on $\delta^{34}S$ values and a chloride/sulfate ratio indicating gypsum dissolution. Chloride/sulfate ratios are low in the BS5-TCF (0.004), KW (0.013), BS2 (0.078), and BS4 (0.055) waters of the study area. These springs with low yields (<1 L/s) have cold waters and discharge from the Kirmir formation. TCF, KW, and BS2 waters are $Ca^{2+}-SO_4^{2-}$ water type (Figs. 8.4 and 8.5). According to the saturation index calculation of all the waters, the waters are oversaturated with respect to aragonite, calcite, and dolomite minerals (Çelmen and Çelik, 2009).

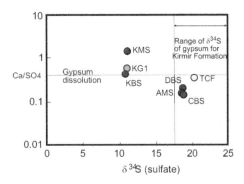

Figure 8.5. Ca/SO₄ *versus* sulfur-34 (sulfate) diagram–cold waters: KG1, KG2: cold springs (discharged from granitoids); TCF, Tahirler cold spring (discharged from gypsiferous formations); SR, Sakarya river; Thermal and/or mineral waters: KMS, Karakoca mineral spring; KBS, Kapullu bath spring; DBS, Dutlu bath spring; AMS, Ayaş mineral spring; CBS, Çoban bath spring; AK1 and AK2: Karakaya bath wells.

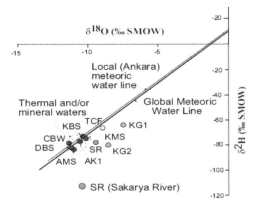

Figure 8.6. Oxygen-18 (%, SMOW) *versus* deuterium (%, SMOW) diagram (see Fig. 8.5 for abbreviations).

In the oxygen-18/deuterium graph, thermal and/or mineral waters (CBW, AMS, DBS, KBS, KMS, and AK1) are on Global Meteoric Water Line and therefore they are meteoric in origin. Some cold waters (KG1, KG2, and SR) were exposed to evaporation (Figs. 8.1 and 8.6). Considering the hydrochemical and isotopic data, the thermal and/or mineral waters represent deep circulating, old waters, and it is derived from waters rising to the surface through the fault zones cutting the granitoids.

In the study area, cold springs (BS-4, BS-3, and BS-5) are affected by seasonal precipitation. The cold springs are located in the recharge area of the thermal waters (Fig. 8.7). According to the graph, geothermal springs of BT-1, BT-2, BT-3, and BT-5 are recharged by precipitation from elevations between 950 and 1150 m (Çelmen and Çelik, 2009); elevations that correspond to the gypsiferous Kirmir formation (Mio-Pliocene units) and to the granitoids (Late Cretaceous units) (Figs. 8.1 and 8.2).

8.3.2 *Sulfur isotope evaluation*

Sulfur isotope analyses were first performed on the thermal and mineral waters in the study area. Hot waters ascending from a geothermal reservoir may cool by mixing in the upflow with

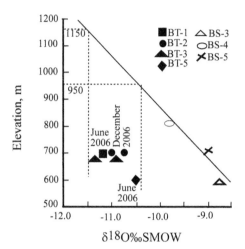

Figure 8.7.   Oxygen-18 *versus* elevation graph (Çelmen and Çelik, 2009).

shallow, relatively cold water. When this is the case, geothermometers may yield misleading results (Arnorsson, 2000). It is thought that mixing processes of the waters, SO_4–H_2O geothermometers are applied rather than cation geothermometer ones. The $\delta^{18}O$ (SO_4–H_2O) geothermometer is based on the oxygen isotope exchange reaction:

$$SO_4 + H_2O \leftrightarrows SO_3O + H_2O$$

Experimental fractionation factors for the sulfate-water system have been determined (Lloyd, 1968) and, are in reasonable agreement between 100 and 200°C (Mizutani and Rafter, 1969). It is reported that an equilibrium fractionation is between dissolved sulfate and water of (Lloyd, 1968)

$$1000 \ln \alpha_{SO_4-H_2O} = 3.251(10^6/T^2) - 5.6$$

but (Mizutani and Rafter, 1969) found

$$1000 \ln \alpha_{SO_4-H_2O} = 2.88(10^6/T^2) - 4.1$$

where T is in K.

The results of SO_4–H_2O geothermometer reveal that the reservoir temperatures of the study area geothermal waters range between 42 and 73°C. According to Adam and Jan (2009), reliable reservoir temperature estimates were obtained while applying the oxygen isotope geothermometer in SO_4–H_2O system. However, attention must be paid to the secondary processes in geothermal systems, which may disqualify the use of this temperature indicator, for example, bacterial reduction of sulfates or considerable CO_2 content (carbonated waters). Reservoir temperature is low in the region. According to Dickson and Fanelli (1996), if the reservoir is connected to a surface recharge area that replenishes all or part of the fluids emerging naturally in springs or extracted in wells, reservoir temperature can be low.

According to oxygen-18 (SO_4) and sulfur-34 (SO_4) contents, the sulfate in the waters of Ayaş and Dutlu baths is similar to that of several other thermal springs in Turkey (Çelik et al., 2005; Pasvanoglu and Chandrasekharam, 2011; Ünsal et al., 2003), and are primarily derived from the gypsum of the Kirmir formation. However, sulfates of the KMS, KCS (KG1), KBS, and SR waters probably originated from the mixing of terrestrial, atmospheric, and Tertiary sources (Clark and Fritz, 1997) (Figs. 8.8 and 8.9). It is thought that the Kapullu bath spring was probably diluted by Sakarya river waters. However, isotopic and hydrochemical evidence of dilution is not clear from the Kapullu bath spring.

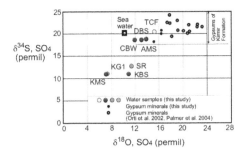

Figure 8.8.   Sulfur-34 (SO$_4$) *versus* oxygen-18 (SO$_4$) diagram (Çelmen and Çelik, 2009).

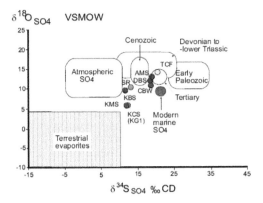

Figure 8.9.   Oxygen-18 (SO$_4$) *versus* sulfur-34 (SO$_4$) diagram (Çelmen and Çelik, 2009; Clark and Fritz, 1997).

8.4   RESULTS

Hydrochemical and isotopic characteristics, discharges, and temperatures of the thermal and/or mineral waters indicate that the waters are deeply circulated and manifest from the fault zones.

Two water types of the cold springs can be distinguished in the region: Ca^{2+}–SO_4^{2-} and Ca^{2+}–HCO_3^- water. Thermal and mineral springs have been formed Na^+, SO_4^{2-}, and Cl^- hydrochemical facies. It was thought that thermal effect resulted from geothermal gradient, radioactive degradation, and up lifting of the underground volcanic rocks.

Kirmir formation gypsum minerals in Mio–Pliocene age have sulfur-34 isotopes between 17% and 25%. According to oxygen-18 (SO$_4$) and sulfur-34 (SO$_4$) contents, sulfate in waters of Ayaş and Dutlu mineral and Çoban bath waters are derived primarily from Kirmir formation gypsum. Sulfates of Kapullu bath water and Karakoca mineral water are located near the atmospheric and terrestrial regions, respectively.

Among the cold groundwater, the sulfate of Tahirler cold spring is derived from gypsum in the Kirmir formation (primary source) and the sulfate in Kapullu cold water has a secondary atmospheric origin. Sakarya river water sulfate probably originates from atmospheric and terrestrial regions. According to SO$_4$–H$_2$O geothermometer, reservoir temperature of the geothermal springs at Kapullu bath is up to 73°C. According to the sulfur isotope results, the open and low-enthalpy Ayaş-Beypazarı geothermal system is not expected to be useful for district heating and electric power generation.

ACKNOWLEDGMENTS

This study was financially supported by TÜBİTAK under grant no. 104Y056 and Scientific Research Projects Unit of the Ankara University under grant no. 2005-07-45-027.

REFERENCES

Adam, P. & Jan, D.: Application of selected geothermometers to exploration of low-enthalpy thermal water: the Sudetic geothermal region in Poland. *Environ. Geol.* 58:8 (2009), pp. 1629–1638.

Arnorsson, S.: Mixing process in up flow zones and mixing models. In: S. Arnorsson (ed): Isotopic and chemical techniques in geothermal exploration, development and use: Sampling methods, data handling, interpretation. International Atomic Energy Agency, Vienna, 2000, p. 351, Austria.

Canik, B.: Geological and hydrogeological investigation of the Ayaş bath. General Directorate of Mineral Research and Exploration, 1970, pp. 137–152 (in Turkish).

Clark, I. & Fritz, P.: *Environmental isotopes in hydrogeology.* Lewis Publishers, New York, 1997.

Çelik, M.: Evaluation of origin of the Ayaş-Beypazarı geothermal waters with sulfur isotopes, Central Anatolia, Turkey. *International Symposium on Advances in Isotope Hydrology and its Role in Sustainable Water Resources Management (IHS-2007),* 21–25 May 2007 Vienna, Austria, 2007.

Çelik, M., Çelmen, O. & Koçbay, G.R.: Hydrogeological investigation of the Çavundur -Çerkeş geothermal field, north-west Turkey. *Proceedings World Geothermal Congress 2005,* Antalya Turkey, 24–29 April 2005, pp. 1–5.

Çelmen, O. & Çelik, M.: Hydrochemistry and environmental isotope study of the geothermal water around Beypazarı granitoids, Ankara, Turkey. *Environ. Geol.* 58:8 (2009), pp.1689–1701.

Çetin, A.: *Hydrogeological investigation of Ayaş (Ankara) Karakaya and Ilıcaköy hot waters.* MSc Thesis, Hacettepe University, Ankara, Turkey, 2006 (in Turkish).

Dickson, M.H. & Fanelli, M.: *Geothermal energy.* John Wiley and Sons, Chichester, 1996, p. 214.

Diker, S.: *Hydrogeological investigation of the Kapullu bath (Beypazarı) and around.* MSc thesis, Ankara University, Graduate School of Sciences, 2005 (in Turkish).

Diker, S., Çelik, M. & Kadıoğlu, Y.K.: Finger prints of the formation of geothermal springs on the granitoids: Beypazarı-Ankara, Turkey. *Environ. Geol.* 51:3 (2006), pp. 365–376.

Helvacı, C. & Bozkur, S.: Geology, mineralogy and petrogenesis of the granitoid of Beypazarı (Ankara). *Geol. Bull. Turkey* 37:2 (1994), pp. 31–42 (in Turkish).

Kadıoğlu, Y.K.: Nature of Beypazarı Granitoids: Ankara – Turkey. *Geochim. Cosmochim. Acta* 68:11 (2004), pp. 687–688.

Kadıoğlu, Y.K., Dilek, Y. & Foland, K.A.: Slab break off and syncollisionalorigin of the Late Cretaceous magmatism in the Central Anatolian crystalline complex, Turkey. *Geol. Soc. Am. Bull.* 409 (2006), pp. 381–415.

Lloyd, R.M.: Oxygen isotope behavior in the sulfate-water system. *J. Geophys. Res.* 73:18 (1968), pp. 6099–6110.

Mayer, B., Fritz, P., Priatzel, J. & Krouse, H.R.: The use of stable sulfur and oxygen isotope ratios for interpreting the mobility of sulfate in aerobic forest soils. *Appl. Geochem.* 10:2 (1995), pp. 161–173.

MTA (General Directorate of Mineral Research and Exploration): 1/500.000 scale Zonguldak geological map, Ankara, Turkey, 1963 (in Turkish).

Mizutani, Y. & Rafter, T.A.: Oxygen isotopic composition of sulfates-3. oxygen isotopic fractionation in the bisulfate ion-water system. *NewZeal. J. Sci.* 12 (1969), pp. 54–59.

Özbek, T.: *Hydrogeological investigation of the Ankara-Beypazarı Dutlu bath around.* PhD Thesis, Ankara University, Graduate School of Natural and Applied Sciences Ankara, Turkey, 1984 (in Turkish).

Özeke, H.: Karakaya bath (Ayaş-Ankara) AK1 and AK2 wells finishing report. Mineral Research and Exploration Report No: 8195, 1987 (in Turkish).

Pasvanoğlu, S. & Chandrasekharam, D.: Hydrochemical and isotopic study of thermal and mineralized waters from the Nevşehir (Kozakli) area, Central Turkey. *J. Volcanol. Geotherm. Res.* 202 (2011), pp. 241–250.

Sacks, L.A., Herman, J.S. & Kauffman, S.J.: Controls on high sulfate concentrations in the Upper Floridan aquifer in southwest Florida. *Water Resour. Res.* 31:10 (1995), pp. 2541–2551.

Ünsal, N., Çelik, M. & Murathan, A.M.: Hydrochemical and isotopic properties of the mineralized thermal waters of Kırşehir province, Turkey. *Journal Geological Society of India* 62 (2003), pp. 455–464.

CHAPTER 9

Geochemistry of thermal waters in eastern Anatolia: a case study from Diyadin (Ağrı) and Erciş-Zilan (Van)

Suzan Pasvanoğlu

9.1 INTRODUCTION

Although in some countries the utilization of geothermal energy has a long history, in Turkey intensive research and exploration activities and evaluation of this potential source have been emphasized only after the establishment of the Geothermal Code, which was issued by the Turkish Assembly in June 2007. The geotectonic position of Turkey between the African-Arabian and Eurasian continents, and within the Alpine orogenic belt, has given rise to tectonic activity resulting in folding and faulting during several geological periods, and is still shown by the active tectonic zones (Canik and Baksan, 1983; Koçyiğit et al., 2001, Kurtman, 1977; Şengör, 1980; Şengör and Kidd, 1979; Şengör and Yilmaz, 1981). Several volcanoes, active in historic times, such as Kula, Erciyes-Hasandağ, Nemrut, Süphan, Tendürek, and Ağrı Mountains are known (Şaroğlu et al., 1980), and these together with the tectonically active belts have given rise to the many hot-water springs that occur throughout Turkey (Fig. 9.1). It is possible to consider four main geothermal provinces, namely, the western Anatolia horst-graben system, the Northern Anatolia Fault Zone (NAF), and the areas affected by Upper Tertiary recent volcanic activity localized mainly in central and eastern Anatolia. Eastern Anatolian contractional province is characterized by E–W trending folds, manifestation of the N–S shortening, and a widespread volcanism with several young volcanoes emitting gasses and aligned in NE–SW direction. The important geothermal fields of the region are concentrated around these young volcanoes (Ağrı, Van, Muş, Bitlis, and Bingöl).

Geothermal activities have been developed significantly for direct use and electricity production especially in western Anatolia, and the thermal waters have been well studied by several authors (Baba and Armannsson, 2006; Bülbül, 2009; Gemici and Gültekin, 2000; Gökgöz, 1998; Güleç et al. 2002; Gültekin and Gemici, 2003; Mutlu, 2007; Pasvanoğlu, 2011; 2012; Şamilgil, 1977; Şimşek, 1997; Vengosh et al., 2002; Yilmazer et al., 2010). However, only a few studies of the thermal springs in eastern Anatolia are available. Geothermal histories and chemical analysis of these thermal waters are rarely discussed in books related to the thermal waters of Turkey (Akkuş et al., 2005; Çağlar, 1950; Yenel, 1976). Thus, the present study aims at filling this gap.

In the eastern Anatolian region, two neighboring towns of Diyadin (Ağrı) and Zilan (Erçiş-Van) have significant thermal and mineral water potential. Thermal waters in these areas are used for district heating, agriculture, industrial process, baths, and treatments. Due to poorly accessible sites, the hydrogeochemical and isotopic characteristics of these two thermal springs have not been investigated. Detailed geological, geochemical, and isotopic investigation on these thermal springs have been carried out in order to understand their chemical evolution, to estimate the reservoir temperatures using chemical geothermometers, and to characterize the hydrothermal activity from known and newly investigated hydrothermal springs and thus to provide a framework for future studies of the two active provinces of eastern Anatolia.

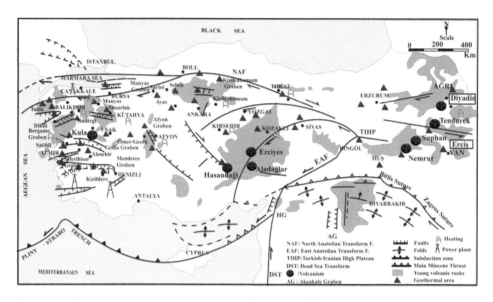

Figure 9.1.   General tectonic and volcanic features of Turkey (modified from Yılmaz *et al.*, 1987).

9.2   SITE DESCRIPTION

9.2.1   *Diyadin (Ağrı) area*

The geothermal area is located 70 km east of the city of Ağrı and 5 km southwest of Diyadin town. The study area occurs at an altitude of 1925 m and is located in the 1:25000 scaled shown in Figure 9.2 (Burcak *et al.*, 1997). The Murat river is the most important water source in the region and flows from the Tendurek and Aladağ mountains in the southeastern part of the region. The thermal activity consists of thermal springs, steaming ground, gas seepage, travertine deposition, widespread and actual sulfur precipitation, and hydrothermal alterations (Figs. 9.3 and 9.4). The distribution of hot-water springs in the Diyadin area roughly parallels to the distribution of the fault systems and young volcanics. In the Diyadin geothermal field, there are several features that discharge thermal, mineralized waters, including the Yilanli Kirecli, Kopru Çermik, Davutlu Çermik, Hidircayir, Tazekent, and Kusburnu Çermik springs are the main emergences. Temperatures of these springs range from 24 to 64°C, with discharge rates of 0.5–10 L/s (Pasvanoğlu and Güler, 2010a). Some of the thermal spring are unused and discharge directly into the Murat river. Others are used for spa, greenhouse, and district heating. The surface location of thermal springs has changed over time due to $CaCO_3$ deposition blocking fracture pathways. The General Directorate of Mineral Research and Exploration (MTA) started the drilling operations in the Diyadin area in 1998. As a result, 6 artesian wells were drilled to a depth of 77–215 m (Keskin, 1998), which now tap 420 L/s of thermal water with a temperature of 62–78°C. Among these wells, the MT-3 well is currently the only production well, while the others were abandoned or closed due to gas flux at high pressure. Hot fluid produced from the MT-3 well is piped to to the distribution center 7 km away, where freshwater is heated to a temperature of 60°C by heat exchangers, and delivered to users in isolated pipes. A factory to produce liquefied CO_2 and dry ice has also been established in Diyadin through a Turkish–German joint venture with production capacity in the past years of 100 ton/day. In addition, 15 km away from Diyadin Kusburnu in Molakir village, there are thermal water springs with temperatures range between 40 and 60°C, with two shallow wells that discharge 37–73°C waters (Pasvanoğlu, 2008). The depth of artesian wells is 65 m (Yilmaz Bakir, 2008, personal communication). The heads of the wells at Molakir were closed due to the intense carbonate scale.

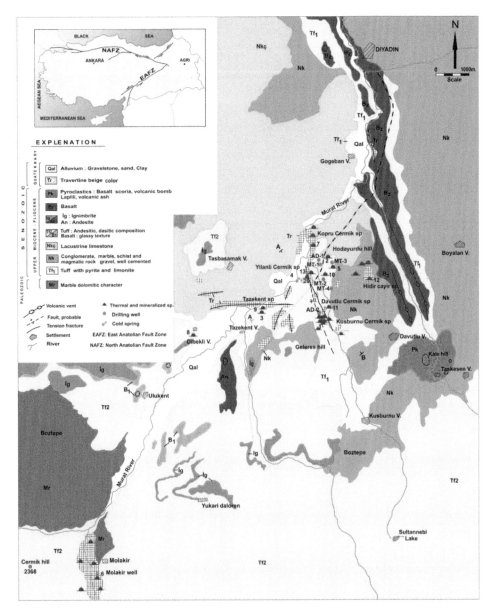

Figure 9.2.   Geological map of the Diyadin geothermal field (modified from Burcak *et al.*, 1997).

9.2.2   *Zilan (Van-Erciş) area*

The Zilan geothermal field is located 30 km north of Erciş of Van province in the Zilan valley in eastern Anatolia. The Zilan geothermal area is located to the south of Diyadin town. The area with an altitude of 1750 m is shown in Figure 9.5. The study area is high and very rugged, with a maximum peak of 3538 m in Hüdavendiger. Due to rapid uplifting, the area has been cut deeply by a drainage network, and consequently, considerable landslides have occurred in some parts of this region (Fig. 9.5). Major creeks of the region are Saman, Ganisipi, Kömler, and Karakaya. They form Zilan creek flowing from the north to south into Van Lake. The discharge of the Zilan creek,

Figure 9.3.   CO$_2$ gas bubbling from thermal water spring at Diyadin.

Figure 9.4.   Diyadin thermal waters and the deposition of CaCO$_3$.

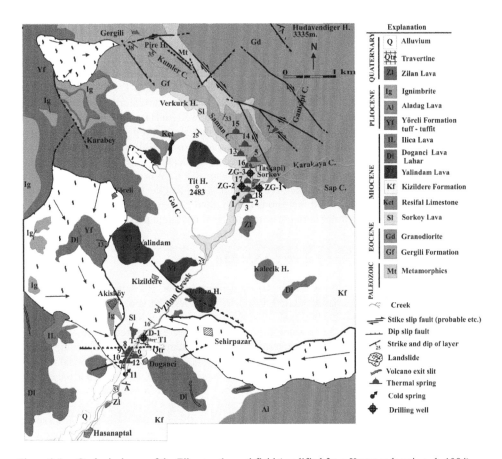

Figure 9.5. Geological map of the Zilan geothermal field (modified from Karamanderesi *et al.,* 1984).

which is fed by springs near Hüdavendiger hill in the northeast, is 1944 L/s (Öztekin *et al.,* 1977). The area has been subjected to frequent and severe earthquakes in the last decades. Therefore, in the investigated area, the locations of thermal water springs have been displaced frequently, and new thermal water springs have emerged.

Due to the October 23 and November 9, 2011 earthquake (also called the Van Earthquake), with magnitude 7.1 and 5.7, respectively, some differences (many of which were temporary) have been observed in the springs and wells in the Erciş geothermal area. The Van earthquake took place near the city of Van province in eastern Turkey, close to the Iranian border (Fig. 9.1). Young faults with vertical movement and E–W, NW–SE direction exists along the Zilan valley.

The thermal activity consists of numerous fumaroles, thermal springs, steaming ground, fluid seepage, and recent hydrothermal alterations, which are present in the area. Thermal water resources inside Zilan valley basically cluster in two different areas: (i) in the vicinity of Şorköy (Taskapi) and Gergili villages which lie in the north of the area; (ii) in the neighborhood of Hasanaptal (Dogaci) thermal springs, located further south in the valley (Fig. 9.5). Zilan waters emerge, at elevations of about 2000 m. Temperatures of the springs range from 22 to 78°C, with discharge rates of 0.5–20 L/s (Pasvanoğlu and Güler, 2010b). Some of the thermal waters are unused, and they discharge directly to the Zilan creek. Others are used as spa waters. The distribution of thermal waters in Şorköy and its vicinity and their general course (N–S trending Zilan valley) in the Zilan valley probably indicate their close association with the Diyadin geothermal waters in the north.

Most of the thermal-mineral waters contain CO_2 and H_2S gases (Yenel, 1976). The geological, hydrogeological, and geophysical studies aimed toward the determination of the geothermal potential of Zilan area were first started by MTA in the 1988 (e.g., Akkuş *et al.*, 1990; Burkay and Şahin, 1986; Doğan, 1986; Karamanderesi *et al.*, 1984; Ölmez and Güner, 1989; Öngür *et al.*, 1974). In 1988, an exploration well was drilled in the area by MTA (ZG-1 well) (Fig. 9.5) where water was found as artesian at 394.20 m depth in granodiorite and quartzdiorite, with a flow rate of 40 L/s and temperature of 80°C. The ZD-1 gradient drilling well (in Hasanaptal region) conducted to research the units in the field in 1989 by MTA, located 500 m north of Hasanaptal thermal area could not be properly investigated because the capacity of the drill rig was not sufficient to reach the desired depth and the drilling was stopped at the depth of 1172.70 m. The measured temperature of the mud flow at 1168 m well depth was 105°C. When the samples taken from the well at 1064–1066 m depth were analyzed, it was detected that the minimum homogenization temperature was 132°C using fluid inclusions (Ölmez and Güner, 1989). All these wells were abandoned or closed due to water high pressure and gas flux at high pressure.

9.3 FIELD SURVEY–METHODOLOGY–ANALYSIS

The study was conducted in three stages: (i) field observations, (ii) sampling, and (iii) data processing and interpretation of results. During the fieldwork, the geological map of the hot springs area was prepared at 1:25000 scale. Coordinates, temperatures, pH, and EC values of a total of 17 water from Diyadin and 12 water springs from Zilan were measured (Tables 9.1 and 9.2). In addition, samples were collected to determine the hydrochemical and isotopic compositions of the thermal waters in May 2008 (Table 9.4). For the Diyadin geothermal field, a total of seven water samples from the thermal pools, bubbling spring (samples 1, 2, 3, and 7) used in the Diyadin spa, and from wells (samples 5 and 6) that are used for heating, and from surface waters (sample 4) representative of groundwater in the area were collected (Tables 9.1, 9.2, and 9.4).

A total of seven water samples were collected from Zilan; five samples from the thermal spring (samples 2, 3, 4, 5, and 6), one sample (sample 1) from cold waters representing the groundwater in the area, and one sample from Zilan creek (ZC) representing surface water.

All water samples were collected in high-density polyethylene containers and 0.2 mL concentrated HNO_3 added to 100 mL samples for cation and trace element analyses. The other sample (1000 mL) taken for anion analyses was untreated. The 100 and 200 mL samples were collected for $\delta^{18}O$–δ^2H analyses, respectively. Chemicals, heavy metals, and tritium (3H) analyses were conducted at the water chemistry laboratory of the Hacettepe University in Ankara, and (^{18}O), (^2H) analyses were conducted at the G.G. Hatch Stable Isotope Laboratory of the Canada University of Ottawa. Heavy metal analyses for waters were carried out by ICPMS (inductively coupled plasma mass spectrometry). Temperature, pH, and EC measurements were carried out both at the sampling sites and in the laboratory. Water chemistry analyses were conducted in accordance with APHA, AWWA, and WPCF (1989) standards. Ca^{2+}, Mg^{2+}, Na^+, K^+, and SiO_2 concentrations were analyzed by atomic absorption spectrophotometry. SO_4 concentrations were determined by spectrophotometer together with alkalinity standard titration. Cl^- was analyzed by the $AgNO_3$ titrimetric method. Minor and trace elements analyses were carried out by the ICPMS. Standard deviation of isotope analyses is 0.15‰ for $\delta^{18}O$ and 2‰ for δ^2H. In addition, results of analyses of MTA (Akkuş *et al.*, 2005) were also utilized for the evaluation of water chemistry data.

9.4 GEOLOGICAL SETTING

9.4.1 *Geology of Diyadin geothermal field*

During the fieldwork, a new geological map of the hot springs area was prepared at 1:25,000 scale (Burçak *et al.*, 1997). Paleozoic metamorphic rocks, characterized by marble and schists, comprise

Table 9.1. Location, elevation, temperature, and discharge data of water sample sites in the Diyadin (Ağri) region.

Nr	Sample name	Date	T (°C)	pH	EC (µS/cm)	Flow rate (L/s)	X	Y	Z (m)	References
1	Kusburnu Cermik spring	2008	64	6.45	3662	5.5	4370152	3084138	1993	Pasvanoğlu (2008)
2	Facility spring	2008	58	6.54	4182	~3	4370002	3084141	1956	Pasvanoğlu (2008)
3	Yilanli Kirecli	2008	30	6.82	2786	0.5	4370488	3082326	1946	Pasvanoğlu (2008)
4	Murat river	2008	10	7.92	222.3	n.m.	4371375	3083500	1935	Pasvanoğlu (2008)
5	MT-3 well	2008	73	6.81	5750	150	4371782	3084473	1960	Pasvanoğlu (2008)
6	Molakir well-1	2008	45	6.26	2751	n.m.	4362704	3077160	2049	Pasvanoğlu (2008)
7	Kopru Cermik spring	2008	52	6.61	4996	10	4372437	3083963	1957	Pasvanoğlu (2008)
8	Dibekli spring	1998	42	7.10	1661	0.5	4369800	3080525	1980	Akkus et al. (2005)
9	Tazekent spring	1998	37	6.40	3120	0.5	4370550	3082750	1970	Akkus et al. 2005
10	MT-2 well	1999	78	6.55	2791	150	4371375	3084300	1950	Akkus et al. (2005)
11	Davutlu Cermik spring	1998	44	7.60	1990	2.5	4370700	3084424	1980	Keskin (1998)
12	Hidir Cayir spring	1998	45	7.36	3200	0.5	4371300	3085575	1980	Keskin (1998)
13	Yilanli Cermik spring	1998	41	6.90	2500	0.5	4371750	3085575	1950	Keskin (1998)
14	Kusburnu spring	2008	55	6.55	2700	<2	4370090	3084137	1994	Pasvanoğlu (2008)
15	Davutlu spring	2008	48	6.88	2995	1	4370217	3084115	1992	Pasvanoğlu (2008)
16	Davutlu spring	2008	41	6.51	2642	0.5	4369968	3084111	1996	Pasvanoğlu (2008)
17	Davutlu spring	2008	54	6.76	3132	0.5	4370182	3084120	1993	Pasvanoğlu (2008)
18	Davutlu spring	2008	57	6.60	3365	0.5	4370207	3084151	1991	Pasvanoğlu (2008)
19	Yilanli (hydrothermal Davutlu spring	2008	25	6.35	2345	2	4370152	3084138	1989	Pasvanoğlu (2008)
20	Yilanli (hydrothermal cracks spring)	2008	33	6.54	3488	2	4375515	3083752	1979	Pasvanoğlu (2008)
21	cracks spring)	2008	36	6.62	3680	0.5–1	4371463	3085185	1974	Pasvanoğlu (2008)
22	Molakir spring	2008	60	6.65	3816	1	4362704	3077160	2049	Pasvanoğlu (2008)
23	Molakir well-2	2008	73	6.31	2744	n.m.	4362271	3077249	2047	Pasvanoğlu (2008)

Table 9.2. Location, elevation, temperature and discharge data of water sample sites in the Erçiş–Zilan region.

Sample name	Nr	Date	T (°C)	pH	EC ($\mu S/cm$)	Discharge (L/s)	X	Y	Z (m)	References
Taskapi cold spring	1	16.05.2008	13.3	7.1	900	~5	4348178	364227	2078	Pasvanoğlu (2008)
Taskapi well spring	2	16.05.2008	52	6.3	7800	~20	4348326	364565	2020	Pasvanoğlu (2008)
Taşkai şikeft spring	3	16.05.2008	67	6.2	5650	~18	4348173	364538	2015	Pasvanoğlu (2008)
Hasanaptal main spring	4	15.05.2008	64	6.2	6600	~10	4343056	360852	1898	Pasvanoğlu (2008)
Sorkoy spring	5	15.05.2008	59	6.9	360	~5	4349192	364539	1873	Pasvanoğlu (2008)
Hasanaptal spring	6	15.05.2008	22	7	4350	–	4343250	360925	1870	Pasvanoğlu (2008)
Ganisipi spring	7	24.07.1981	78	5.1	6400	~4	4350150	365350	2620	Karamanderesi et al. (1984)
Hasanaptal kuzey spring	8	17.07.1981	64	6.3	7200	–	4343200	360900	1860	Karamanderesi et al. (1984)
Hasanaptal kadinlar hamami	9	17.07.1981	65	6.5	6400	–	4343175	360875	1870	Karamanderesi et al. (1984)
Hasanaptal erkekler Hamami	10	17.07.1981	65	6.5	6400	–	4343150	360875	1870	Karamanderesi et al. (1984)
Zilan creek spring	11	02.07.1974	20	6.4	–	3	4342800	360850	–	Karamanderesi et al. (1984)
ZG-1 well	ZG-1	25.11.1988	80	7.92	3086	40	4348360	364647	2002	Akkuş et al. (2005)
ZG-2 well	ZG-2	2000	92	7.5	4450	4	4348459	364586	1997	Akkuş et al. (2005)
ZG-3 well	ZG-3	2000	98	7.7	4450	22	4348677	364568	2000	Akkuş et al. (2005)
Hasanaptal spring	12	15.05.2008	63.7	6.15	9560	0;5	4343056	360852	1898	Pasvanoğlu (2008)
Taskapi germe Mahmut spring	13	15.05.2008	41.8	6.63	3479	~1	4349379	364344	2040	Pasvanoğlu (2008)
Taskapi havuz	14	15.05.2008	49	6.28	6810	~2	4349501	364373	2039	Pasvanoğlu (2008)
Şepha Pazar spring	15	15.05.2008	42	6.35	4735	~0,2	4350065	363876	2077	Pasvanoğlu (2008)
Nüsret havuzu	16	15.05.2008	22	6.5	3536	–	4348975	364504	2046	Pasvanoğlu 2008
Taskapi well spring	17	16.05.2008	32.8	5.68	2024	–	4348543	364545	2024	Pasvanoğlu (2008)
Taskapi cave	18	16.05.2008	71.2	6.35	8410	~0, 5	4348234	364540	2016	Pasvanoğlu (2008)
Zilan creek	ZC	16.05.2008	14	8	127					

the basement rocks in the study area. Marbles, with a thickness of about 200 m, are white, grayish, and black in color, with a dolomitic composition, and contain calcite veins (Keskin, 1998).

The metamorphic rocks are unconformably overlain by a thick and widely distributed sequence of Upper Miocene-Pliocene volcanic and lacustrine units (Keskin, 1998; Şaroğlu *et al.,* 1980). Metamorphic units exposed in the vicinity of Ulukent village are overlain by the Upper Miocene lacustrine sediments, with an angular unconformity marked by a claret conglomerate of 20–30 m thickness. The Upper Miocene formations are composed of pyroclastic, lava, ignimbrites, and lacustrine units (Keskin, 1998). Silicification is common in veins that cut the volcanics. The pyroclastic rocks consist of agglomerate, volcanic breccias, lapilli, tuff, and ash, with about 0.5– 5 m thick. Products of the Tendurek volcanism, which started in Upper Miocene and continued through the Quaternary, are widely distributed (Pamir, 1951; Şaroğlu *et al.,* 1980).

The radiometric age (K–Ar) of the volcanics is $430,000 \pm 150,000$ to 6.2 ± 0.2 Ma (Ercan *et al.*, 1990). It is also suggested that the source of alkaline type volcanism (Tarhan, 1992) in the region is not derived only from upper mantle, but also has a signature of crustal contamination via post collision. Ignimbrites in the area are exposed around Tasbasamak and Dibekli-Ulukent. It is believed that the ignimbrites are of crustal origin (Keskin, 1998), with the other pyroclastic rocks in lateral and horizontal transition to the lacustrine units. Pliocene units (volcanic and lacustrine sediments) and travertine deposits unconformably overlie the Upper Miocene volcanics and lacustrine units (Keskin, 1998). The Upper Miocene lacustrine units are composed of plant-bearing tuffite, silty marl, lacustrine limestone, tuffaceous-sandy mudstone, and conglomerates. The Pliocene lacustrine units are overlain, with an angular discordance, by Pliocene volcanics, Diyadin lava and pyroclastic rocks, younger lacustrine sediments, Quaternary travertine, and alluvium deposits. The Diyadin lava is 20–30 m thick and is exposed along the Murat river between Davutlu village and Diyadin town (Fig. 9.2). The lava has a basaltic-andesite composition. Pliocene volcanics occur intercalated with lacustrine units. Quaternary alluvium and travertine deposits in the bed of the Murat river are the youngest units in the area.

9.4.2 *Geology of Zilan geothermal field*

During the fieldwork, the geological maps of the Zilan geothermal area were prepared at 1:25,000 scales with data from the field work conducted in 2008 and from previous studies (Karamanderesi *et al.*, 1984; Fig. 9.5). Paleozoic metamorphic rocks characterized by marble and schists constitute the basement in the study area. Metamorphic rocks having been subject to a metamorphism in the green schist facies (biotite, chlorite, quartz, actinolite schist, feldspar, sericite schist, quartz chlorite schist) are exposed in a thin band form along Saman creek from Şorköy village to Gergili in the north (Akkuş *et al.*, 1990; Ölmez and Güner, 1989; Oruç, 2008; Pamir; 1951; Şaroğlu *et al.*, 1980). The plutonic mass is granodioritic to quartz dioritic in character. It has a stock shape forming an irregular contact zone with the surrounding rocks. Since metamorphic rocks outcrop around the granodioritic stock, it can be inferred that these metamorphic rocks have been uplifted by the intrusion of this pluton (Karamenderesi *et al.*, 1984). Following a considerably large stratigraphic gap, the metamorphic rocks have been overlain by the Eocene aged marine deposition of Gergili formation with angular unconformity. This formation has a thickness of 200–250 m; it is composed of sandstone, limestone, claystone, marl, and ultrabasic units, and it has been uplifted and exposed together with metamorphic rocks (Karamanderesi *et al.*, 1984; Fig. 9.5).

During the period from early Neogene up to now, including the Quaternary, an extensive volcanism has occurred in the area (Karamanderesi *et al.*, 1984). As a result of this volcanism, the marine and continental sediments of the area have acquired a volcano/sedimentary character. Neogene is represented by the Lower and Middle Miocene aged Kizildere and Upper Miocene, Lower Pliocene aged Yöreli formations (Akkuş *et al.*, 1990). Kizildere formation, which developed in a shallow marine facies during the Lower and Middle Miocene, consists of alternating beds of tuffites, tuffs, agglomerates, sandstones, clayey limestones, and marls, and it is characterized by a decreasing ratio of volcanic material in its upper levels (Karamanderesi *et al.*, 1984). The formation is approximately 574 m (Akdoğan, 1987) thick and overlies the Gergili formation with unconformity.

In the small lakes that have formed as a result of the uplifting of the region starting from Middle Miocene–Lower Pliocene aged sediments of Yöreli formation have been deposited on Kizildere formation with an angular unconformity. The Yöreli formation consists of yellowish color of marl, claystone, sandstone, lava flows, and tuffs (Akkuş *et al.*, 1990). This formation is approximately 150 m thick and it is overlain by lava flows, pyroclastic material of Quaternary, and alluvium depositions. Quaternary alluvium and travertine deposits in the bed of the Zilan creek are the youngest units in the area. The travertine is mostly $CaCO_3$ composition, beige in color, and deposited from thermal waters that rise from fractures to the north of Erciş. The volcanic activity that began from Lower Miocene continued in the Quaternary. As plutonic rocks, the Eocene aged granodiorite intrusions in the region as well as Miocene aged Şorköy lava, Yalındam lava, Doğanci lava, Ilica lava and Pliocene related Aladağ lava, ignimbrites, and Zilan lava are representatives of magmatic event. The volcanism here is generally andesitic character. However, the Quaternary aged Zilan lavas, which are the products of the latest volcanic activity, are basaltic in character (Karamanderesi *et al.*, 1984; Türkunal, 1980). Şaroğlu *et al.* (1980) indicated that volcanism in the region started in the Pliocene and continued to the end of the Quaternary. Neotectonics in the region began in the Middle Miocene. As a result of compression, thrusts trending in the direction of E–W, dipping to the north or south, and E–W trending folds were formed.

The NE–SW trending left lateral strike-slip faults, in the NW–SE trending, right lateral strike-slip faults, cracks, and volcanics are documented (Şaroğlu *et al.*, 1980). The ongoing regional uplift in the region started after Pliocene (Öngür *et al.*, 1974). At present, gradual steps of valleys according to the regional uplifts and subsidences, E–W and NW–SE striking faults have resulted. Near Doğanci village along the Hasanaptal resources, there are two faults striking ENE–WSW. Along these faults, travertine and silica deposition are found throughout the thermal water outlets. These faults have been detected in the coal drillings between 90 and 100 m depths (Ağrali, 1966; Has, 1973).

9.5 HYDRGEOLOGY

9.5.1 *Diyadin (Ağri) geothermal field*

The Diyadin geothermal field is located in an area that has been extensively affected by Neo-tectonic activities. Several strike-slip faults and tensional fractures developed in association with a N–S regional compression. Most tension fractures are oriented in a N–S direction. Extensional structures formed by multiple-fault systems tend to be E–W trending and are conducive for controlling flow of thermal fluids. Thermal water and gas are expelled from most of tensional fractures, associated with travertine deposition. As a result of intensive tectonism, the metamorphic rocks are extensively deformed, and they reveal a well-developed fracture network.

Due to their permeability, Paleozoic marbles and limestones underlying the Neogene units can be regarded as potential reservoir rocks for the Diyadin geothermal system (Fig. 9.6; Pasvanoğlu and Güler, 2010a). Schists, altered tuffs, and ignimbrites are the cap rocks (Table 9.3). The Pliocene lacustrine units provide a weak cap rock, but do not have noticeable thickness and distribution within the Diyadin geothermal field (Keskin, 1998). Due to their permeability, these units can host reservoir fluids. Despite their limited extent and thickness, metamorphic schists beneath the reservoir rocks of marbles are assumed to form an impermeable basement. However, due to N–S compression and intense, recently active tectonism, aquifers may be formed within the overlying fractured volcanic rocks. The Tendurek volcanism that lasted from Upper Miocene to Quaternary is the inferred heat source for the Diyadin thermal waters. The Tendurek volcano is located near the Turkey–Iran border and is composed of two cones (Ercan *et al.*, 1990). The first cone is inactive and expels only gas, while the other hosts a lake, and steam-heated fluids discharge along fault associated with the volcano (Pamir, 1951).

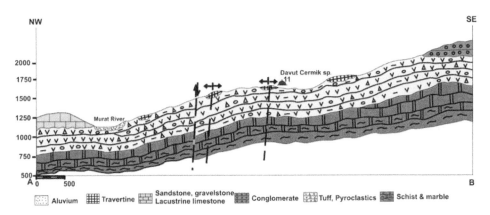

Figure 9.6.   Hydrogeological cross-section of the Diyadin geothermal field A–B section line on Figure 9.2, sample numbers correspond to locality shown in Figure 9.2.

9.5.2   *Zilan (Erçiş) geothermal field*

The distribution of hot water springs in the Zilan region roughly parallels to the distribution of the fault systems and young volcanism. Because of intensive tectonism, the metamorphic rocks are extensively deformed, and they reveal a well-developed fracture network. The Gergili formation, overlying the metamorphics, has acquired secondary permeability because of the intense faulting tectonism and thus became a reservoir rock. Therefore, aquifer rocks of Erciş thermal springs waters are intrusives (granodiorite, quartzdiorite), sandstones, and limestones (Table 9.3, Pasvanoğlu and Güler, 2010b). Şorköy lava, Kizildere formation, Yöreli formation, and ignimbrite, which are characterized with thick horizons of altered tuffs, tuffite, and marl, have a low porosity and therefore act as a cap rock containing heat and maintaining water pressure. These concerns are very typical, especially around Hasanaptal resources. Thermal water supplies at the bottom of the deep Zilan valley are very good examples for outcropping of very young faults and Tertiary recent volcanics in the edge zones. The propagation of these faults is thought to be caused by Sehirpazar landslides that have been observed in the vicinity of Yöreli, Akişköy, Yalindam, and Kizildere. These landslides that have slid on a cover plane are parallel to this cap rock (Karamanderesi *et al.*, 1984, Fig. 9.7). Metamorphic schists beneath the reservoir rocks of marbles are thought to be the impermeable basement. Resistivity, gravity, and magnetic measurements were conducted in the region (Akdoğan, 1987; Burkay and Şahin, 1986), and despite the positive geothermal aspects of the northern parts of Şorköy field it is determined that no impermeable blanket of sufficient thickness exists. In addition, a cutting fault line in the northeast of Şehirpazari and in the northwest–southeast direction of Zilan creek was identified. The recent volcanism and the active tectonism form the heat source in the geothermal area.

9.6   RESULTS AND DISCUSSION

9.6.1   *Water chemistry*

The analyses show that Ca and HCO_3 are the dominant cation and anion in Diyadin waters, respectively. Sodium and potassium are also significant. Based on their chemistry, pH, and EC, the Dyadin waters can be divided into four groups. The first group consists of waters from high elevation (1946–1996 m) and includes samples 1, 2, 5, 7, 8, 10, 13 thermal waters. These waters have EC in the 1661–5750 μS/cm range, and pH from 6.45 to 7.9. Bicarbonate is the dominant anions in this group; major cations are Ca and Na. The waters in this group have the highest measured spring temperatures (41–78°C). The second group (sample no 3, 6, 11, 12 in Table 9.4)

Table 9.3.   Spring and well informations of Ağri (Diyadin) and Van (Erçiş) geothermal systems.

Geothermal system	NWD	MMDD	TDR (L/s)	Reservoir rocks	Temperature Cap rocks	Present use	(°C)	Heat source
Diyadin geothermal system	8	77–215	Spring: 0.5–10 Well: 560	Paleozoic marbles and limestones	Schists, altered tuffs, and ignimbrites	Bath and district heating	Spring: 24–64 Well: 37–78	Tendurek volcanism
Erçiş geothermal system	4	264–1172	Spring: 0.1–18 Well: 4–40	Intrusives (granodiorite, quartzdiorite), sandstones, and limestones	Şorköy lava, Kizildere, and Yöreli formation (ignimbrite, alterated tuffs, tuffite and marl	bath and treatment	Spring: 22–78°C Well: 80–105°C	Tendurek volcanism

NMD: number of well drilled (m), MMDD: maximum and minimum drilled depth (m), and TDR: Total discharge rate (L/s).

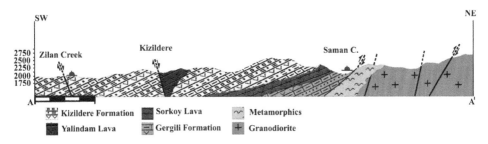

Figure 9.7.   Hydrogeological cross-section of the Zilan geothermal field (modified from Karamanderesi *et al.*, 1984) A–A′ section line on Figure 9.5.

consists of thermal waters from elevation between 1980 and 2049 m. EC and pH of these waters range from 1990 to 3200 μS/cm and pH range from 6.26 to 7.90. As in the first group, bicarbonate is the dominant anion in this group; major concentrations of cations are Na and Ca. The third group (sample 9 in Table 9.4) consists of a single thermal spring on the elevation at about 1970 m. The water has EC of 3120 μS/cm and pH of 6.40. The main cation and anion in this spring is Ca and SO$_4$. The last group of waters (sample 4 with $T = 10°C$) is cold waters, occurring at lower elevation (about 1935 m). Their EC and pH are 222.3 μS/cm and 7.92, respectively. As in the first group, the main anion is HCO$_3$; the main cations are Ca and Mg.

Results of the major ion analyses of thermal and cold waters from the Zilan area are presented in Table 9.4. The pH values for Zilan thermal waters range between 5.1 and 7.9 from high elevation (1860–2620 m), but for cold water springs these range between 6.4 and 7.1 (samples 1 and 11). The TDS (total dissolved solids) content of thermal waters range from 1975 to 4992 mg/L, with the cold water having a maximum TDS value between 81 and 576 mg/L (samples 1 and ZC). Electrical conductivities range from 360 to 9560 μS/cm for thermal waters and from 127 to 900 μS/cm for cold waters (Tables 9.2 and 9.4). The TDS concentration of Zilan thermal water is higher than Diyadin geothermal water. These values probably reflect longer circulation and residence times. The dominant cation is Na$^+$ and the dominant anion is HCO$_3$ and Cl in thermal water. Cold waters (sample 1 and ZC) are mainly dominated by Ca, Mg, and HCO$_3$, and their ion contents are low. The Cl concentration is relatively rich in the Zilan water samples taken from fields to reservoirs, which can be explained by dissolution of rock units or as a result of circulation for a long residence time with rocks. The thermal waters from wells and springs show temporal variation of Cl and HCO$_3$ anions concentrations. The variation is mainly due to a large incidence of HCO$_3$ over SO$_4$ and Cl in the samples. The thermal water in the Zilan geothermal area has concentration ratios of Na/Cl, Ca/Mg, and HCO$_3$/Cl that all exceed a value of 1. It has high Na, K, B, and SiO$_2$, which indicates strongly that the water is in contact with silicic rocks. In addition, the thermal waters were mixed with high ratio of cold water during their ascend to the surface. K-gain during the rise of fluid could be related to K-alkaline volcanism in the region (Pasvanoğlu, 2008).

The ion contents of deep groundwater in Diyadin are more variable than those in Zilan (Fig. 9.9). This is probably due to the greater variety of mixing extents between cold and thermal groundwater in the field.

In summary, the Diyadin samples generally have lower EC, Cl, Na, K, SiO$_2$, B, and Li concentrations than the Zilan samples, as shown in Table 9.4. However, the Ca and Mg concentrations of the Diyadin samples are higher than those of the Zilan samples. In contrast to the geothermal fluids below Diyadin, those below Diyadin have likely mixed with more cold groundwaters enriched in Ca and Mg, but depleted in Cl, Na, K, SiO$_2$, and B.

In both geothermal areas, Na in the waters is derived from dissolution of Na-bearing salts or alteration of feldspars in the volcanic rocks and schists in the area:

$$2NaALSi_3O_8 + 2H^+ + 9H_2O \rightarrow Al_2Si_2O_5(OH)_4 + 4H_4SiO_4 + 2Na^+ \tag{9.1}$$

Table 9.4. Results of chemical composition of waters from the study area.

Nr	T (°C)	pH	EC (µS/cm)	Ca^{2+}	Mg^{2+}	Na^+	K^+	Cl^-	SO_4^{2-}	HCO_3^-	SiO_2	TDS (mg/L)	Li	B	Water type
DIYADIN															
1	64	6.45	3662	171.6	76.8	145.9	68.9	116	122	891	63.3	1683	1.8	25.5	Ca–Na–HCO₃
2	58	6.54	4182	363.6	66.7	179	66.2	155	189	1811	59	2922	1.3	30.8	Ca–Na–HCO₃
3	30	6.82	2786	142.7	75.6	168.4	65.8	136	253	861	44	1777	1.2	28.6	Na–Ca–HCO₃
4	10	7.92	222.3	44.2	10.4	11.6	4.4	7	48	176	22.7	325	0.1	0.7	Ca–Na–HCO₃
5	73	6.81	5750	242.2	58.4	182.8	84.9	160	106	1667	138	2676	1.4	35	Ca–Na–HCO₃
6	45	6.26	2751	177.2	35.1	182.7	82.8	231	150	1000	86.9	1971	2.7	21.9	Na–Ca–HCO₃
7	52	6.61	4996	532.2	83.4	124.4	48.9	111	149	1922	53.4	3042	1.4	16	Ca–Na–HCO₃
8	42	7.1	1661	133	98.2	118	97	114	261	957	40	1846	1.5	27	Ca–Na–HCO₃
9	37	6.4	3120	762	104	59.6	39	134	2367	60	45.0	3601	1.9	29	Ca–Mg–SO₄
10	78	6.55	2791	341	50.3	216	73.5	182	189	1673	42	2803	1.5	34.9	Ca–Na–HCO₃
11	44	7.6	1990	132	65	212.1	75	164	137	976	54	1852	1.4	35	Na–Ca–HCO₃
12	45	7.36	3200	360	65	363.6	59	192	141	2013	65	3295	1.4	35	Na–Ca–HCO₃
13	41	6.9	2500	208	71.2	173	58	151	144	1141	42	2027	1.7	37	Ca–Na–HCO₃
ERÇIŞ															
1	13	7.1	900	39.27	10.6	3,2	1,2	6,3	50	95	24	230	0	0.13	Ca–Na–HCO₃
2	52	6.3	7800	207.22	70.47	997.5	91.24	764	250	1755	122	4325	4.66	62.45	Na–Ca–HCO₃–Cl
3	67	6.2	5650	115.84	47.77	765.99	69.31	593	225	1342	120	3323	3.45	40.57	Na–Ca–HCO₃–Cl
4	64	6.2	6600	144.21	23.69	1062	93.55	1026	245	1800	79	4572	4.24	94.46	Na–Ca–HCO₃–Cl
5	59	6.9	360	110	90.1	220	105	90	175	1400	128	2315	1.31	31.91	Na–Ca–HCO₃
6	22	7	4350	120.6	19	738	105	878	185	1154	31	3370	3	76.9	Na–Ca–HCO₃–Cl
7	78	5.1	6400	160	8	750	68	750	540	875	90	3276	0,4	34*	Na–Ca–HCO₃–Cl
8	64	6.3	7200	135	14	838	99	1075	250	1075	90	3637	2,4	58*	Na–Ca–HCO₃–Cl
9	65	6.5	6400	150	12	850	88	950	208	1000	80	3382	3,6	40*	Na–Ca–HCO₃–Cl
10	65	6.5	6400	150	12	875	86	975	219	1000	80	3489	3,2	88*	Na–Ca–HCO₃–Cl
11	20	6.40	244	32	12	70	8,4	15,6	62	251,8	27	484		4	Na–HCO₃ ZG-1
12	80	7.9	3086	96	56	830	74	715	565	994	109	3504	4,1°	60°	Na–HCO₃Cl ZG-2
13	92	7.5	4450	36.9	54.6	773	110	543	470	897	95	3038	4,2	53.9°	Na–HCO₃–Cl ZG-3
14	98	7.7	4450	29.5	47	858	108	560	491	779	118	3039	4	44.1°	Na–HCO₃3–Cl
ZC	14	8	127	17	9.7	20	3.2	11	48	80	15	208			Na–HCO₃

Diyadin samples (Nr 8, 9, 10, 11, 12, and 13) are results of 1998 analyses of MTA (Akkuş et al., 2005) Erciş: *Karamanderesi et al. 1984; °Akkuş et al., 2005; sample numbers correspond to locality shown in Figures 9.2 and 9.5; TDS total dissolved solids; cation and anion concentration are in mg/L; Zilan samples (Nr 1, 2, 3, 4, 5, 6 and ZC) are results of present study.

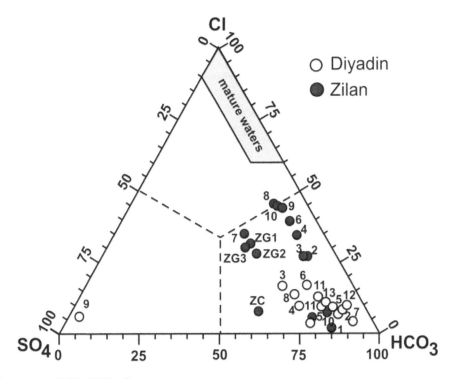

Figure 9.8.   Cl–SO$_4$–HCO$_3$ diagram.

Low Cl$^-$ concentration is attributed to mixing of the ascending thermal waters with cold waters. Low chloride concentrations in hot waters may indicate that water circulation in Diyadin geothermal field is shallow.

High bicarbonate concentrations are due to the reaction of the circulating meteoric waters with limestone and marbles forming CO$_2$-rich waters, and possibly from magmatic fluid input as well (Browne, 1991).

Although the TDS values in Diyadin is high, the concentration of silica is higher in Zilan thermal waters than Diyadin thermal waters. This can be attributed to flow through different lithologies, with different amounts of Si-bearing minerals, and different temperatures in the two fields (Han *et al.*, 2010). The Si content in Zilan thermal groundwater is roughly double than that of Diyadin thermal groundwater, suggesting more extensive silicate mineral dissolution. The lithology around Zilan is made exclusively of intrusive and metamorphic rocks, while in Diyadin some recharge water travels through marbles (karstic), contacting carbonate minerals and mixing with fissure water during through flow. Hotter reservoir temperatures in Zilan field (see "geothermometers") may also contribute to more extensive silicate mineral weathering.

Silica increase in waters is due to alteration of volcanic glass and silicate minerals or dissolution of quartz in alkaline conditions. Indeed, Ercan *et al.* (1994) state that silica concentration of volcanic ash, lava flow, and tuffs are high, with SiO$_2$ content of rocks near spring sites being around 60 wt%. Thermal and cold waters from the Diyadin and Zilan are classified on Figures 9.8 and 9.9. According to Giggenbach (1988), Cl–SO$_4$–HCO$_3$ geothermal waters that are in equilibrium with reservoir rocks and are least contaminated by near surface waters have chloride concentration greater than bicarbonate and sulfate concentrations (Fig. 9.8). The pH of such waters is neutral due to phyllosilicates buffering reactions (Giggenbach, 1988). Hence, the majority of geothermal waters that emerges to the surface with minimum contamination with groundwaters or by H$_2$S containing waters of volcanic origin fall within the mature thermal waters field proposed by Giggenbach (1988) in Figure 9.8.

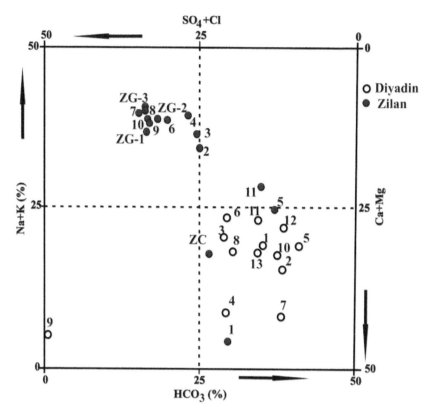

Figure 9.9.   Langelier–Ludwig diagram (Langelier and Ludwig, 1942).

For the Diyadin geothermal field, two types of waters are identified: (i) water (sample no 9), which is located close to the SO_4 corner, and acid sulfate waters, which occur in the geothermal area. Sample 9 has low pH value (6.40) and is likely to be volcanic fluids in which hydrogen sulfide has been oxidized to sulfides (pyrite), sulfates, or elemental sulfur; and (ii) peripheral water that plots in the HCO_3 corner is not sufficiently mature and HCO_3 could be mostly derived from CO_2-rich groundwater and possibly from magmatic activity. Bicarbonate waters are typically found on the margins of the geothermal system and can occur in umbrella shaped perched aquifers overlying the geothermal system. Common surface features are 'soda springs' and warm to hot springs with a near neutral pH. Surface deposits of $CaCO_3$ (travertine) are common around bicarbonate springs (Browne, 1991) and may indicate subsurface temperatures $<150°C$. Rare aragonite may form if surface discharge cooling is rapid (Nicholson 1993). All these mineral forms occur in the vicinity of the springs in Diyadin (Demirel et al., 2000; Ertürk and Manav, 1999). As a result, the hydrochemical types of geothermal waters from the Diyadin geothermal area include Ca–Na–HCO_3, Na–Ca–HCO_3, and Ca–Mg–SO_4 type (Fig. 9.8).

In the case of thermal waters of Zilan, from the local geological setting, enrichment in SO_4 can be due to the oxidation of H_2S derived from the cooling magma in the Miocene volcanics giving rise to SO_4 ions ($H_2S + 4H_2O \rightarrow SO_4^{2-} + 10H^+$). In such reaction, a significant amount of hydrogen is released, which increases the acidity of the fluid. Sulfur-rich minerals of copper, arsenic, and other transition metals in the Miocene lithological units are also the main sources of sulfate in the groundwaters as reported by Gemici et al. (2007) and by Pasvanoglu and Chandrasekharam (2011).

High bicarbonate concentrations are due to reaction of circulating meteoric waters with limestone and marbles, forming CO_2-rich waters. Thermal waters from wells (ZG-1, ZG-2, and ZG-3)

Table 9.5.   Results of trace element (μg/L) and isotope analyses of the waters from the study area.

Nr	Sr	Cs	Rb	As	Cr	Cu	Al	Se	Fe	δ^2H (‰SMOW)	$\delta^{18}O$ (‰SMOW)	^3H (TU)
						DIYADIN						
1	2255	975.2	444.6	76.5	14	0.9	7	7	<10	−92	−12.1	0.64 ± 0.25
2	2816	486.7	357.8	107	27.14	1.5	35	11.40	153	−91	−11.8	0.01 ± 0.23
3	1678	568.5	347.6	50.3	27.1	0.7	5	7.8	<10	−90	−12.2	2.22 ± 0.27
4	189	22.6	17.7	14.6	4.6	8.2	365	<0.5	252	−76	−11.4	3.00 ± 0.24
5	1923	1047	584.7	56.4	3.4	1.1	18	6.2	<10	−94	−12.1	0.09 ± 0.24
6	1855	2368	679	5068	16.2	0.8	5	4	<10	−93	−12.8	0.17 ± 0.24
7	2.257	617.4	339	60.9	22.2	4.8	693	<5	616	−94	−12.6	0.21 ± 0.24
						ERÇIŞ						
1	29	0.28	2.43	0.6	4.8	0.2	<1	<0.5	<10	−77	−11.7	7.94 ± 0.38
2	3706	869	610	387	59	3.1	<0	6.2	<100	−77	−10.3	0.06 ± 0.23
3	2167	704	494	588	49	2	<10	6	<100	−80	−11.1	1.07 ± 0.26
4	2292	833	536	36	75	1.5	<10	10.2	<100	−79	−10.5	2.02 ± 0.29
5	1209	957	562	61	32	0.5	47	2.5	<10	−79	−10.2	0.66 ± 2.25
6	3700	866	601	63	58	3.1	<10	6.2	<100	−79	−10.4	2.03 ± 0.29
9				138*			340*		870*			
10				250*			400*		875*			
ZC	28	0.28	2.42	0.59	4.2	0.1	<1	<0.5	<10	−84	−11.93	8.9 ± 0.23

Sample numbers correspond to locality shown in Figures 9.2 and 9.5; trace element concentrations are in μg/L; the isotopic results are expressed as ‰ deviation from standard Mean Ocean Water (SMOW) with uncertainty ranges of ±0.15‰ for $\delta^{18}O$ and ±2 for δ^2H, respectively. Tritium results are reported in tritium units (TU), where 1 TU has a ratio of $^3H/^1H + 10^{-18}$. The average analytical error is ±0.56 TU. *Yenel (1976).

and spring (sample 7) are waters with high SO_4 and SiO_2 contents. Waters of sample 2, 3, 4, and 6 are a mixture of chloride and bicarbonate end-member waters as are spring samples 8, 9, and 10 from the area. The concentration of Cl and HCO_3 ions in the sample 8, 9, and 10 are almost the same. Except for the thermal waters from wells (ZG-1, ZG-2, and ZG-3), all thermal springs have low pH value (5.1–7) and are likely to be volcanic fluids in which hydrogen sulfide was oxidized to sulfate by oxidation in the groundwater. For the Zilan area, the cold springs (sample 1) plot in the Ca+Mg HCO_3 corner pertaining to typical groundwaters. The thermal waters from springs and wells have much higher TDS (>400 mg/L) relative to the groundwaters. All the thermal waters from the wells and springs have different types of waters mixed during their ascent. These waters would be classified as Na–Cl and Na–HCO_3 water types or mixtures thereof (Fig. 9.9). Waters of samples 5 and 11 of Zilan have values with low EC (360–244 μS/cm). These two samples are located between cold groundwater springs and thermal waters suggesting mixing with two types of waters with different composition during their ascent to the surface. Therefore, the diagram shows the direction of mixing between alkaline Cl–SO_4 type waters and Ca–HCO_3 type groundwater (Fig. 9.9). The trend from thermal waters to sample 1 indicates the direction of the increasing rate of the contamination by the local groundwater.

9.6.2   *Trace element contents of Diyadin waters*

Trace metals and non-metals were analyzed in thermal and cold waters from the study area (Table 9.5). Rare elements such as Li, Rb, and Cs are unaffected by the dilution processes at shallower levels or steam loss during ascent and retain their parentage (Chandrasekharam and Bundschuh, 2002). These elements provide information on the geochemical processes influencing the chemical signature of the thermal waters and their origin. These elements enter the liquid phase during water-rock interaction at deeper depths due to initial dissolution process and then enter

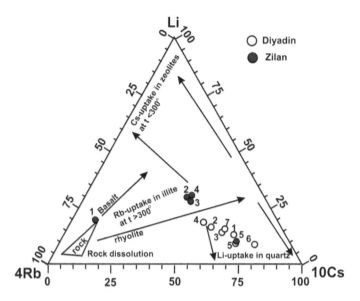

Figure 9.10.   Li–Rb–Cs diagram.

the solid phases again when the thermal water and rock attain new chemical equilibrium. The new solid phase equilibrium conditions result in precipitation of the secondary minerals, which are commonly encountered along the fluid paths. To attain such new equilibrium conditions, the water and the rock should be in contact for a very long period of time.

Relative concentrations of Rb, Cs, and Li of the thermal waters from wells, springs, and surface water from the study areas are plotted in the Li–Rb–Cs ternary diagram (after Giggenbach, 1991) as shown in Figure 9.10. The mineralization trends and water composition paths depicted in Figure 9.10 are discussed in Giggenbach and Goguel (1989). Thermal waters in which these elements enter due to initial dissolution fall in the "rock dissolution" field in Figure 9.10. Initial dissolution takes place when the pH of the reacting waters is very low.

The Li, Rb, and Cs are higher in thermal waters than cold waters (Table 9.5). Concentrations of these elements at the surface decrease with increasing migration and lateral flow. The relatively higher Li content of Zilan spring waters compared to the Diyadin spring waters may be a result of Li release from clays or other minerals due to shallow processes. Mixing with meteoric water would lower Li concentrations, and they are almost zero in both Diyadin and Zilan cold water springs. The thermal waters show the compositional range from basalt to rhyolite.

The position of Diyadin waters indicate that these waters have already lost some amount of Rb to minerals like illite and K-feldspar, which accommodate Rb in their structure at temperature >300°C. Precipitation of quartz results in depletion of Li from the thermal waters during their ascent. The Li/Cs ratios of the Zilan thermal waters (i.e., excluding cold waters) range from 1.37 to 5.36, reflecting fluid–rock interactions (Giggenbach, 1991). This ratio is close to that of intermediate volcanic rocks, suggesting that Li and Cs in the spring waters are the result of water-rock interactions (Giggenbach and Glover, 1992).

Strontium concentrations of the thermal waters are very high in Diyadin and Zilan. According to Pisarskii et al. (1998), Li, Rb, and Sr may come from a magma chamber linked to the Tendurek. The very high strontium concentration of thermal waters reflects the interaction between the ascending thermal waters and volcanic rocks (Delalande et al., 2011 (Table 9.5). Sr concentrations increase with increasing Cl$^-$, indicating accumulation during water-rock interaction along groundwater flow paths. The different extents of this increase in the two fields again highlight differences in lithology. A plot of the relative concentrations of chloride, lithium, and boron is displayed in Figure 9.11. The position of all the Diyadin and Zilan waters in this diagram suggests the

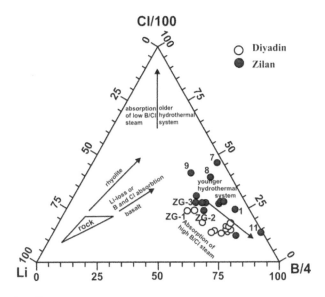

Figure 9.11.   Cl–Li–B diagram.

absorption of high B/Cl steam after neutralization of HCl. The low B/Cl ratio of near surface water is also possibly due to the dilution of deep Cl waters (Reyes *et al.*, 1993). The relatively high concentrations of boron in the thermal waters (16–37 mg/L in Diyadin and 4–94.46 mg/L in Zilan) also indicate a relatively young geothermal system because B would likely be expelled during the early heating-up stages of the geothermal system (Giggenbach and Goguel, 1989). Concentrations of B and Cl cannot be controlled by temperature-dependent chemical equilibrium; both could have been extracted from the country rocks by the circulating thermal waters (Ellis, 1970), in which case their relative concentrations would be similar. Because of the high ratio of Cl/B, in the range of about 4.5–10.5 in Diyadin and 4.5–23.75 in Zilan, it is more probable that they do not have the same origin. It is more likely that the Cl is contributed by magmatic waters and that the B is extracted from the wall rock (Shakeri *et al.*, 2008).

Lithophilic element concentrations of the water are variable: the maximum aluminum (Al) content in water is 693 μg/L (in Diyadin, sample 7), with high Al and Fe, and in some cases, Se concentrations of the Diyadin and Zilan waters are probably due to acidic alteration (i.e., low pH) and association with sulfide minerals in skarn areas along contacts between granodiorite and metamorphic rock.

Iron input may be due to the oxidation of pyrite and other sulfides during water-rock interaction in the reservoir:

$$2FeS_2 \text{ (pyrite)} + 7O_2 + 2H_2O \rightleftarrows 2Fe^{2+} + 4SO_4^{2-} + 4H^+ \qquad (9.2)$$

High boron concentrations in Diyadin and Zilan thermal waters are also accompanied by relatively high concentrations of some other volatiles such as arsenic. As and Cu are chalcophilic elements and their concentrations range from 50.3 to 5068 μg/L (sample 6), 0.7–4.8 μg/L (sample 7) in Diyadin, and between 36 and 588 μg/L (sample 3), and 0.5–3.1 μg/L (sample 2) in Zilan, respectively. Concentrations of As in the thermal water are very high compared to the cold groundwater. Arsenic is usually found in metallic mineralization such as arsenate (AsO_4^{4-}) and in apatite, taking the place of phosphate. The degradation of volcanic rocks allows As to pass more easily into the groundwater. Therefore the high concentration of As are possibly related to the leaching of metal sulfides in the rocks.

The dissolution of volcanic rocks makes arsenic to pass more easily to the groundwater. The increase of As concentration is related to interaction between hydrothermal decomposition zone

Table 9.6.   Results of geothermometers application in the study area.

Thermal waters	Discharge temperature (°C)	TK–Mg[4]	TLi–Mg[3]	T Qtz[1]	Ağri-Diyadin T Chalcedony[2]
1	64	84	84	113	84
2	58	90	78	113	84
3	30	88	74	96	66
5	73	97	81	150	132
6	45	104	105	129	102
7	52	79	77	105	75
8	42	94	77	92	61
9	37	71	82	97	67
10	78	95	85	94	64
11	44	93	80	106	76
12	45	86	80	115	86
13	41	86	80	114	86
	Erciş-Zilan				
2	52	97	111	143	123
3	67	95	108	141	122
4	64	113	125	122	96
5	59	97	81	145	127
6	22	119	117	84	49
7	78	119	75	128	104
8	64	122	115	128	104
9	65	121	128	123	97
10	65	120	127	123	97
ZG-1 well	80	94	111	137	116
ZG-2 well	92	105	112	130	107
ZG-3 well	98	107	113	141	121

[1,2]Fournier (1973); [3]Kharaka and Mariner (1988); [4]Giggenbach (1988); sample numbers and names are the same as in Tables 9.1 and 9.2.

and water. Zilan thermal waters are used in spas at present. Considering As content is much higher than the prescribed limit for drinking water standards of 10 μg/L (As) by WHO (2003), and caution should be exercised in using cold springs water for drinking purposes. As given in Table 9.4 and Figure 9.5, mixing thermal water and cold spring water takes place at near surface environment and thus there is every chance of contamination of groundwater that may be used for domestic purposes. Finally, the high levels of As in solution can be considered as normal in this volcanic district, and are possibly related to the leaching of metal sulfides in the rocks. Moreover, the Li and As concentrations of the thermal springs are higher compared to cold waters (Tables 9.4 and 9.5). Cr concentration in Zilan thermal waters is 31.7–74.8 μg/L. Hasanaptal main spring (sample 4) has a high Cr concentration due to the oxidation of pyrite.

9.6.3   Geothermometers

Chemical geothermometers, which are used to estimate the reservoir temperatures for geothermal systems, are based on the temperature-dependent geochemical equilibria, which exists between water and the minerals within the rock at the temperature of the deep thermal reservoir. The principal assumption during the use of these geothermometers is that the chemical composition of water attained through the equilibration process at depth is preserved during its rise to the site of sampling at the surface.

In this study, various silica and cation geothermometers were used to infer the reservoir temperatures of the Diyadin and Zilan geothermal system (Table 9.6). However, estimates for reservoir temperatures calculated for thermal waters from both fields are affected by chemical processes

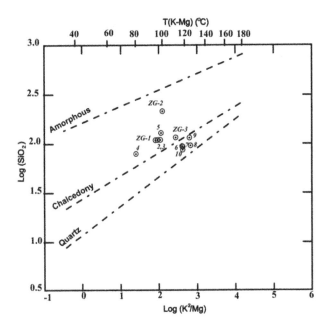

Figure 9.12.   Plot of log (K²/Mg) *vs.* log(SiO₂), concentrations in mg/L. The lines represent simultaneous attainment of equilibrium for the systems involving silica and K–Mg (Giggenbach and Glover, 1992).

such as mixing and evaporation. Using the quartz and chalcedony geothermometers (Fournier, 1973), a range of reservoir temperatures from 92 to 150°C and 61 to 132°C; from 84 to 145°C and 49 to 127°C were obtained for thermal springs in the Diyadin and Zilan, respectively (Table 9.6).

The estimates obtained from the Chalcedony temperatures are approximately 18°C lower than the quartz temperatures in both fields. For Diyadin geothermal areas, chalcedony geothermometer does not yield reasonable reservoir temperatures, because the estimated temperatures are lower than the (sample 10, $T = 78°C$) discharge temperatures. The K–Mg geothermometers of Giggenbach (1988) can be used, assuming that deeper waters, which are characterized by low Mg values, are in equilibrium with the mineral assemblages as described by Giggenbach. For the Diyadin thermal waters, the K–Mg geothermometer (proposed by Giggenbach, 1988) and the Li–Mg geothermometer of Kharaka and Mariner (1988) yield similar temperature estimates, ranging from 71°C to 104°C and 74 to 105°C, respectively (Table 9.6). Using the K–Mg and Li–Mg geothermometers, a range of reservoir temperatures from 75 to 128°C and 94 to 122°C were obtained for thermal waters in the Zilan, respectively (Table 9.6).

These temperatures are similar to those obtained from the chalcedony geothermometer and to discharge temperatures of some thermal waters in the Zilan geothermal area; however, they are lower than those from quartz geothermometers.

These differences suggest that quartz geothermometer reflects the reservoir temperatures whereas K–Mg geothermometer records intermediate temperatures between the reservoir values and the spring outlet values as the geothermal waters re-equilibrate upon conductive cooling or mixing with cooler Mg-rich waters. Further inspection indicates that the discrepancies between the temperatures obtained using K–Mg geothermometer and quartz geothermometer are much larger for the geothermal springs at Diyadin and Zilan geothermal areas. It means that conductive cooling or mixing with cooler Mg-rich waters has much more influence on temperature decrease of geothermal water when it flows upward to the spring vents, and as a result, their K–Mg temperatures and quartz temperatures have significant differences.

To assess the appropriate solid SiO₂ species that is involved in the equilibration of the Zilan thermal waters, the K, Mg, and SiO₂ data from Table 9.6 are plotted in the log(K²/Mg) *vs.*

log(SiO$_2$) diagram together with the solubility curves for amorphous silica, chalcedony, and quartz (Fig. 9.12). The figure shows that thermal waters of Taskapi (samples 2, 3, 4, 5, 7, ZG-1, ZG-2, and ZG-3) plot along the chalcedony curve; the waters of Doganci (samples 6, 8, 9, and 10) lies on the quartz curve. The preliminary check of data in Figure 9.12 has shown that the quartz geothermometer of Fournier (1973) can be used to assess the equilibrium temperature at Doganci (samples 6, 8, 9, and 10), and the chalcedony geothermometer at Taskapi (samples 2, 3, 4, 5, 7, ZG-1, ZG-2, and ZG-3). The rather fast equilibrating K–Mg geothermometer (Giggenbach, 1988) can be used for all samples. The resulting equilibrium temperatures are listed in Table 9.6 which shows that the SiO$_2$ and the K–Mg equilibrium temperature are rather close, namely between 119 and 122°C for the Doganci (samples 6, 8, 9, and 10), and between 94 and 120°C for the Taskapi samples (samples 2, 3, 4, 5, 7, ZG-1, ZG-2, and ZG-3).

For the Diyadin and Zilan geothermal waters, the real reservoir temperature cannot be estimated by geothermometers without considering mixing. Therefore, silica enthalpy mixing model suggested by Truesdell and Fournier (1977) was used to estimate the temperature for the Diyadin and Zilan mixed water. This model is based on the assumption that no silica deposition took place before or after mixing and that quartz controls the solubility of silica in the thermal water component. Figure 9.13a and b show the results of the silica mixing model applied to the Diyadin and Zilan data. Samples 1, 4, and ZC are used as a reference cold water for the Diyadin and Zilan region, respectively. The silica enthalpy mixing models suggest a reservoir temperature of 110°C to 161°C for Diyadin thermal waters, and between 145°C and 178°C for Zilan thermal waters.

The geothermal gradient in the ZG-1 well is found to be 10 m/2.25°C (Ölmez and Güner, 1989). The presence of Eocene granodiorite stock and Miocene volcanism in and around the Zilan area caused the increase of the geothermic gradient value. Homogenization temperatures inferred from fluid inclusions can also represent the equilibrium temperatures of geothermal systems (e.g., González-Partida et al., 2000). As we mentioned previously, the minimum homogenization temperature of 132°C in the samples taken from the ZD-1 well at 1064–1066 m depth is detected (Ölmez and Güner, 1989). The Zilan-Doğanci samples have the highest estimated quartz geothermometric reservoir temperatures, which close to the homogenization temperature of 132°C. The Zilan-Taskapi samples have maximum reservoir temperature of 121°C (sample ZG-3 well) estimated by chalcedony geothermometer.

9.6.4 Isotopic composition of waters

Results of the $\delta^{18}O$–δD ratios of Diyadin and Zilan thermal waters are presented in Table 9.5. In Figure 9.14, stable isotopic compositions of Zilan waters are plotted in $\delta^{18}O$–δD diagram with Van meteoric ($\delta^2H = 8\delta^{18}O + 16.4$; Aydin et al., 2009) and Global meteoric ($\delta^2H = 8\delta^{18}O + 10$; Craig, 1961) water lines. The Mediterranean meteoric water line of Gat and Carmi (1970) is also plotted for comparison.

On the ^{18}O–^2H diagram (Fig. 9.14), all the thermal waters from Diyadin and Zilan plot close to the global meteoric water line indicate that they are likely to be of meteoric origin. It is suggested that the rainwaters percolated downward through faults and fractures are heated by an intrusive-cupola and then rise to the surface along permeable zones that act a hydrothermal conduits.

Both thermal waters from Diyadin and Zilan show ^{18}O shift from the meteoric water line indicating that their isotopic composition is affected by water-rock interaction process, evaporation and different rock types of aquifer. It is noticeable that Diyadin thermal waters represent more negative oxygen-18 values than Zilan thermal waters. It suggests that thermal waters below Diyadin have likely mixed with more cold groundwaters ratio enriched than Zilan thermal waters. This is also confirmed by water chemistry and temperatures of these waters. However, they are affected by fluid–rock interaction and the differences in the ratios of Cl-B, Li, and Rb concentrations between Diyadin and Zilan imply that there are different subsurface temperatures and host rock types in both areas. Cold waters from Diyadin and Zilan areas are characterized by low chloride. In addition, tritium values suggest that water comes from shallow circulating and are recharged by recent, low-altitude precipitation (Figs. 9.15 and 9.16).

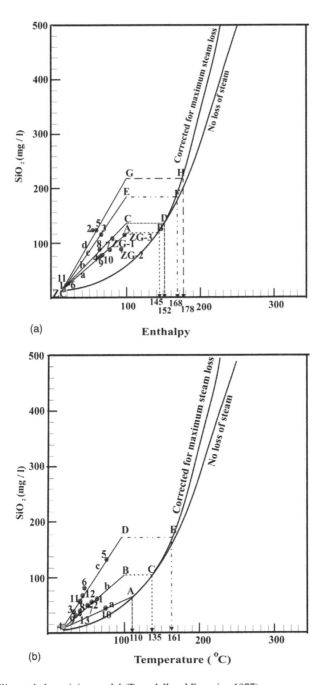

Figure 9.13. Silica enthalpy mixing model (Truesdall and Fournier, 1977).

Both Diyadin and Zilan thermal waters constitute different groups with different concentrations of tritium and chloride. It suggests that thermal waters that are discharging from two different fields are cooled and mixed with different quantities of shallow cold groundwater during their ascent to the surface. Therefore, thermal waters from two different areas represent different flow path systems. Moreover, spring waters that discharge from Zilan geothermal area have tritium

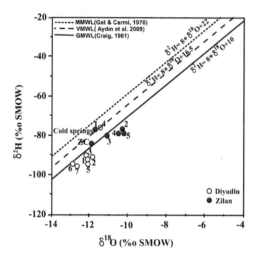

Figure 9.14.   Oxygen-18–deuterium diagram.

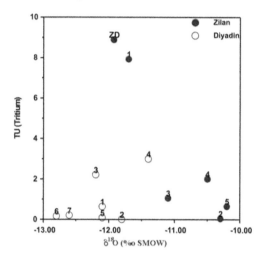

Figure 9.15.   Tritium–oxygen-18 diagram.

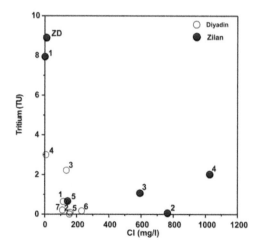

Figure 9.16.   Tritium–chloride diagram.

contents of 0.06–0.66 TU and are represented by high electrical conductivity and oxygen-18 values. Their isotopic characteristics indicate that the water is derived from relatively old (>50 years), deeply circulated waters inferred to have a long residence time in the aquifer. This is also confirmed by water chemistry as well. Higher concentrations of some minor elements in Zilan thermal waters, such as Si, B, Rb, Sr, and Cs, are probably derived from extended water-rock interaction, and these elements can also be regarded as indicators of flow paths and residence times.

9.7   CONCLUSION

The Diyadin (Ağri) and Erçiş (Zilan-Van) geothermal systems occur in a volcanic terrain, with interconnected fault/fracture networks providing conduits for water flow.

Physical, chemical, and isotopic investigation of thermal and cold groundwaters made it possible to reconstruct the thermal water evolution from infiltration to discharge, including water-rock interaction and shallow water mixing (Imbach, 1997).

Meteoric waters infiltrate on Mount Tendurek (Ölmez *et al.*, 1994) and recharge the thermal waters at Diyadin and Zilan. In addition, tectonic activities in association with volcanism transfer the magma heat to the surface, and therefore, the geothermal gradient in these fields is high. These waters, which are heated at depth, fill the fractures, cracks, and, rising along the permeable zones, form the thermal waters of Diyadin and Zilan. During their residence times spanning at least 50 years, the waters of the two different systems were mineralized by water-rock and gas interactions in the metamorphic rocks of Mount Tendurek. Calcite dissolution and hydrolysis of silicates are the dominant water-rock interactions. These thermal waters follow separate flow paths. The upflow of the two waters occurs in the western part of Mount Tendurek along faults that are related to the neotectonic stress regime of the region. Within each thermal water district, a travertine complex created by the thermal waters covers the upflow zone.

Both Diyadin and Zilan geothermal fields is a liquid-dominated system with different reservoir temperature. For both thermal provinces, thermal waters have been cooled and mixed with different quantities of shallow cold groundwater during their ascent to the surface. Therefore, they are different in hydrochemical type and compositions. Based on the main constituents, Diyadin thermal waters can be classified as $CaHCO_3$ and $NaHCO_3$, and Zilan thermal waters can be classified as NaCl and $CaHCO_3$ type, with high concentrations of Si, B, Li, Rb, Sr, and Cs. Conservative elements indicate that the types of analyzed waters have similar origin, and the difference in concentration is due to the dilution of thermal water with shallow groundwater. Long circulation of meteoric waters within the basement rocks is indicated by low tritium values in the thermal waters. Surface and cold waters are of Ca–Na–HCO_3 type and represent shallow circulating groundwaters with low TDS and EC.

REFERENCES

Ağrali, B.: Erciş Zilan bölgesinin jeolojisi ve linyit. General Directorate of Mineral Research and Exploration (MTA) report, No: 2766 (Yayinlanmamiş), Ankara, Turkey, 1966.

Akdoğan, N.: Van-Erçiş Zilan deresi sahas*i* jeotermal Enerji aramalari gravite etüd raporu. General Directorate of Mineral Research and Exploration (MTA), No: 8223, Ankara, Turkey, 1987.

Akkuş, I., Güner, A. Demir, A. & Islamoğlu, T.: Van-Erçiş-Zilan jeotermal enerji derin arama sondaj*i* (ZD-1) kuyu bitirme ve saha değerlendirme raporu. General Directorate of Mineral Research and Exploration (MTA), No: 3026, Ankara, Turkey, 1990.

Akkus, I., Akilli, H., Ceyhan, S., Dilemre, A. & Tekin, Z.: Turkiye jeotermal kaynaklari envanteri serisi-201. ISBN975-8964-36-4, Ankara, Turkey, 2005.

APHA, AWWA, WPCF: *Standad method for the determination of water and waste water.* 15th Edition: American Public Health Association, Washington, DC, 1989.

Aydin, H., Ekmekçi, M., Tezcan, L., Dişli, E., Aksoy, L., Yalçin, M.P. & Özcan, G.: Assesment of water resources potential of Gürpinar (Van) karst springs with regard to sustainable management. The Scientific and Technical Research Council of Turkey. (TUBITAK Project) No. 106Y040) Final report, 2009 (In Turkish).

Baba, A. & Ármannsson, H.: Environmental impact of the utilization of a geothermal area in Turkey. *Energy Sources*, Part B: *Economics, Planning, and Policy* 1:3 (2006), pp. 267–278.

Browne, P.R.L.: Minerological guides to interpreting the shallow paleohydrology of epitermal mineral depositing enviroments. *Proceedings 13th New Zealand Geothermal Workshop*, Auckland, New Zealand, 1991, pp. 263–270.

Bulbul, A.: *Hydrogeological and hydrogeochemical investigation of hot and cold water systems of Alasehir (Manisa)*. PhD Thesis, University of 9 Eylul, Izmir, Turkey, 2009 (in Turkish).

Burkay, I. & Şahin, H.: Van-Erciş-Zilan deresi sahasi jeotermal enerji aramalari jeoelektrik etüd raporu. General Directorate of Mineral Research and Exploration (MTA) Rap no 8137, 1986.

Burçak, M., Yildirim, T. & Yücel, M.: Ağri - Diyadin - Cermik sahasi jeotermal-jeofizik etüt raporu General Directorate of Mineral Research and Exploration (MTA), No. 10020, Ankara, Turkey, 1997.

Çağlar, K.Ö.: Türkiye maden sulari ve kaplicalari. General Directorate of Mineral Research and Exploration (MTA) Publised No:11 Vol 4, 1950, pp. 772–778.

Canik, B. & Baskan, E.M.: IAH map of mineral and thermal waters of Turkey Aegean region. General Directorate of Mineral Research and Exploration (MTA) report no. 189. Ankara, Turkey, 1983. Craig, H.: Isotopic variations in meteoric waters. *Science* 133 (1961), pp. 1702–1703.

Chandrasekharam, D. & Bundschuh, J.: Geochemistry of thermal waters and thermal gases. In: D. Chandrasekharam & J. Bundschuh (eds): *Geothermal energy resources for developing countries*. Balkema, Leiden, The Netherlands, 2002, pp. 253–267.

Delalande, M., Bergonzini, L., Gherardi, F., Guidi, M., Andre, L., Abdallah, I. & Williamson, D.: Fluid geochemistry of natural manifestations from the southern Poroto-Rungwe hydrothermal system (Tanzania): preliminary conceptual model. *J. Volcanol. Geoth. Res.* 199 (2011), pp. 127–141.

Demirel, V. & Özkan, H.: Ağri-diyadin MT-2, 3 ve 4 jeotermal sondajlari kuyu bitirme raporu. General Directorate of Mineral Research and Exploration (MTA) Report No 10451, Ankara, Turkey, 2000.

Ellis, A.: Quantitative interpretation of chemical characteristics of hydrothermal systems. *Geothermics* 2 (1970), pp. 516–528.

Ercan, T., Fujitani, T., Matsuda, J.I., Notsu, K. & Tokel, S.: Doğu ve Güneydoğu Anadolu Neojen – Kuvaterner volkanitlerine ilişkin yeni jeokimyasal, radyometrik ve izotopik verilerin yorumu. General Directorate of Mineral Research and Exploration (MTA) publ. No 110, Ankara, Turkey, 1990, pp. 143–164.

Ercan, T., Olmez, E., Matsuda, J.I., Nagao, K. & Kita, I.: Kuzey ve Bati Anadolu'da sicak ve mineralize sular ile içerdikleri gazlarin kimyasal ve izotopik ozellikleri. *Turkiye Enerji Bülteni* 1:2 (1994), pp. 10–21.

Ertürk, I. & Manav, E.: Ağri-Diyadin AD-1 ve AD-2 jeotermal sondajlarinin kuyu bitirme raporu. General Directorate of Mineral Research and Exploration (MTA), Ankara, Turkey, 1999.

Fournier, R.O.: Silica in thermal waters: laboratory and field investigations. *International Symposium on Hydrogeochemistry and Biogeochemistry*, Tokyo, 1970, vol.1, *Hydrochemistry*, Washington, DC, 1973, pp. 122–139.

Fournier, R.O.: A revised equation for the Na-K geothermometer. *Geotherm. Resour. Council Trans.* 3 (1979), pp. 221–224.

Fournier, R.O.: Water geothermometers applied to geothermal energy. In: F. D'Amore (ed): *Applications of geochemistry in geotherma reservoir development*. UNITAR/UNDP publication, Rome, Italy, 1991, pp. 37–69.

Fournier, R.O. & Truesdell, A.H.: An empirical Na-K-Ca geothermometer for natural waters. *Geochim. Cosmochim. Acta.* 37 (1973), pp. 1255–1275.

Gat, J.R. & Carmi, I.: Evolution in the isotopic composition of atmospheric waters in the Mediterranean Sea area. *J. Geophys. Res.* 75 (1970), pp. 3039–3048.

Gemici, U. & Tarcan, G.: Hydrogeochemistry of the Simav geothermal field, western Anatolia, Turkey. *J. Volcanol. Geoth. Res.* 116 (2002), pp. 215–233.

Gemici, U., Tarcan, G., Helvaci, C. & Çolak, M.: Distribution of As, B and other elements in Locacay River catchment (Emet-Turkey). In: T.D. Bullen & Yanxin Wang (eds): *Water-Rock Interaction 2007*, 2007, pp. 1047–1051.

Giggenbach, W.F.: Geothermal solute equilibrium. Derivation of Na-K-Mg-Ca geoindicators. *Geochim. Cosmochim. Acta* 52 (1988), pp. 2749–2765. NITAR/UNDP Publication, Rome, pp. 119–142.

Giggenbach, W.F.: Chemical techniques in geothermal exploration. In: F. D'AMORE (ed): *Application of geochemistry in geothermal reservoir development*. UNITAR, USA, 1991, pp. 119–144.

Giggenbach, W.F. & Glover, R.B.: Tectonic regime and major processes govering the chemistry of water and gas discharges from the Rotorua geothermal field, NZ, New Zealand. *Geothermics* 21:1–2 (1992), pp. 121–140.

Giggenbach, W.F. & Goguel, R.L.: Collection and analysis of geothermal and volcanic water and gas discharges. Report no. CD 2401. Chemistry division, DSIR, Petone, New Zealand, 1989.

González-Fernández, A., Martín-Atienza, B. & Paz-López, S.: Identificación de fallamiento en la Península de Punta Banda, B.C., a partir de datos de topografia, magnetometría y gravimetría. *GEOS* 20 (2000), pp. 98–106.

Gökgöz, A.: Geochemistry of the Kizildere-Tekkehamam-Buldan-Pamukkale geothermal fields, Turkey. In: L.S. Georgsson (ed): *Geothermal training in Iceland 1998*. United Nations University Geothermal Training Programme, Reykjavik, Iceland, 1989, pp. 115–156.

Güleç, N. Hilton, D.R. & Mutlu, H.: Helium isotope variations in Turkey: relationship to tectonics, volcanism and recent seismic activities. *Chem. Geol.* 187 (2002), pp. 129–142.

Gultekin, T. & Gemici, U.: Water geochemistry of the Seferihisar geothermal area, Izmir, Turkey. *J. Volcanol. Geoth. Res.* 126 (2003), pp. 225–242.

Han, D.M., Liang, X., Jin, M.G., Currell , M.J. Song , X.F. & Liu, C.M.: Evaluation of groundwater hydrochemical characteristics and mixing behavior in the Daying and Qicun geothermal systems, Xinzhou Basin. *J. Volcanol. Geoth. Res.* 189 (2010), pp. 92–104.

Has, F.: Van-Erciş-Zilan linyit havzasina ait işletme raporu. General Directorate of Mineral Research and Exploration (MTA) report No. 5151, Ankara, Turkey, 1973.

Henley, R.W. & Ellisa, A.: Geothermal systems ancient and modern: A geochemical review. *Earth-Sci. Rev.* 19:1 (1983), pp. 1–50.

Imbach, T..: Geology of Mount Uludag with emphasis on the genesis of the Bursa thermal waters, northwestern Anatolia, Turkey. In: C. Schindler & M. Pfister (eds): *Active tectonics of northwestern Anatolia — the MARMARA Poly-Project: a multidisciplinary approach by space- geodesy, geology, hydrogeology, geothermics and seismology*. VDF Hochschulverlag ETH, Zurich, Switzerland, 1997, pp. 239–266.

Karamanderesi, I.H., Coşkun, B., Çağlav, F., Güner, A. & Polat, Z.: Van Erçiş Zilan deresi jeoloji ve jeotermal enerji olanaklari. General Directorate of Mineral Research and Exploration (MTA) report No. 7793, Ankara, Turkey, 1984.

Keskin, B.: Agri-Diyadin jeotermal alani jeolojik etüd raporu ve jeotermal potansiyeli. Doğan Jeotermal, Ankara, Turkey, 1998.

Kharaka, Y.K. & Mariner, R.H.: Chemical geothermometers and their application to formation waters from sedimentary basins. In: ND Naeser, & T. McCulloh (eds): *Thermal history of sedimentary basins*. Springer Verlag, New York, 1988, pp. 99–117.

Koçyiğit, A., Yilmaz, A., Adamia, S. & Kuloshvili, S.: Neotectonics of east Anatolian Plateau (Turkey) and Lesser Caucasus: implication for transition from thrusting to strike-slip faulting. *Geodyn. Acta* 14 (2001), pp. 1–19.

Kurtman, F.: Geothermal energy investigation in Turkey. *Symposium on Geothermal Energy*, Ankara, Turkey, 1977, pp. 34–44.

Langelier, W. & Ludwig, H.: Graphical methods for indicating the mineral character of natural water. *J. Am. Water Assoc.* 34 (1942), pp. 335–352.

Mutlu, H.: Constraints on the origin of the Balikesir thermal waters (Turkey) Stable Isotope (^{18}O, D, ^{13}C, ^{34}S) and major-trace element compositions. *Turkish J. Earth Sci.* 16 (2007), pp. 13–32.

Nicholson K.: *Geothermal fluids chemistry and exploration techniques*. Springer-Verlag, Berlin, Heidelberg, Germany, 1993.

Ölmez, E. & Güner, A.: Van-Erciş-Zilan gradyan sondaji (ZG-1) kuyu bitirme raporu no 8724. General Directorate of Mineral Research and Exploration (MTA), Ankara, Turkey, 1989.

Ölmez, E., Ercan, T. & Yildirim, T.: Tendurek (Doğu Anadolu) jeotermal alaninin (Diyadin, Zilan, Çaldiran) volkanolojisi ve jeotermal enerji olanaklari. *TJK bülteni* 9 (1994), pp. 48–55.

Öngür, T., Kahramanderesi, H.I., Ünlü, M.R., Suludere, A. & Yoğurtçuoğlu, A.: Diyadin-Erciş jeotermal araştirma sahasinin, jeolojisine ilşkin ön rapor. General Directorate of Mineral Research and Exploration (MTA) report, Ankara (Yayinlanmamiş), Ankara, Turkey, 1974.

Oruç, N.: Occurrence and problems of high fluoride waters in Turkey: an overview. *Environ. Geochem. Health* 30 (2008), pp. 315–323.

Öztekin, M., Metin, N. & Çuhadar, G.: Van Erciş plain. Hydrogeological investigation report, DSI, Ankara, Turkey, 1977, pp. 7–10.

Pamir, H.N.: Tendurek daği. *Istanbul Üniversitesi Fen Fakültesi Mec.* B 16 (1951), pp. 83–88.

Pasvanoğlu, S.: Doğu Anadolu Ağri (Tutak-Diyadin) Van (Erçiş-Zilan), Bitlis (Nemrut) jeotermal enerji olanaklari. RARIK-Turkison Ltd., unpublished report, Ankara, Turkey, 2008.

Pasvanoglu, S. & Chandrasekharam, D.: hydrogeochemical and isotopic study of thermal and mineralized waters from The Nevşehir (Kozakli) area, central Turkey. *J. Volcanol. Geoth. Res.* 202 (2011), pp. 241–250.

Pasvanoğlu, S. & Güler, S.: Hydrogeological and geothermal features of hot and minerlized waters of the Ağri-Diyadin (Turkey). *WGC2010 World Geothermal Congress*, 24–30 April, Bali, Indonesia, 2010a.

Pasvanoğlu, S. & Güler, S.: Zilan (Van-Erciş) Sicak ve mineralli su kaynaklarinin hidrojeokimyasal incelemesi; *IV. Ulusal Jeokimya Semposyumu*, 26–28 May, Elaziğ, Turkey, 2010b, p. 129.

Pasvanoğlu, S.: Hydrogeochemical and isotopic investigation of the Bursa-Oylat thermal waters, Turkey. *Environ. Earth Sci.* 64:4 (2011), pp. 1157–1167.

Pasvanoğlu, S.: Hydrogeochemical study of the thermal and mineralized waters of Banaz (Hamambogazi) area, western Anatolia, Turkey. *Environ. Earth Sci.* 65:3 (2012), pp. 741–752.

Pisarskii, B.A., Konev, A.A., Levi, K.G. & Delvaux, D.: Carbon dioxide-bearing alkaline hydrotherms and strontium-bearing travertines in the Songwe River valley (Tanzania). *Russ. Geol. Geophys.* 39:7 (1998), pp. 941–948.

Reyes, A.G., Giggenbach, W.F., Saleras, J.R., Salonga, N.D. & Vergara, M.C.: Petrology and geochemistry of Alto Peak, a vapour-cored hydrothermal system, Leyte, Philippines. *Geothermics* 22 (1993), pp. 479–519.

Şamilgil, E.: Geothermal situation in the Aegean region and its continental shelf. *Symposium on Geothermal Energy*, Ankara, Turkey, 1977, pp. 47–60.

Şaroğlu, F., Güner, Y., Kidd, W.S.F. & Şengör, A.M.C.: Neotectonics of eastern Turkey: new evidence for crustal shortening and thickening in a collisional zone: *EOS Trans. Am. Geophys. Union* 61:17 (1980), p. 360.

Şengör, C. & Kidd, W.S.F.: Post-collisional tectonics of the Turkish-Iranian plateau and a comparison with Tibet. *Tectonophysics* 55 (1979), pp. 361–376.

Şengör, C. & Yilmaz, Y.: Tethyan evolution of Turkey; A plate tectonic approach. *Tectonophysics* 75 (1981), pp. 181–241.

Shakeri, A., Moore, F. & Zare, M.K.: Geochemistry of the thermal springs of mount tatfan, southeastern Iran. *J. Volcanol. Geoth. Res.* 178 (2008), pp. 829–836.

Şimşek, N.: Geothermal potential in northwestern Turkey. In: C. Schindler & M. Pfister (eds): *Active tectonics of northwestern Anatolia — the MARMARA Poly-Project: a multidisciplinary approach by space-geodesy, geology, hydrogeology, geothermics and seismology*. VDF Hochschulverlag, ETH, Zurich, 1997, pp. 111–136.

Tarhan, N.: Hinis-Varto-Karliova (Erzurum-Muş-Bingöl) dolayinin Neojen volkanitlerinin jeolojisi ve petrolojisi. *General Directorate of Mineral Research and Exploration (MTA) Magazine* 113, 1992, pp. 45–60.

Truesdell, A.H. & Fournier, R.O.: Procedure for estimating the temperature of a hot water component in a mixed water using a plot of dissolved silica *vs.* enthalpy. *U.S. Geol. Survey J. Res.* 5 (1979), pp. 49–52.

Vengosh, A., Helvaci, C. & Karamanderesi, I.H.: Geochemical constraints fort he origin of thermal waters from western Turkey. *Appl. Geochem.* 17 (2002), pp. 163–183.

WHO: Guidelines for Drinking Water Quality. 3rd ed., Radiological Quality of Drinking Water. World Health Organization, Geneva, Switzerland, 2003.

Yenel, O.: Turkiye maden sulari, Akdeniz, Karadeniz Doğu ve Güney–Doğu Anadolu Bölgeleri. Istanbul University, Istanbul, Turkey, 1976, pp. 202–205.

Yilmaz, Y., Şaroğlu, F. & Güner, Y.: Initiation of the neomagmatism in East Anatolia. *Tectonophysics* 134 (1987), pp. 177–199.

Yilmazer, S., Pasvanoğlu, S. & Vural, S.: Geological features of Geven (Kütahya) field and evaluation of its geothermal resources. *Istanbul Earth Sciences Review* 23 (2010), pp. 73–85 (in Turkish).

CHAPTER 10

Balçova geothermal field district heating system: lessons learned from 16 years of application

Mahmut Parlaktuna

10.1 GEOGRAPHICAL SETTING, GEOLOGY, AND GEOCHEMISTRY OF THE FIELD

Balçova geothermal field is situated 11 km southwest of the city of İzmir in western Anatolia (38.2° latitude, 27.0° longitude) (Fig. 10.1). It is located along the E–W trending Izmir Fault Zone. The oldest and common stratigraphic unit around the geothermal field is Upper Cretaceous

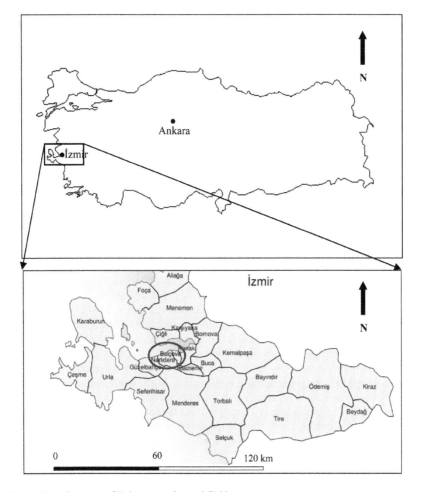

Figure 10.1.   Location map of Balçova geothermal field.

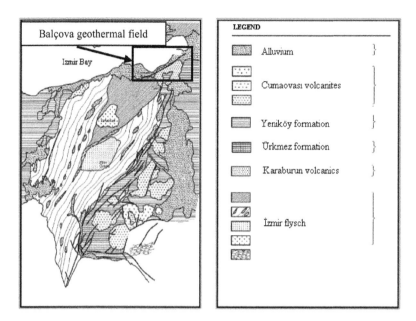

Figure 10.2.   Regional geological map of Balçova geothermal field (Öngür, 1972).

Izmir Flysch. In the eastern part of the Balçova, Miocene age Yeniköy formation exposing an angular unconformity. Pliocene volcanics are widely exposed in the northeast and southeast side of the Balçova. Other units are Quaternary talus breccia and alluvium (Fig. 10.2). The thickness of the alluvium is between 25 and 200 m. Fractured and jointed rocks of the Upper Cretaceous Izmir Flysch unit are the reservoir rocks in the Balçova geothermal system (Öngür, 2001). The cap rock of the system is the clay-rich zones of the interbedded sandstone-shale units because of their impermeable feature.

The Upper Cretaceous Izmir Flysch, is exposed from Seferihisar-Doğanbey in south to Izmir Bay in north–northeast direction (Fig. 10.2). It is composed of an intercalation of a variety of rocks such as metasandstones, clayey schists, phyllites, limestones, limestone olistoliths, granodiorites, and the altered products of submarine volcanism such as serpentinites and diabases (Fig. 10.3). In the field, sedimentation period is accompanied by tectonic activities such as continental rise and compression. Thus, in sedimentary basin, pieces and blocks of old geologic formations can be observed. The formation overlies the Menderes metamorphics (Erdoğan, 1990). The wells in the Balçova geothermal field are mainly completed in lightly metamorphosed sandstones, clays, and siltstones of the Izmir Flysch sequence (Öngür, 2001; Serpen, 2004). The thickness of the Izmir Flysch exceeds 2000 m in some regions. Fractured metasandstones and fault zones existing in limestone and granodiorites are permeable. Other zones existing in the Izmir Flysch formation can be thought as impermeable (Serpen, 2004).

Yeniköy formation, exposed in the east and southern east of the geothermal field, consist of sandstone, claystone, conglomerate and lignite units. At the lower boundary of the formation poorly sorted, blocky and coarse-grained conglomerates are observed and overlain by intercalation of limestone, claystone and sandstone. The upper boundary of formation is represented by clayey limestone, claystone, sandstone and volcanic sedimentary units of tuff and tuffites. The thickness of it varies between 800 and 1000 m. Lower boundary unconformably overlies Izmir Flysch and is mostly a faulted unit (Aksoy, 2001).

Cumaovası volcanites outcrop at the southern part of the field. Cumaovası volcanites are composed of rhyolite, latile-andesite, and volcanic breccia-tuffs. Pliocene units are unconformably overlying the Yeniköy formation.

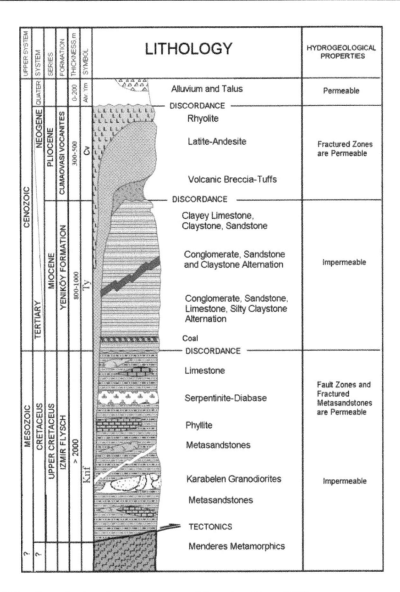

					LITHOLOGY	HYDROGEOLOGICAL PROPERTIES

Figure 10.3. Generalized stratigraphic sequence of the region (Aksoy, 2001).

Quaternary talus breccia and alluvium consists of younger clay and sand deposits all through the active stream beds. Alluvium existing over the field has good porosity and some permeability, while talus breccia serves as a cap rock.

The most important tectonic feature in the region is the E–W oriented Agamemnon-I Fault, which extends over 30 km. (Yılmazer, 1989; Öngür, 2001). Other than this main fault, series of E–W and S–N oriented faults and fractures can be observed in the region, with the NE–SW oriented Agamemnon-II Fault being the most pronounced (Fig. 10.4). The hot waters recharging in the Balçova region circulate through the major, about 2 km long, fracture zone associated with the Agamemnon-I Fault.

Faults and fractures within the Izmir Flysch formation provide a hydrothermal system to Balçova region. The meteoric precipitations in recharge area infiltrate through faults, fracture into deeper parts of the region (deeper than 2000 m), and are heated by an undefined heat source,

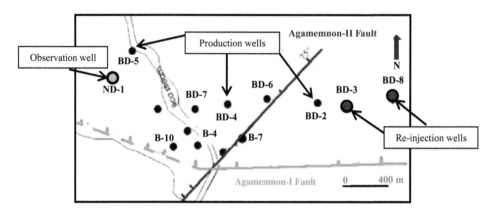

Figure 10.4.   Location map of Agamemnon-I and Agamemnon-II faults (after Yılmazer, 1989).

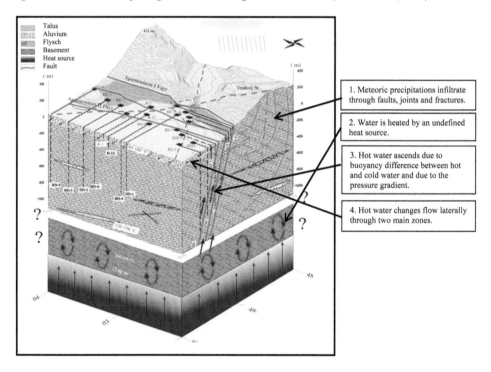

Figure 10.5.   Hydrogeological model of Balçova geothermal field (modified after Aksoy, 2001).

and rise along the Agamemnon Fault (Fig. 10.5). From this zone, the thermal waters flow mainly into two permeable horizons, one in the alluvium located in the upper 100 m of the system, and the other more permeable layers of the Izmir Flysch formation between 300 and 1100 m depth (Satman *et al.*, 2001).

Mineral concentrations of Balçova geothermal fluids are relatively low. The amount of total dissolved solids does not exceed 2000 mg/L and the amount of non-condensable gas content is lower than 1000 ppm. All waters taken from the field have nearly neutral pH. The main cation is Na^+ and the anion is HCO_3^- (Tables 10.1a and b) (Aksoy *et al.*, 2009). High Na^+ values can be due to the presence of clay minerals, and high HCO_3^- values can be due to carbonate rocks. The geothermal waters can be classified as sodium bicarbonate thermal water (Alacalı, 2006).

Table 10.1a.   Chemical analysis results for geothermal water and surface water samples (major ions) (Aksoy et al., 2009).

Sample #	X	Y	Well depth (m)	T (°C)	pH	EC (µS/cm)	Ca^{2+} (mg/L)	K$^+$ (mg/L)	Mg^{2+} (mg/L)	Na^{2+} (mg/L)	HCO$_3^-$ (mg/L)	SO$_4^{2-}$ (mg/L)	Cl$^-$ (mg/L)
Geothermal fluid samples													
B1	503.115	4248.997	104	115	7.2	1925	23.3	26.2	4.9	345.1	550	219	190
B4	503.462	4248.980	125	117	7.4	1950	21.2	24.8	5.1	347.3	661	186	172
B5	503.015	4248.996	110	124	7.4	1813	31.4	21.6	6.2	323.1	610	165	178
B7	503.172	4249.033	100	115	7.0	1948	22.7	27.6	7.6	402.0	611	198	216
B10	502.765	4249.032	125	114	7.5	1948	26.9	26.5	8.0	400.5	650	192	205
BD1	502.864	4249.196	564	135	8.2	1632	19.1	29.7	2.7	417.1	698	168	194
BD2	503.482	4249.219	677	137	8.4	2010	15.2	30.1	2.5	508.5	664	186	260
BD3	503.706	4249.187	750	128	8.2	1926	22.2	27.9	3.0	435.5	680	205	241
BD4	503.065	4249.240	630	132	8.3	2020	20.1	32.5	2.4	460.3	682	231	250
BD5	502.599	4249.492	1100	117	8.2	1853	17.6	31.8	2.5	399.1	690	189	227
BD7	502.804	4249.221	600	115	7.0	2190	26.8	32.0	9.1	382.9	575	213	220
BD9	504.219	4249.217	776	138	8.9	2020	14.5	35.1	4.9	476.4	390	228	276
BD10	502.724	4248.998	750	104	7.6	1473	22.3	27.7	9.0	345.4	690	159	138
Surface water samples													
SW-1	502.832	4248.526	–	19	7.9	684	64.1	1.6	33.6	10.3	225	117	17
SW-2	502.962	4249.038	–	40.1	7.8	2236	22.1	35.5	5.4	392.8	595	296	180
SW-3	502.463	4251.118	–	23.1	9.8	1384	34.3	35.4	13.6	284.3	413	164	208
SW-5*	506.200	4250.880	–	23	8.2	55600	570.4	415.0	1565.1	14420.5	180	2450	30406

*Seawater

Table 10.1b. Chemical analysis results for geothermal water and surface water samples (minor ions) (Aksoy et al., 2009).

Sample #	Al	As	B	Ba	Be	Br	Cd	Cr	Cu	Fe	Li	Mn	Ni	Pb	Sb	Zn
(μg/L)																
Geothermal fluid samples																
B1	51	197.7	9950	129.13	0.75	273	0.32	2.7	6.9	570	1126.0	64.38	3.3	2.3	124.8	20.7
B4	21	173.2	9224	100.19	0.42	242	0.27	0.9	15.7	138	1080.0	43.62	14.7	4.7	38.9	39.6
B5	29	242.7	9804	528.27	0.74	262	0.24	1.3	6.5	419	999.0	44.68	<.2	1.2	688.5	9.1
B7	20	384.2	15904	125.92	0.66	270	<.05	28.2	13.0	146	1465.0	44.58	<.2	0.6	234.0	5.3
B10	40	363.7	15079	128.91	0.70	262	<.05	30.1	23.8	126	1442.4	45.70	3.7	0.6	52.6	14.7
BD1	30	298.4	13756	124.48	0.89	266	<.05	25.5	10.3	99	1417.2	31.10	<.2	0.3	26.0	7.1
BD2	42	1419.8	21333	102.16	0.70	352	<.05	30.6	26.6	24	1873.9	15.01	<.2	0.5	104.4	7.6
BD3	53	674.6	20483	116.54	1.09	318	<.05	30.3	13.0	31	1652.8	21.69	<.2	0.4	109.0	8.3
BD4	68	776.8	20822	125.59	0.90	326	<.05	31.0	11.9	24	1706.4	19.33	0.4	0.5	114.8	7.6
BD5	40	163.5	12935	109.50	1.00	276	0.19	1.5	4.3	262	1101.4	21.53	0.8	0.8	193.9	6.8
BD7	30	357.2	18270	135.35	0.55	291	<.05	31.2	12.9	95	1555.9	23.16	<.2	0.4	152.3	7.2
BD9	5050	278.1	20453	125.48	1.38	364	<.05	24.2	19.2	4886	1615.2	83.28	14.0	9.5	58.0	14.0
BD10	127	261.6	7806	107.25	0.68	211	<.05	1.5	58.8	747	851.9	47.69	40.8	2.5	169.3	37.0
Surface water samples																
SW-1	21	1.5	<20	20.7	<.05	22	<.05	0.9	6.3	10	5.3	15.57	5.3	0.2	0.8	4.9
SW-2	26	182.4	9499	121.5	0.82	271	<.05	0.6	1.7	323	1014.2	52.38	<.2	0.3	23.7	5.8
SW-3	126	63.7	3354	35.3	<.05	278	<.05	1.7	5.1	563	281.6	99.50	20.1	2.8	14.6	10.1
SW-5*	1034	231.0	5046	87.7	<.05	83061	2.26	15.2	184.2	1876	0.2	2.39	<.2	3.5	0.1	0.8

*Seawater

The high Mg and HCO₃ concentrations are the main evidence of mixing thermal water with meteoric water close to surface.

Among several chemical geothermometers applied for Balçova field, Na–K geothermometer resulted with an estimate of $189\pm11°C$ and Na–K–Ca geothermometer gave value $179 \pm 18°C$ for the temperature of the reservoir rock in Balçova region (Aksoy, 2001).

10.2 DEVELOPMENT OF THE FIELD

The General Directorate of Mineral Research and Exploration of Turkey (MTA) started the first geothermal drilling studies in the region in 1963. Resistivity, thermal probing, and self-potential surveys were conducted. Three wells were drilled including the first geothermal exploratory well in Turkey. The first well (S-1) drew a mixture of hot water and steam at 124°C at a depth of 40 m. S-2 and S-3/A were drilled to 100 and 140 m, with downhole temperatures of 102°C and 101°C, respectively. S-3/A did not flow. From 1981 to 1983, 16 wells, including 7 thermal gradient and 9 production wells (100–150 m), were drilled. They encountered temperatures of 50°C to 126°C with flow rates of 4–20 kg/s. In 1982, system of geothermally heated hotels, curing center, swimming pools, and piped hot water began operation. Nine wells produced 4,500,000 kcal/h (~18,828 MJ/h) for surrounding hotels, buildings, and greenhouses. A district heating system with a total capacity of 2.2 MW$_t$ began operation in 1983 for heating offices, hospital, and dormitories of Dokuz Eylül University (~30,000 m²). Heating for Turkey's largest indoor swimming pool, which has a capacity of 1,600,000 kcal/h (~6694 MJ/h), began operation in February 1987. In 1989, two new wells (B-10 and B-11) were drilled to 125 m that encountered fluids with temperatures of 109°C and 114°C and flow rates of 5 and 3 kg/s, respectively. Geothermal heating of an 11,000 m² curing center became operational with a capacity of 1,200,000 kcal/h (~5020 MJ/h) on September 1989. Heating system for an additional 110,000 m² (1100 residence equivalent (RE), 1 RE = 100 m² heated area) plus hot water for the Hospital of Faculty of Medicine at Dokuz Eylül University was installed on February 1992. An additional system with a capacity of 6,900,000 kcal/h (~28,869 MJ/h) (8.0 MW$_t$) began running on November 1992. The most important stage was realized by starting the operation of the Balçova District Heating System (BDHS) in 1996.

There are more than 40 wells drilled in the Balçova geothermal field. Some of those wells are gradient wells aiming to get information on the geology as well as geothermal gradient of the region and indicated by G. The wells indicated by B or BG are shallow wells, while the wells with BD are deep wells (Fig. 10.6). The depths of shallow wells are in the range of 50–150 m, while the

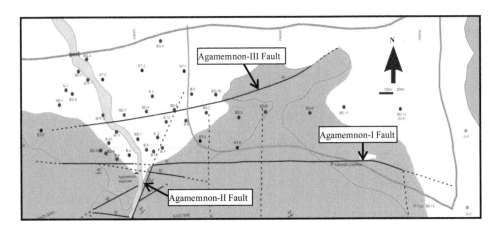

Figure 10.6. Well locations of Balçova geothermal field.

Table 10.2.   List of active wells.

No.	Production well	Maximum rate (m^3/h)	Injection well	Average reinjection rate (m^3/h)
1	B-5	117	BD-3	180
2	B-7	41	BD-8	800
3	B-10	220	BD-10	93
4	BD-2	130	BD-15	134
5	BD-4	209	BD-1	11
6	BD-5	55	B-4	70
7	BD-6	199	B-1	160
8	BD-7	69	B-7	12
9	BD-9	360	BH-1	10
10	BD-11	225	BTF-2	7
11	BD-12	256		
12	BD-14	125		
	Total rate (m^3/h)	2006		1478

deep wells have an average depth of 700 m. Twelve of these wellbores are utilized as production wells and four of them are used for reinjection purposes (Table 10.2).The wells shaded in Table 10.2 are seldom used as reinjectors during peak times. As indicated in Table 10.2, the maximum capacity of the production wells is over 2000 m³/h, which was not tested yet. The producing temperatures of the wells are in the range of 97–140°C.

10.3   UTILIZATION OF THE FIELD

Balçova geothermal field is located in a densely populated area (Fig. 10.7) which makes direct heat applications very efficient and economical. The heat produced from Balçova geothermal field is utilized for three main purposes: greenhouse heating, balneology, and residential heating. Among these three applications, the latter one is the main application throughout the BDHS (Fig. 10.8). There are three main flow loops within BDHS:

1. *Geothermal water loop*: This is the loop in which the produced geothermal fluid at an average temperature of 120°C is sent to heating centers to transfer its heat energy to the closed loop of city water with the help of heat exchangers. Geothermal fluid after heating centers is reinjected into ground at an average temperature of 60°C.
2. *City water loop*: In this loop, the city water is circulated in a distribution network between the heating centers and residences. The city water is heated to a temperature of 90°C at the heating centers and headed to residences in which each residence has its own heat exchanger to heat its radiator water.
3. *Residences loop*: This is the loop within a single residence through which the heat energy of city water is transferred to radiator system.

There are eight heating centers within the system to cover the residential area of Balçova–Narlıdere districts. Each heating center serves the residences to their close vicinity. Heating centers do not serve with their full installed capacities since some of the residences do not subscribe for the BDHS. On average, 80% of the installed capacity is used by subscribers (Fig. 11.7). As seen in Figure 10.9, both installed capacity and subscription increases with time and a project is under way for an estimated subscription of 3500 RE (residence equivalent, 1 RE $= 100$ m² heated area). In addition to heating centers, BDHS has two pumping stations and more than 350 km pipeline network. Individual houses, governmental institutes, and private firms such as Dokuz Eylül

Figure 10.7. Residential view of Balçova district with well locations.

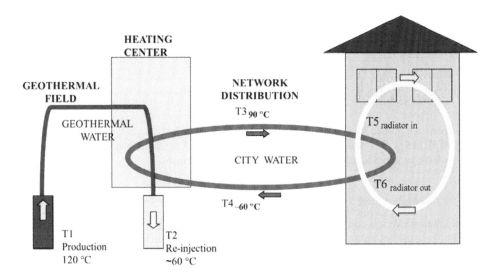

Figure 10.8. Schematic plan of BDHS.

University Hospital, University Dormitories, Izmir Economy University Campus, one shopping center, and a hotel are the subscribers of BDHS. In addition to BDHS, Balçova geothermal field supplies geothermal water to two health centers hotel for balneological purposes, namely Balçova Thermal hotel and Kaya hotel.

BDHS was managed by a private geothermal company (1996–2000) until the establishment of Balçova Geothermal Inc. In 2005, the status of the company as well as its name was changed, becaming Izmir Geothermal Inc. and the owner of the company became Izmir Governorship and

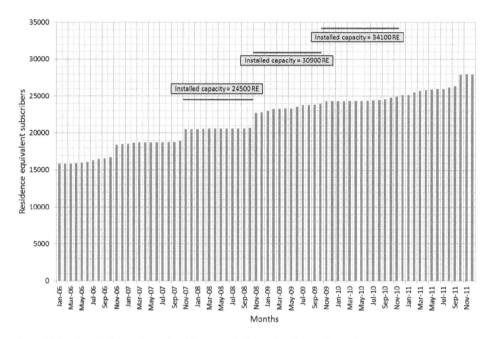

Figure 10.9.   Installed capacity and residence equivalent subscribers of BDHS.

Municipality of Izmir with 50% share. Since then the company is run by professional employees but reports to the Directors of Board formed by the representatives from the shareholders.

10.4   LESSONS LEARNED

The experience gained by Izmir Geothermal Inc. during the last 16 years of operation (1996–2012) of BDHS will be shared in the following paragraphs. Each incidence will be discussed as observations, problem identification, and the corrective measures.

10.4.1   *Pipeline network*

The pipeline network is one of the crucial parts of a district heating system. It is used to transport and distribute the hot water. The pipeline network of BDHS was initially constructed from buried steel pipes. Buried pipes started to give problems, such as leakage and broken pipes, after few years of operation (Fig. 10.10). The main cause of malfunctioning was determined to be the outside corrosion of pipes, since there was no preventive action although the pipeline was subjected to the wet environment of soil. Izmir Geothermal decided to change the pipes from steel to PPR or PEX-A plastic pipes smaller than 160 mm in diameter, cathodic protection is applied for larger size steel pipes, and main pipes started to be installed above ground (not buried) wherever possible (Fig. 10.11).

The obvious consequence of broken pipes was the heavy leakage of water from the pipeline network. In order to overcome the loss of water from the system, geothermal water was injected into city loop for long periods of time which reached a value of 140 m^3 during the heating season of 2004 (Fig. 10.12). A leakage detection campaign was started in 2005, resulting in a gradual decrease in the amount of water to be added to the loop. The amount of leakage from the system was reduced to 10–15 m^3/h although it is aimed to reduce the leakage to zero, which is difficult

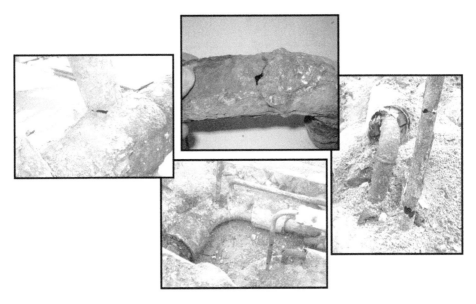

Figure 10.10.   Corroded, broken pieces of pipes in BDHS.

Figure 10.11.   Repairment of pipe network.

to achieve in a 350 km long network of pipes having t-junctions and branches almost every 50 m. As the result of this corrective action, addition of geothermal fluid into the city loop was stopped.

10.4.2   *Decline in reservoir pressure*

Balçova geothermal reservoir is a water-dominated hydrothermal reservoir, which is prone to relatively higher pressure drop response to fluid production. The main precaution for the decline

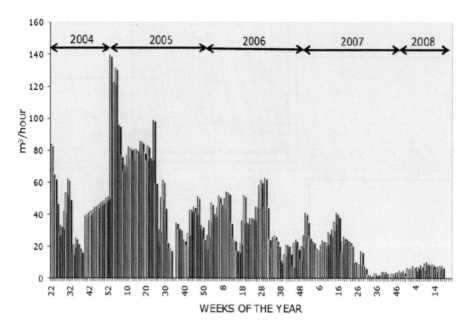

Figure 10.12.   Water added to the city loop because of leakage.

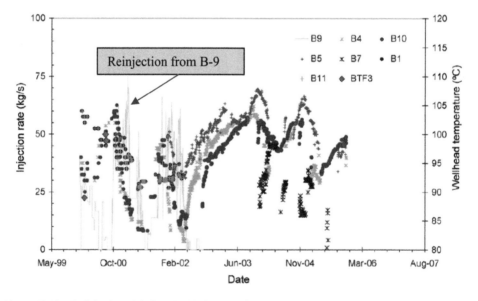

Figure 10.13.   Reinjection trials from B-9 (Aksoy *et al.*, 2008).

in reservoir pressure of geothermal reservoirs is the reinjection of produced fluid. At the early years of operation, reinjection into a shallow well (B-9) was tried but resulted with a rapid decline in fluid temperatures (10–15°C) of nearby shallow wells (Fig. 10.13) (Aksoy *et al.*, 2008).

In order to find a permanent solution to the reinjection practice of the field, MTA drilled a deep well (BD-8) in the eastern part of the field in 2002, which turned out to be a very powerful wellbore in terms of reinjection ($>700 \, m^3/h$). Although the reinjection capacity of the field increased after

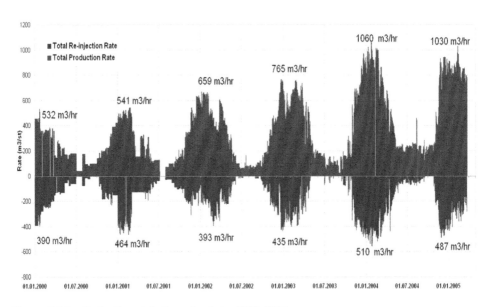

Figure 10.14. Production reinjection rates during 2000–2005.

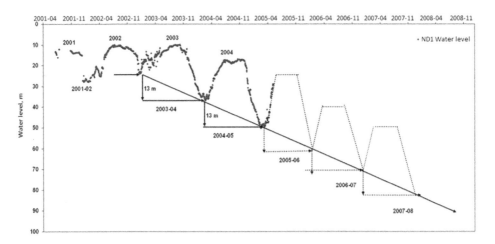

Figure 10.15. Forecast of reservoir pressure decline.

the drilling of BD-8, adding the geothermal fluid into the city loop decreased the amount of hot water to be reinjected and the ratio of reinjection to production continued to decrease over time during the period of 2000–2005 (Fig. 10.14) (Aksoy, 2005). This observation was interpreted as continuous pressure decline in reservoir (Fig. 10.15). A decline in water level of observation well ND-1 was interpreted as a total water level decline of more than 80 m for the 2007–2008 heating season. Fortunately remediation of the pipeline network after 2005 decreased the production rate for the same heating capacity, since there was no need to add geothermal water into city loop which is actually a permanent loss. This resulted in an increase in reinjection/production ratio (Fig. 10.16) and an obvious recovery of reservoir pressure (Fig. 10.17). The decreasing trend of water level of ND-1 during 2004–2005 reversed its direction, leading to a recovery after 2006 because of an increase in reinjection/production ratio.

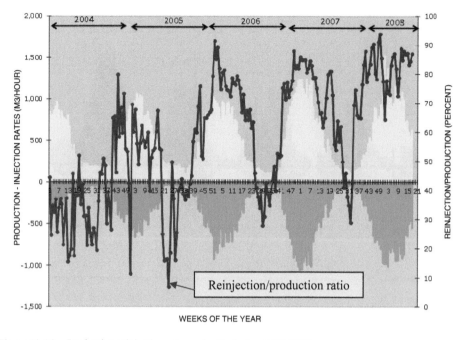

Figure 10.16.   Production reinjection rates and ratio during 2004–2008.

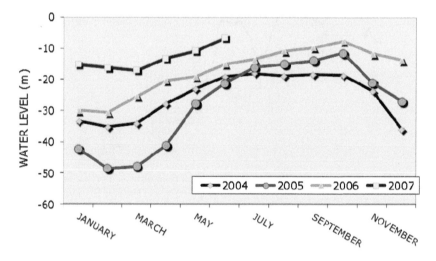

Figure 10.17.   Water level measurements at ND-1 during 2004–2007.

10.4.3   *Pricing policy*

Every subscriber of the BDHS needs to pay for the service that they receive. The very first payment to the BDHS is known as "Area Based Payment." The subscriber pays a fixed price per month depending on the area to be heated. This practice do not take into account any energy efficiency or energy saving. Since the payment is not dependent on the amount of energy used, in general, the users heat their houses to relatively higher temperatures, which results in an inefficient use of energy. This observation forced the Izmir Geothermal to take an action and by

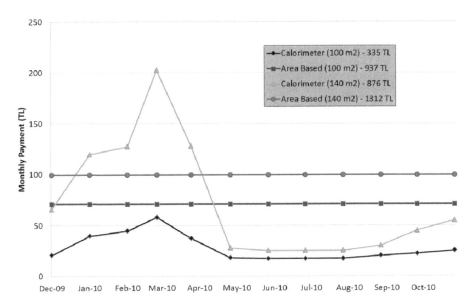

Figure 10.18. Annual payments by subscribers in 2010.

2006, all new subscriptions are made as "Energy Based Payment" by using calorimeters. Those subscribers had to pay the price of the energy they were supplied with, and who started to pay attention on energy efficiency measures such as insulation applications, thermostatic radiator valves, turning off radiators in unused space, or lowering the comfort temperature in the living area. All these measures resulted in a decrease in energy use for the same area. Average energy consumption was 14,150,000 kcal/RE/year (~59,204 MJ/RE/year) and 7,560,000 kcal/RE/year (~31,631 MJ/RE/year) for area-based and energy-based subscribers, respectively, in 2009. As indicated, area-based subscribers consume about twice the energy compared to the energy-based subscribers. Another consequence of the change in pricing is the decrease in annual payment by users (Fig. 10.18). Energy efficiency law of Turkey became compulsory since May 2012; therefore, all area-based users of BDHS need to convert to energy-based subscription.

10.5 CURRENT STATUS OF THE FIELD

Sixteen years of operation in a district heating system have equipped the management of Izmir Geothermal Inc. with enough experience to take the necessary precautions to prevent the long-term possible problems during operation. Some of these measures will be discussed in the following paragraphs.

As mentioned earlier, reinjection of the geothermal fluid is crucial for the continuation of a successful operation. The key parameter for the success of reinjection operation is the site selection for the reinjection well. A previous trial to reinject the fluid into the shallow reservoir (B-9) was unsuccessful resulting with immediate cooling of the nearby shallow wells (Fig. 10.13), since then this practice is not applied except for a few days a year during peak heating times (Table 10.2). This protective measure is obvious from the performance of shallow well B-10 (Fig. 10.19). This wellbore is a 125 m deep well with an initial producing temperature of 105°C. It is the flagship well of the field producing throughout the year, from delivering heat in the winter to supplying water for spas and hotels during the summer. Its temperature is still about 100°C with no sign of decrease in its productivity. It should be mentioned that during the reinjection trial into B-9, the temperature of B-10 dropped below 85°C.

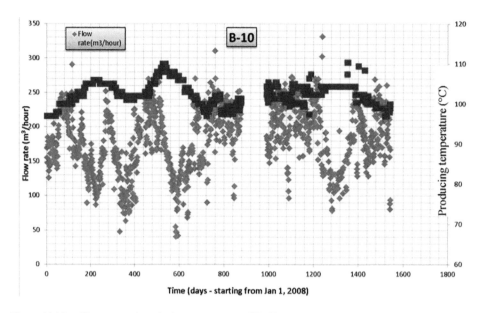

Figure 10.19. Flow rate and producing temperature of B-10.

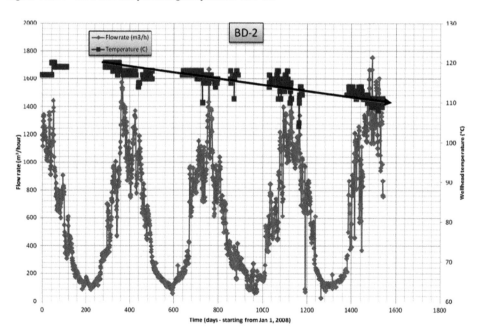

Figure 10.20. Total flow rate of the field and producing temperature of BD-2.

Two wellbores, BD-2 and BD-7, in the field were negatively affected by long-term operation. BD-2 is relatively old wellbore drilled in 1995. The producing temperature of this wellbore decreased from an initial 132°C to 112°C during the heating period of 2012 (20°C drop). BD-7 has more dramatic changes in temperature, from an initial 115°C [some kind error preventing comments, I just want to ask whether this number is correct, given that it is corrected HNdJ] to the current temperature of 82°C (33°C drop) (Figs. 10.20 and 10.21). The decrease in temperature

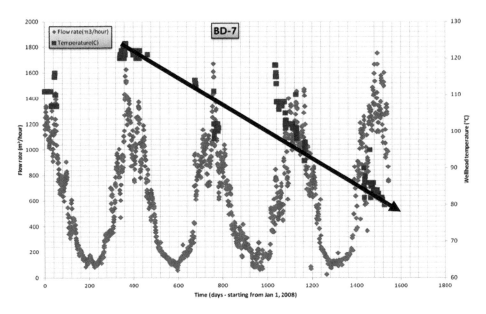

Figure 10.21. Total flow rate of the field and producing temperature of BD-7.

Figure 10.22. Velocity profiles of tracer (NaCl) injected from BD-10 (red = fast, yellow = medium, blue = slow).

of these two wellbores can be interpreted by using the results from the tracer test carried out in the field in 2009–2010 heating season. Two different tracers (NaCl from BD-10 and rhodamine from BD-8) were injected simultaneously into the field, and the arrival times of those tracers at the producing wells were observed (Figs. 10.22 and 10.23). Both wellbores (indicated by green circles in figures) are either in fast or medium velocity profiles obtained from tracer test data. The temperature decrease in those two wellbores could be affected by the injected fluids.

Figure 10.23.   Velocity profiles of tracer (rhodamine) injected from BD-8 (red = fast, yellow = medium, blue = slow).

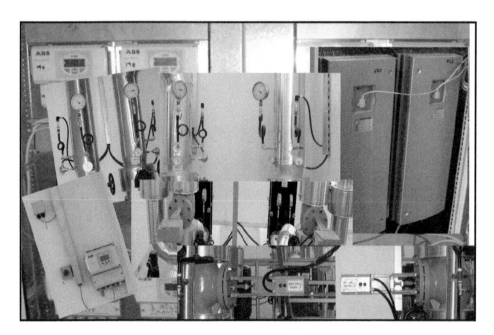

Figure 10.24.   Automation components of the field.

One very important achievement in the field is the automation of the facilities of the system. All heating centers, pumping stations, main pipelines, and some wellbores were already connected to a central data collection/recording system which enables to remotely control some of the components, such as pumps and valves (Figs. 10.24 and 10.25).

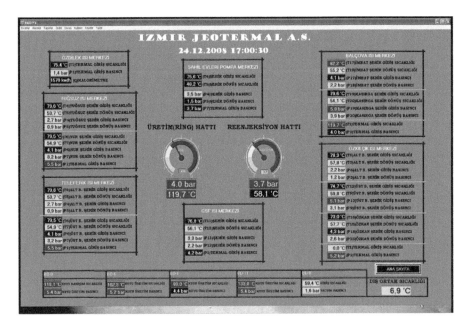

Figure 10.25.   Snapshot of the screen in the control room of automation system.

ACKNOWLEDGMENTS

The author thanks to Izmir Geothermal Inc. to permit the use of data presented in this chapter. Contributions of Ayşe Merve Turanlı during literature review (Research Assistant, Middle East Technical University) are highly appreciated.

REFERENCES

Aksoy, N.: *Monitoring Balcova-Narlidere geothermal system with tracers*. PhD Thesis, Dokuz Eylul University Graduate School, Izmir, Turkey, 2001 (in Turkish).

Aksoy, N. & Filiz, S.: Investigation of Balcova-Narlidere Geothermal Field by Isotopes. *Proceedings, 1st Environment and Geology Symposium (ÇEVJEO'2001)*, Izmir, 2001.

Aksoy, N., Serpen, U. & Filiz, S.: Management of the Balcova – Narlidere Geothermal Reservoir, Turkey. *Geothermics* 37 (2008), pp. 444–466.

Aksoy, N., Şimşek, N. & Gunduz, O.: Groundwater contamination mechanism in a geothermal field: a case study of Balcova, Turkey. *J. Contam. Hydrol.* 103 (2009), pp. 13–28.

Alacalı, M.: İzmir Balçova geothermal field, proceedings, *Thirty-First Workshop on Geothermal Reservoir Engineering*, 30 January–1 February 2006, Stanford University, Stanford, CA, 2006, SGP-TR-179.

Erdoğan, B.: Tectonic relations between İzmir-Ankara zone and Karaburun Belt. *Bulletin of the Mineral Research and Exploration of Turkey* 110 (1990), pp. 1–15.

Öngür, T.: Geological report on Izmir – Urla geothermal exploration area. General Directorate of Mineral Research and Exploration (MTA) report No. 4835, Ankara, Turkey, 1972.

Öngür, T.: Geology of Izmir Agamemnon hot springs – Balcova geothermal area and new conceptual geological model. Report for Balcova Geothermal Ltd., Izmir, Turkey, 2001 (in Turkish).

Satman, A., Serpen, U. & Onur, M.: Reservoir and production performance of Izmir Balcova-Narlıdere geothermal field. Project report, Balcova Jeotermal Ltd., Izmir, Turkey, 2001 (in Turkish).

Serpen, U.: Hydrogeological investigations on Balçova geothermal system in Turkey. *Geothermics* 33:3 (2004), pp. 309–335.

Yilmazer, S.: *Geochemical features of Balcova hot springs and geothermal energy possibilities for the area*. PhD Thesis, Akdeniz University Graduate School, Isparta, Turkey, 1989.

CHAPTER 11

Rapid development of geothermal power generation in Turkey

Murat Karadaş & Gülden Gökçen Akkurt

11.1 INTRODUCTION

Population and industrialization in Turkey have increased as the country develops, which has also resulted in a tremendous increase in energy demand. According to the information provided by the Ministry of Energy and Natural Resources, electricity consumption of Turkey, which was 229.3 billion kWh in 2011, is expected to reach 499 TWh by 2020, with an annual increase of around 8% according to the higher demand scenario, or 406 TWh with an annual increase of 6.1% according to the lower demand scenario (MENR, 2012). Currently, over 80% of the energy supply comes from fossil fuels in Turkey. According to Aneke *et al.* (2011), since fossil fuels are exhaustible, there is a need for their conservation. Because of the over-dependence on fossil fuels for energy supply, fossil fuels have resulted in the release of large quantity of anthropogenic CO_2 (greenhouse gas) into the environment. They cause environmental pollution and global warming. This associated environmental danger is caused by the burning of fossil fuels and has resulted in a clamor by the world leaders to develop better and more efficient means of meeting the world energy demand at the least possible environmental impact. Recently, there has been a gradual shift from over-dependence on fossil fuels to the use of renewable and cleaner energy sources such as wind and geothermal energy.

In Turkey, geothermal energy is a renewable heat energy, with fluid production temperatures varying from 50 to 287°C. It occurs mainly in the form of water and water-steam mixture. Geothermal heat energy has been identified as a good source of power generation. Geothermal power plants differ from thermal power plants in that they have boilers underground and they are less efficient because of low resource temperatures.

Geothermal exploration studies in Turkey started at the beginning of 1960s in Balçova, İzmir (1963), and in Kizildere, Denizli (1968). Kizildere geothermal field was the first geothermal field discovered at high enthalpy. Some medium enthalpy fields such as Seferihisar, Salavatlı, and Simav were discovered in 1970s and 1980s. The Kizildere geothermal power plant, which was the first geothermal power plant of Turkey, was installed in 1984. Kizildere geothermal power plant with a 17.4 MW$_e$ electricity generation capacity has had to deal with some problems such as precipitation of calcium carbonate in the wellbores, pipelines, and the condenser.

Investment in new geothermal power plants was delayed for many years due to the lack of experience in geothermal power plant operation and regulatory framework. After many years of the Kizildere experience, Dora-1 binary geothermal power plant was installed in 2006 in Salavatlı, Aydın. Beliefs in the success of the geothermal power plant investments increased after installation of Bereket (Sarayköy, Denizli) and Gurmat (Ömerbeyli, Aydın) geothermal power plants in 2007 and 2009, respectively.

In recent years, investments on geothermal power plants have increased with the enactment of regulations and incentives for electricity generation from renewable energy sources. The government provides tax and customs duty exemption for investors who want to generate electricity from renewable energy by the Encouragement of Investments and Employment Law (TGNA_1). Also, it gives 10.5 US$cent/kWh purchase guarantee from geothermal energy for a

Table 11.1.   Geothermal power plants of Turkey as of June 2013 (in operation).

Location	Power plant	Types of geothermal powerplants	Startup date	Maximum resource temperature (°C)	Average resource temperature (°C)	Gross power capacity (MW$_e$)
Denizli						
Kizildere	Zorlu-Kizildere	Single flash	1984	242	217	17.4
Sarayköy	Bereket	Binary cycle	2007	–	145	7.5
Aydın/Sultanhisar						
Salavatlı	Dora-1	Binary cycle	2006	172	168	7.35
Salavatlı	Dora-2	Binary cycle	2010	176	175	11.2
Aydın/Germencik						
Ömerbeyli	Gurmat	Double flash	2009	232	220	47.4
Hıdırbeyli	Irem	Binary cycle	2012	190	170	20
Çanakkale						
Tuzla	Tuzla	Binary cycle	2010	174	160	7.5
Total						**118.35**

period of 10 years by another law enacted (TGNA_2). If turbine, generator, power electronics, vacuum pumps, or compressors are manufactured in the country, an increase ranging from 0.7 to 1.7 US$cent/kWh is added to the current purchase price of electricity.

Currently, the total installed capacity of geothermal power plants in Turkey is 166.35 MW$_e$, which is expected to increase rapidly with new investments. In this chapter, development of geothermal power generation of Turkey is evaluated briefly and geothermal power plants in the Aegean region are introduced in detailed.

11.2   PRESENT STATUS OF GEOTHERMAL POWER PLANTS IN TURKEY (2013)

High temperature geothermal fields suitable for conventional electricity generation in Turkey are Denizli-Kizildere (242°C), Aydın-Ömerbeyli (232°C), Aydın-Salavatlı (176°C), Canakkale-Tuzla (174°C), Kutahya-Simav (184°C), Izmir-Seferihisar (153°C), and Manisa-Kavaklıdere (215°C). The other high temperature fields with electricity generation potential are Manisa-Caferbeyli (168°C), Aydın-Yilmazkoy (142°C), Aydın-Umurlu (130°C), Izmir-Dikili (120°C), and Izmir-Balcova (125°C). The operating geothermal power plants of Turkey are listed in Table 11.1.

Table 11.1 indicates that 77% of geothermal power plants are binary power plants. The operating temperatures of the all plants range between 145 and 242°C. By June 2013, the total installed capacity of geothermal power plants reached to 166.35 MW$_e$.

Increase in installed geothermal power capacity of Turkey is exhibited in Figure 11.1 by year. Kizildere geothermal power plant, which was the first geothermal power plant of Turkey, installed in 1984. Then, the second geothermal power plant had to wait for 22 years to be installed. Thereafter, investments of power plants have boomed along with the enactment of laws and regulations on power generation from renewables. The geothermal power plant (planned and under construction) locations and installed capacities are given in Table 11.2.

By the end of 2013, Salavatlı–Dora-3 unit 1, Germencik–Gümü köy, and Sarayköy–Kizildere GPP projects are expected to be in operation, and the total installed capacity of Turkey is expected to be 283.48 MW$_e$. Also installed capacity of all geothermal plants is expected to reach up to 500 MW$_e$ with additional GPPs by the year of 2016 (EMRA, 2013).

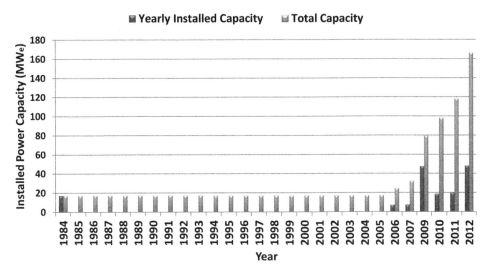

Figure 11.1.   Installed geothermal power plant capacity of Turkey by year.

Table 11.2.   Geothermal power plants as of June 2013 (planned and under construction).

Location	Planned capacity (MW$_e$)	Maximum temperature (°C)
Manisa		
Alaşehir-Piyadeler	24	–
Alaşehir-Erenköy	30	–
Salihli-Caferbeyli	15	168
Denizli		
Sarayköy-Seyitler	2.52	–
Sarayköy-Kizildere	86.93	242
Aydın		
Salavatlı: Dora-3	34	180
Salavatlı: Dora-4	17	–
Sultanhisar	9.9	–
Nazilli-Atça	9.5	124
Nazilli-Gedik	20	–
Köşk-Umurlu	4.85	131
Germencik-Gümüşköy	15	–
Germencik-Bozköy	24	–
Germencik-Hıdırbeyli	24	–
Kuyucak-Pamukören	61.72	–
Total	**378.42**	

11.3   CHARACTERISTICS OF GEOTHERMAL RESOURCES IN AEGEAN REGION

All discovered geothermal fields in Turkey are liquid dominated. Geothermal reservoirs are classified as low-, medium-, and high-enthalpy resources, and the temperature ranges are 100–160, 160–190, and over 190°C, respectively (Bronicki, 1995). Although some higher enthalpy wells exist in the Kizildere and Germencik geothermal fields, generally the low- and medium-enthalpy resources have been encountered in the Aegean region, which contains high amounts of non-condensable gases (NCGs).

Presence of NCGs plays an important role in the driving mechanism of the reservoirs because their partial pressure is quite high (Kaplan and Serpen, 2010). Since most of the fields in Aegean region have high amounts of NCGs; pressure and mass flow rate of geothermal fluids are consequently high as well, and the wells are artesian. On the other hand, NCGs cause scaling problems in the wellbores, pipelines, and all interacted plant equipment.

A common problem of the geothermal wells with two-phase flow is that the deposition of calcium carbonate above the flash point in the wellbores. CO_2 causes flashing of geothermal fluid; calcium carbonate becomes oversaturated and is precipitated above the flash horizon. To prevent scaling, inhibitor according to the content of the geothermal fluid is pumped to the flash point by a pipeline.

In addition, silica (SiO_2), which is in the structural forms of amorphous silica and quartz, is always found in the geothermal fluids. When the temperature and pressure of geothermal fluid is decreased until undersaturation of amorphous silica and quartz, silica precipitation is possible in the plant system or reservoir (Dipippo, 2005). Besides NCGs, some ions and chemical compounds are dissolved in the geothermal fluids, such as CI, SO_4, and CO_2, and calcite and aragonite (Mutlu and Gulec, 1998).

A power plant generates electricity between temperature of high heat source (geothermal reservoir) and temperature of low heat source (air or water). Therefore, the design of a power plant is essentially determined by thermophysical properties of high heat and low heat sources such as temperature, pressure, and mass flow rate. In addition, the contents of geothermal fluid, NCGs rate, and the other characteristics of the field are vital during the design stage of geothermal power plants.

11.4   TYPES OF GEOTHERMAL POWER PLANTS FOR RESERVOIR CHARACTERISTICS OF AEGEAN REGION

Flash and binary cycles are mostly applied to liquid-dominated resources. Since efficiencies of geothermal power plants are lower compared with fossil-fuelled power plants, most effective and economical plant type should be selected for geothermal resources (Shokouhmand and Atashkadi, 1997). In the world, some combined cycles such as flash-binary combined cycle have been used to increase efficiency and power generation.

11.4.1   *Single and double flash geothermal power plants*

Flash cycle geothermal power plants are generally installed on the high-enthalpy geothermal fields. A single flash plant mainly consists of a separator, a turbine-generator unit, a condenser, a cooling tower, a circulation water pump, and a gas removal system (Fig. 11.2). Geothermal fluid, which comes from the wellbore at high pressure, can flash in the reservoir, production well, or inlet of the separator (Dipippo, 2005). In this type of plant, the mixture firstly enters a separator, and then two-phase flow is physically separated as steam and liquid phases with a minimum pressure loss in the separator. Separated liquid phase (brine) returns into reinjection line. Steam phase, including NCGs, passes through a ball check valve, enters the turbine at high pressure, and drives the blades of turbine to produce electricity. After that steam expands, its pressure drops, and flows to the condenser. Steam passes liquid phase in the condenser by using cooling water coming from cooling tower. NCGs are extracted from the condenser by a gas removal system such as compressor or jet ejector (Gokcen et al., 2004a). Condensed water is pumped down a reinjection well with separated liquid phase to sustain production.

Double flash geothermal power plants are different from the single flash in that they have additional low-pressure separator and turbine. After the first separation, separated liquid phase enters the low-pressure separator producing low-pressure steam. Steam and liquid phases are separated and then steam passes through the low-pressure turbine. It enters the condenser with

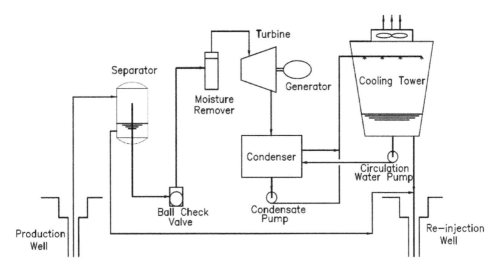

Figure 11.2.   Flow diagram of a single flash geothermal power plant (adopted from Dipippo, 2005).

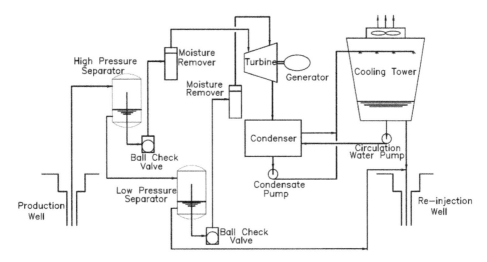

Figure 11.3.   Simplified flow diagram of a double flash geothermal power plant (adopted from Dipippo, 2005).

steam of high-pressure turbine. The rest of the cycle is similar to single flash cycle. A simple schematic diagram of a double flash geothermal power plant is shown in Figure 11.3. The double flash plant can produce more power than single flash under the same conditions (Swandaru, 2009).

11.4.2   *Binary cycle geothermal power plants*

In binary geothermal power plants, the heat of geothermal fluid is transferred to an organic working fluid in heat exchangers such as preheater and evaporator. Geothermal fluid is used for the heat transfer and passes only in the heat exchangers. The heat of geothermal fluid is transferred to an organic working fluid that has a low boiling point, and then returns to the ground

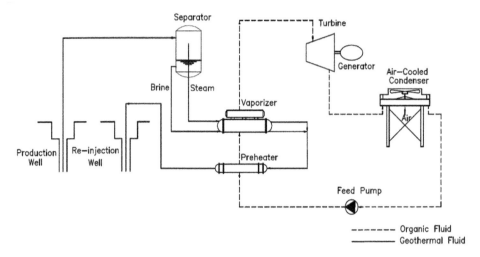

Figure 11.4.   Flow diagram of a two-phase binary geothermal power plant (from Kaplan and Serpen, 2010).

by the reinjection wells to recharge reservoir (Dipippo, 2005). Working fluid becomes saturated or superheated vapor in the heat exchangers. The working fluid then passes through the turbine and electricity is generated. After that, it enters the condenser and is cooled by a cooling fluid until it becomes liquid. Finally, it is pumped to the preheater and the evaporator by a feed pump. The simple schematic of binary cycle power plants is shown in Figure 11.4.

The cooling system is more important in binary plants. If there is an unlimited resource of cooling water in the field, wet cooling system may be used to condensate working fluid by cooling water. Dry air-cooling systems are more suitable for locations with water shortage because in this system, heat of working fluids is rejected directly into air. Another advantage of the air-cooling systems is that there is no emission to the atmosphere. However, parasitic load of cooling fans is excessive because of a need of a large heat transfer surface for condensation. Therefore, the efficiency of the plant significantly decreases with increasing ambient temperatures for air-cooling systems.

In addition, thermo-physical properties of working fluid directly affect the performance of the plant. The main criteria for the selection of the working fluid are evaporation at atmospheric pressures and boiling at low temperatures. The working fluids are generally selected from hydrocarbons such as pentane and isobutane.

Binary systems have more environmental benefits than flash systems because binary geothermal power plants operate in closed circuits with no loss of working fluid to the environment. Equipment of the binary cycle plants have economically a longer life than the flash type plants because geothermal fluid does not have contact with the turbine or other moving mechanical components of the plant (Franco, 2011).

Currently in Turkey, single flash, double flash and binary cycle power plants are in operation. Reservoir characteristics and environmental conditions directly affect the determination of appropriate option of power plant. Since binary cycle is more feasible for low- and medium-enthalpy geothermal fields because of high concentration of calcium carbonate and NCG content of geothermal fluids, it is the most common cycle applied in Turkey (Kaplan and Serpen, 2010). Integrated two level unit (ITLU) type binary cycle power plants, consisting of two organic Rankine cycles (ORCs) and CO_2, are separated at the wellhead by a separator and passed to the heat exchanger at one of the cycles, and are the ones mostly applied in Turkey.

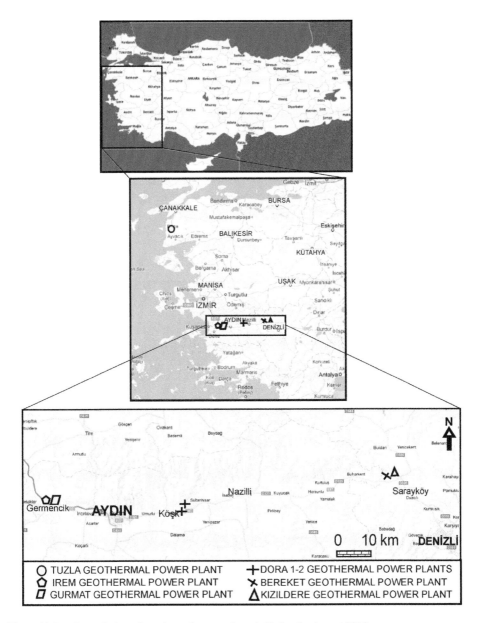

Figure 11.5. General view of geothermal power plants in Turkey by August 2012.

11.5 GEOTHERMAL POWER PLANTS IN TURKEY

Detailed information about operating geothermal power plants (Fig. 11.5) of Turkey is summarized in this section.

11.5.1 *Kizildere geothermal power plant*

Kizildere geothermal power plant is the first geothermal power plant of Turkey, therefore quite a number of studies exist in the literature.

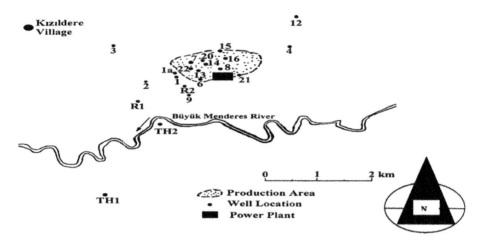

Figure 11.6.   Well locations of the Denizli-Kizildere geothermal field (Gokcen *et al.*, 2004a).

Figure 11.7.   General view of Kizildere geothermal power plant (Simsek *et al.*, 2005).

11.5.1.1 *Brief historical development of Denizli-Kizildere geothermal field*

The Denizli-Kizildere geothermal field is located 30 km away from the province of Denizli, western Anatolia, Turkey. The field was discovered in the mid-1960s and the United Nations Development Program (UNDP) supported the first geological, geochemical, and geophysical studies in 1966. The first well (KD-1), which has 196°C temperature, was drilled in 1968 (Gokcen *et al.*, 2004b). A total of 17 wells were drilled until mid-1970s within the scope of field development. Construction of a single flash geothermal power plant with a 17.4 MW$_e$ gross, 15 MW$_e$ net power generation capacity was initiated in 1982, and the plant was started to generate electricity by using six production wells (KD-6, 7, 13, 14, 15, and 16) in 1984. Two years later, three production wells (KD-20, 21, and 22) were drilled to increase the steam production, and the total number of the production wells reached to nine (Sarikurt, 1983; Serpen and Satman, 2000). The TH-2, R-1, and R-2 wells were drilled for reinjection purposes from 1997 to 2000 by MTA (Zoren, 2012, *pers. commun.*). Figures 11.6 and 11.7 illustrate the well locations of the

Table 11.3. General characteristics of the Denizli-Kizildere geothermal field (Gokcen *et al.*, 2004a).

Description	Unit	Value
Reservoir temperature	°C	200–242
Wellhead steam fraction	%	10–12
CO_2 partial pressure	bar/MPa	30–50/3–5
Total dissolved solid	mg/kg	2500–3200
NCG content in steam	%	10–21
CO_2 content	%	96–99
H_2S content	mg/kg	100–200

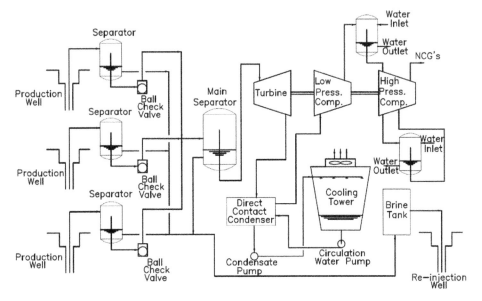

Figure 11.8. Flow diagram of the Kizildere geothermal power plant (from Gokcen *et al.*, 2004a; Serpen and Turkmen, 2005).

field and a general view of the Kizildere geothermal power plant, respectively. A CO_2 and dry ice production plant was installed in the field in 1986. The capacity of the plant increased from 40,000 tons/year to 120,000 tons/year in 1999 (Gokcen *et al.*, 2004b).

At first, geothermal fluid was not fully reinjected into the underground; some of the geothermal fluid was discharged to the Great Menderes river. Full reinjection of the geothermal fluid, except the evaporated fluid from the cooling tower to the atmosphere, was started in October 2009.

The Denizli-Kizildere field is a liquid-dominated reservoir with high amount of NCGs, which is 2.5% in the reservoir, 5% by volume of steam, 10–21% by weight of steam, and average 13% by weight of steam at the turbine inlet with a CO_2 content of 96–99%. The specific steam consumption of the plant is 10.96 kg/kWh. The general characteristics of the field are given in Table 11.3 (Gokcen *et al.*, 2004a).

11.5.1.2 *Power generation*
Kizildere geothermal power plant is a conventional single flash type plant. The plant, of which the flow diagram is shown in Figure 11.8, can be divided into two sections: steam field and power generation unit. The steam field consists of several wellbores, separators, silencers, and the

Table 11.4. General characteristics of surface installations and the power plant
(Serpen and Turkmen, 2005).

Description	Unit	Value
Wellhead pressures	bar/MPa	11–29/1.1–2.9
Separator pressures	bar/MPa	3.5–4.5/0.35–0.45
Separator temperature	°C	147
Turbine inlet pressure	bar/MPa	3.5/0.35
Turbine inlet temperature	°C	147
Turbine outlet pressure	bar/MPa	0.09/0.009
Turbine outlet temperature	°C	51
Steam dryness at turbine exhaust	%	85
Installed capacity of the power plant	MW_e	17.4
Rated capacity of the power plant	MW_e	15
Compressor rating	MW_e	2.38
Compressor capacity	$1000\,m^3/h$	293.5
Auxiliary power	MW_e	0.472
Condenser inlet temperature	°C	28
Condenser outlet temperature	°C	39

steam transmission and the reinjection system. A wellhead contains master and service valves. The master valve is used to close the well for maintenance purposes while the service valve is used for regulation of flow. The two-phase geothermal fluid coming from wellbore passes through wellhead with minimum pressure drop to prevent precipitation of $CaCO_3$ in the wellbores. Wellhead pressures are kept as high as 11–29 bar (1.1–2.9 MPa). The geothermal fluid is sent to the cyclone-type separator and separated as steam and liquid (brine) phases at 3.5–4.5 bar (0.35–0.45 MPa) for each well. In the separators, approximately 87% of geothermal fluid is separated as brine, then it enters to a brine tank and reinjected to underground at 110°C. A ball check valve is used as a safety unit. The steam from the separator passes through the ball check valve to prevent water entrance to the steam line (Gokcen *et al.*, 2004a; Serpen and Turkmen, 2005; Zoren, 2012, *pers. commun.*).

Power generation unit of the plant mainly consists of a scrubber, a turbine-generator unit, a condenser, a gas removal system, and a cooling tower. Scrubber is the main moisture separator and used to remove the impurities and condensate of the steam. After scrubber, steam goes to the demister to remove the moisture and splits into two branches that are sent to the double flow turbine at 147°C. The double flow turbine has two turbines on the same shaft with steam flowing in opposite direction and seven reaction stages on both sides. The steam, including NCGs, expands in the turbine and flows to the condenser. The condenser, which is located under the turbine, is direct contact type. The steam condenses by using cooling water, which is sucked, from a wet cooling tower by vacuum in the condenser at 29°C. Condensate water is pumped to a height of 8.5 m with a temperature of 36.6°C by two centrifugal pumps and it cools by using four motor-driven fans in the wet cooling tower. In the plant, two compressors (high and low pressure) are used to extract NCGs from the condenser (Gokcen *et al.*, 2004a; Serpen and Turkmen, 2005).

Exergetic efficiency of Kizildere geothermal power plant with assuming temperatures of 217°C geothermal sources at 20°C environmental conditions is 25%. The main characteristics of surface installations and the power plant are shown at Table 11.4.

11.5.1.3 *Scaling problems in Kizildere geothermal power plant*
Energy generation in the Kizildere geothermal power plant was very low during the first three years because of the lack of experience. One of the severe problems in the field was the $CaCO_3$ scaling in the production wells and high boron concentration in the geothermal fluid. Calcite scaling

Figure 11.9.   Calcite scaling in a pipe at Kizildere (Simsek *et al.*, 2005).

in a pipe at Kizildere is shown in Figure 11.9. Over many years, calcite scaling problem has been solved at interim by mechanical cleaning operations using a rig equipped with rotating head blowout preventer (Serpen and Turkmen, 2005). In 1988–1989, Dequest 2006 inhibitor injection test was conducted for 10 months. Although the results of the inhibitor test were successful, periodical mechanical cleaning was preferred for once a year due to the high operation cost of inhibitor injection (Gokcen *et al.*, 2004b). Fully inhibitor injection for all production wells started in 2009.

Silica scaling is another important operational problem of the Kizildere geothermal power plant came along with the reinjection trials in 2009. The geothermal fluid contains about 350–400 mg/kg SiO_2 (Yildirim and Olmez, 1999). After the separator, the separated liquid becomes saturated with silica. Nevertheless, silica precipitation is prevented by keeping the reinjection temperature at 110°C since 2009.

11.5.1.4   *Future of Denizli-Kizildere geothermal field*

A new power plant investment with 86.93 MW_e gross power capacity started by The Zorlu Energy Group since 2010 and the power plant is expected to be in operation by summer of 2013. Ten production and 9 reinjection wells were drilled for this plant between years 2009 and 2012. Construction of the plant is shown in Figure 11.10.

The plant is designed as combination of triple flash cycle and two binary units by POWER Engineers, Inc. of the USA. Total seven separators, which are two high-pressure, two medium-pressure, and three low-pressure separators, will be used in the plant. High-pressure steam will first pass through high-pressure steam turbine, after that it will be separated equally and sent to two binary cycle power units. R-134a will be used as secondary fluid for both binary units. Medium- and low-pressure steam will enter the two- stage steam turbine. High-pressure and two-stage steam turbines are planned to have a capacity of 66.13 MW_e while each binary unit will have a 10.4 MW_e gross capacity (Zoren, 2012, *pers. commun.*).

Figure 11.10.   Construction of the new plant of The Zorlu Energy Group.

Table 11.5.   Some characteristics of the Dora-1 and Dora-2 wells.

Powerplant and well no.	Date	Well type	Depth (m)	Reservoir temperature (°C)	Wellhead temperature (°C)	Wellhead pressure (bar/MPa)	Mass flow rate (ton/hour)
Dora-1							
AS-1	1987	Production	1519	167	159.5	9/0.9	350
ASR-2	2004	Production	1300	165	158	6.6/0.66	240
AS-2	1988	Reinjection	962	172	–	–	–
Dora-2							
AS-3	2008	Production	1325	175	170.5	14/1.4	430
AS-4	2008	Production	1275	176	171	15/1.5	460
ASR-4	2008	Reinjection	1923	148	–	–	–
ASR-5	2009	Reinjection	1218	157	–	–	–

11.5.2   *Dora geothermal power plants*

11.5.2.1   *Brief historical development of Salavatlı geothermal field*

Salavatlı geothermal field is located approximately 24 km away from the province of Aydın on the western Turkey. The field was discovered in early 1980s by the General Directorate of Mineral Research and Exploration (MTA) and two geothermal wells (AS-1 and AS-2) were drilled in 1987 and 1988, respectively. After 16 years of discovery, the third well (ASR-2) was drilled in 2004 and the first binary geothermal power plant of Turkey, Dora-1, was installed in 2006 by Menderes Geothermal Inc. (MEGE) with a capacity of 7.35 MW$_e$. Between 2008 and 2009, additional production (AS-3 and AS-4) and reinjection wells (ASR-4 and ASR-5) were drilled in the field for construction of second unit Dora-2 geothermal power plant. Dora-2 was installed in 2010 with a capacity of 11.2 MW$_e$. Reservoir temperatures, depths, and production capacities of the wells of Dora-1 and Dora-2 are listed in Table 11.5.

Following the successful operation of Dora-1 and 2, third and fourth units were planned and seven new wells (ASR-3, AS-6, 7, 8; ASR-6, 7, 8) were drilled for Dora-3 unit-1, and 10 new

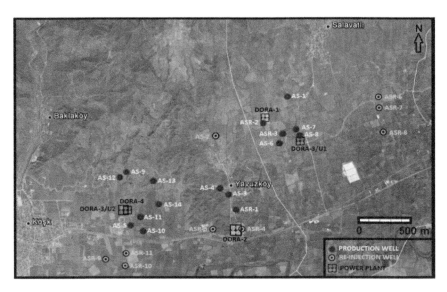

Figure 11.11. Locations of wells and power plants in the Salavatlı geothermal field (Google Earth, August 2012; MEGE, 2012, *pers. commun.*).

wells (AS-9, 10, 11, 12, 13, 14; ASR-5, 9, 10, 11) were drilled for Dora-3 unit-2 and Dora-4 from November 2010 to September 2012. The locations of the wells and the power plants are illustrated in Figure 11.11.

All production wells have low or medium enthalpies with 1% NCG content in the liquid-dominated field except the AS-5 well. The AS-5, which is the deepest well of Turkey with 3224 m, has high enthalpy with a reservoir temperature of 211°C.

11.5.2.2 *Dora-1 geothermal power plant*

The Dora-1 geothermal power plant was put in operation on May 2006 as the first binary power plant of Turkey. The plant type is the ITLU air-cooled binary cycle manufactured by Ormat. A general view of Dora-1 is shown in Figure 11.12.

The geothermal fluid is produced from AS-1 and ASR-2 productions wells, and reinjected to the AS-2 well. The enthalpies of the reservoir range between 660 and 680 kJ/kg with 1% NCG of the total flow. NCGs mainly consist of CO_2 and small amount of H_2S. The two-phase geothermal fluid is physically separated by the separators at each wellhead. After the separation, brine (geothermal fluid) firstly enters a tank and then is pumped to the power plant by brine pumps. Separated steam and NCGs spontaneously go to the power plant by individual pipelines. A general view and flow diagram of wellhead equipment are provided in Figures 11.13 and 11.14, respectively.

11.5.2.3 *Power generation*

The flow diagram of Dora-1 geothermal power plant is shown in Figure 11.15. The plant is divided into two Rankine cycle levels: Level 1 and Level 2. Although the working principles of these cycles are the same, their working pressures are different. N-pentane (C5H12) is the working fluid for both levels. First, the brine enters the vaporizer of Level 1 where working fluid of Level 1 is heated. Then, it flows to the vaporizer of Level 2 to transfer its heat to the working fluid in this heat exchanger. Afterwards, the temperature of brine decreases and is divided equally into two for preheaters of Levels 1 and 2. In the preheaters, the brine temperature is further decreased and leaves the plant to be reinjected to AS-2 well. While the working fluid of Level 2 is vaporized by steam and NCGs (through water vapor and NCGs tube section), which comes from the separator,

Figure 11.12.   General view of Dora-1 geothermal power plant.

Figure 11.13.   Wellhead equipment of the ASR-2.

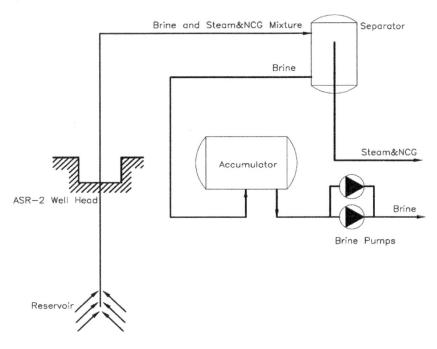

Figure 11.14. Flow diagram of the ASR-2.

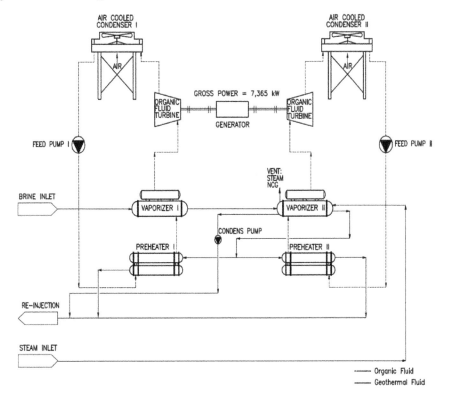

Figure 11.15. Flow diagram of Dora-1 geothermal power plant.

Table 11.6. Design parameters of the Dora-1 plant (Ormat, 2005; Toksoy *et al.*, 2007).

Design parameters and ranges

Design parameters		Design ranges	
Ambient air temperature	17.1°C	Ambient air temperature range	0–40°C
Brine flow rate	542.65 tons/h	Brine flow rate range	80–110%
Vapor flow rate	22.45 tons/h	Vapor flow rate range	80–110%
Brine temperature	157.9°C	Brine temperature range	150–166°C
Steam NCG content	33.6%	NCG content range	29–39%

the steam is condensed and pumped to the reinjection line by a condensate pump. NCGs are vented from the vaporizer, sent to the liquid CO_2 plant, which is located next to the plant.

In the vaporizers, the working fluid (*n*-pentane) is heated up to the boiling point. It evaporates, and then the superheated vapor enters the turbine, expands, and its pressure decreases. The low-pressure vapor flows to an air-cooled condenser, where the vapor is condensed and then pumped back into the preheaters. Both levels have closed cycles and are independent from each other. Design parameters of Dora-1 and their ranges are shown in Table 11.6.

The electricity is generated from the geothermal source at a temperature ranging from 157.9°C (plant inlet) to 78.9°C (reinjection). The total mass flow rate of the geothermal fluid, which comes from two production wells, is 565 tons/h. The geothermal fluid has 22.45 tons/h steam that includes 33.6% NCG content. The Dora-1 geothermal power plant generates 7.35 MW_e gross, 6.5 MW_e net power at 17.1°C ambient air temperature and design conditions. After the electricity generation, NCGs are vented to produce liquid CO_2 at another facility nearby the plant. In this facility, approximately 30,000 tons/year of CO_2 is liquefied and gained for beneficial uses (Linde Gas, 2012, *pers. commun.*). The flow rate and temperature of the reinjection well is 557.55 tons/h (brine+condensate) and 78.9°C, respectively.

11.5.2.4 *Performance assessment of the plant*
The first and the second law efficiencies of Dora-1 are calculated as 12.1% and 47% at 157.9°C geothermal resource and 17.1°C dead-state temperatures. A thermodynamic model of an ITLU binary power plant was conducted by using design parameters of Dora-1. According to this study, net power generation of the plant increases with an increase in brine temperature, and mass flow rates of brine and steam; decreases with an increase of ambient air temperature and NCGs content of the steam. Ambient air temperature is the most effective parameter on electricity generation, because the efficiency of a plant mainly depends on the difference between the temperatures of the source (geothermal fluid) and the sink (air). Efficiency of the plant increases with increasing the source temperature and decreasing the sink temperature. Source temperature is constant while the air temperature changes throughout the year (Karadas, 2012).

The design parameters and their effect on power generation are displayed in Figure 11.16.

11.5.3 *Germencik double flash geothermal power plant*

11.5.3.1 *Brief historical development of Germencik-Ömerbeyli geothermal field*
The Germencik-Ömerbeyli geothermal field is located approximately 40 km from the Aegean Sea and 25 km away from the province of Aydın, western Turkey. This field is the largest high-temperature geothermal field in Turkey with 14.7 km^2 area. The first studies were initiated in the field in 1967 and the nine deep geothermal wells (OB-1, 2, 3, 4, 5, 6, 7, 8, 9) were drilled from 1982 to 1986 by MTA. After 25 years, another nine wells (OB-10, 11, 14, 17, 19; AG-22, 24, 25, 26) were drilled between 2007 and 2008, followed by OB-16 and AG-28 in 2010 and 2011.

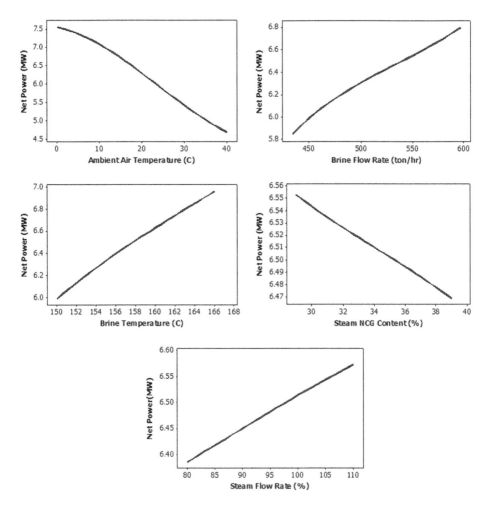

Figure 11.16.   Net power production trends of an ITLU binary power plant model by changing design parameters of Dora-1.

The temperature of the geothermal fluid ranges between 191 and 232°C, with steam content by weight of steam of 10–12% (Gurmat, 2012, *pers. commun.*). The location of the field and the wells are given in Figure 11.17. Well characteristics of the field are listed in Table 11.7.

The construction of the Germencik double flash geothermal power plant is completed in spring 2009 by Guriş's subsidiary Gurmat. The plant is designed by the POWER Engineers, Inc. of the USA (Wallace *et al.*, 2009) and operated since spring 2009 with 47.4 MW$_e$ capacities. The general view of the plant is shown in Figure 11.18.

11.5.3.2   *Power generation of the plant*
The Germencik geothermal power plant is a conventional double flash type plant. The plant consists of eight production wells, two high-pressure vertical cyclone separators, two low-pressure horizontal separators, a high-pressure demister, a low-pressure demister, a dual pressure turbine, a gas removal system, an advanced direct contact condenser (ADDC), and a reinjection system.

Geothermal fluid comes from the production wells at a flow rate of approximately 2500 tons/h and is first directed to the high-pressure separators and it is physically separated as liquid and steam

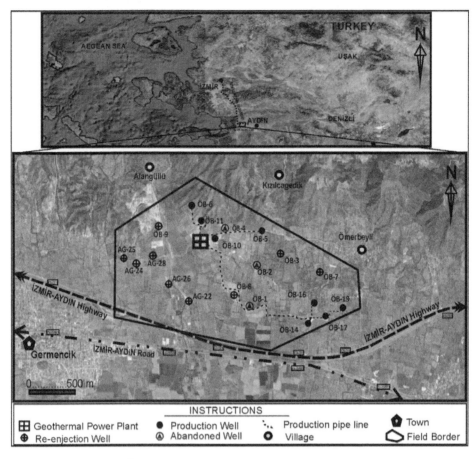

Figure 11.17.   Germencik-Ömerbeyli geothermal field map showing existing well locations (Google Earth, August 2012; Gurmat, 2012, *pers. commun.*).

Table 11.7.   All geothermal wells of Germencik-Ömerbeyli geothermal fields.

Well No.	Date	Depth (m)	Reservoir temperature (°C)	Well No.	Date	Depth (m)	Reservoir temperature (°C)
OB-1	1982	1001	203	OB-11	2007	965	210
OB-2	1982	975.5	232	OB-14	2007	1205	228
OB-3	1983	1195	232	OB-16	2010	–	–
OB-4	1984	285	217	OB-17	2007	1706	228
OB-5	1984	1302	219	OB-19	2008	1651	227
OB-6	1984	1100	221	AG-22	2008	2260	205
OB-7	1985	2398	227	AG-24	2008	1252	199
OB-8	1986	2000	221	AG-25	2008	1838	191
OB-9	1986	1466	213	AG-26	2008	2432	195
OB-10	2007	1524	224	AG-28	2011	–	–

Figure 11.18.   General view of the Germencik double flash geothermal power plant (Wallace *et al.*, 2009).

phases at 6.40 bar (0.64 MPa). Steam goes to the high-pressure demister to remove the moisture and splits into two branches that are sent to the high-pressure section of the dual pressure turbine. The liquid phase enters to low-pressure separators and it is separated again at 1.5 bar (0.15 MPa). During the separation processes, approximately 80% of total geothermal fluid is separated as brine and is directly reinjected by using seven hot reinjection wells at 110°C. Low-pressure steam passes through the low-pressure demister and then it enters the low-pressure section of the turbine. After that, steam condenses in the advanced direct contact condenser by using cooling water. A small amount of steam loses from the cooling tower to atmosphere and condensate water is reinjected by a cold reinjection well. NCGs are extracted with three stage steam-jet ejectors followed by four vacuum pumps (Gurmat, 2012, *pers. commun.*). The flow diagram of the plant is shown in Figure 11.19.

Dipippo (2012) calculated the second law efficiency of Germencik double flash geothermal power plant as 32% for a total geothermal fluid flow rate of 2331.72 tons/h and a reservoir temperature of 218°C.

11.5.4   *Tuzla geothermal power plant*

11.5.4.1   *Brief historical development of Tuzla geothermal field*
The Tuzla geothermal field is located 80 km south of Canakkale city center and 5 km away from the Aegean Sea, northwest Turkey. Geothermal exploration studies in the field started in 1966 by MTA and 10 gradient wells (with a depth of 50–100 m) and two deep exploration wells (with a depth of 814–1020 m) were drilled in 1974 and 1982–1983, respectively. The reservoir temperature of the field is 173°C in the depth of 333–553 m (Baba *et al.*, 2005). The total four wells (T-1, 2, 3, and 4), which are suitable for power generation, were drilled by MTA.

The Tuzla geothermal power plant project was initiated in May 2004 with a capacity of 7.5 MW$_e$. Eight new wells (T-5, 7, 8, 9, 10, 11, 15, and 16) were drilled in the field in 2008 and 2009. The field and well locations are shown in Figure 11.20.

The Tuzla geothermal power plant is an air-cooled binary cycle geothermal power plant manufactured by Ormat. It has been generating electricity since January 2010. The geothermal fluid comes from two artesian production wells (T-9 and T-16) at a temperature of 174–176°C, and reinjected by using two reinjection wells (T-15 and T-10) (Inanli and Atilla, 2011). A general view of the plant is illustrated in Figure 11.21.

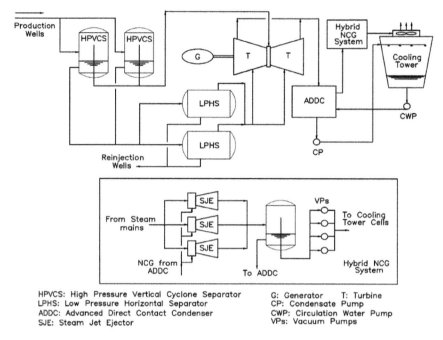

HPVCS: High Pressure Vertical Cyclone Separator
LPHS: Low Pressure Horizontal Separator
ADDC: Advanced Direct Contact Condenser
SJE: Steam Jet Ejector

G: Generator   T: Turbine
CP: Condensate Pump
CWP: Circulation Water Pump
VPs: Vacuum Pumps

Figure 11.19.   Flow diagram of Germencik double flash geothermal power plant (adopted from Dipippo, 2012).

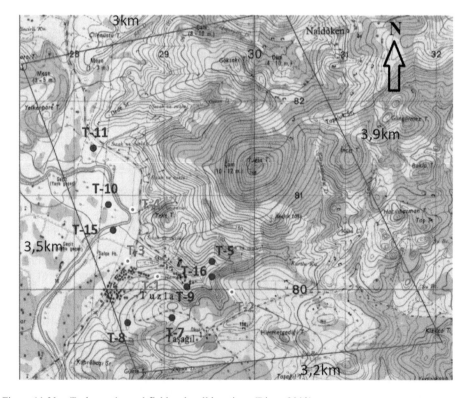

Figure 11.20.   Tuzla geothermal field and well locations (Diner, 2010).

Figure 11.21.   General view of Tuzla geothermal power plant (Inanli and Atilla, 2011).

11.5.4.2   *Power generation and performance assessment of the plant*

The Tuzla geothermal power plant operates on an ORC with a working fluid of isopentane. The plant consists of two main sections: BOP (Balance of Plant) and OEC (Ormat Energy Converter). BOP consists of two production wells, two reinjection wells, two horizontal separators, brine pumps, block, and check valves, reinjection pumps, steam, and brine pipelines. The two-phase geothermal fluid comes from the production wells and is separated as brine and steam by using horizontal separators at each wellhead. The separated brine is pumped to the plant by brine pumps and the steam spontaneously directed to the plant.

Figure 11.22 shows the OEC diagram of the plant. OEC consists of a preheater, a vaporizer, a turbine-generator unit, a recuperator, an air-cooled condenser, and a feed pump. The brine firstly passes through the tube section of the vaporizer and some of the heat of the brine is transferred to the isopentane. In addition, the separated steam and the NCGs pass through the different tubes of the vaporizer. The NCGs are vented into the atmosphere; steam condenses and confuses with brine at the end of the vaporizer. Then, the brine and condensate mixture pass through the tubes of the preheaters and finally the cooled geothermal fluid moves to the reinjection wells. The tubing material of heat exchangers is stainless steel to prevent corrosion of the highly saline brine.

Isopentane is heated in the preheater and vaporizer, becomes superheated at the exit of the vaporizer. The superheated vapor expands in the turbine, and the mechanical power of the turbine is converted into electricity. Afterwards, isopentane is cooled by a recuperator and an air-cooled condenser. Recuperator works as a condenser for the superheated working fluid and as a preheater for the cooled working fluid. Condensate isopentane is pumped to the recuperator by a feed pump for heating and then it reenters the preheater. Circulation of the isopentane is a closed cycle (Coskun *et al.*, 2011a; Inanli and Atilla, 2011).

Tuzla geothermal power plant generates 7.5 MW$_e$ gross, 6.48 MW$_e$ net power at an air temperature of 25°C, a brine flow rate of 693 tons/h, and a steam flow rate of 48 tons/h at 143°C. The NCGs are vented to the atmosphere at a flow rate of 2 tons/h along with a steam flow rate of 3 tons/h. Total reinjection flow rate is 736 tons/h at 97°C (Diner, 2010). The average first and second law efficiencies are 9.47 and 45.2%, respectively (Coskun *et al.*, 2011b).

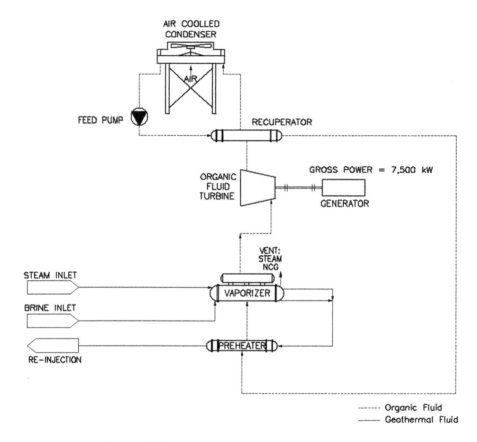

Figure 11.22.   OEC diagram of Tuzla geothermal power plant (utilized from Diner, 2010).

11.5.4.3   *Scaling problems of the plant*

The Tuzla geothermal field has hot saline geothermal fluid with a high-scaling tendency (Tarcan, 2005). The total salt concentration is approximately twice the concentration of seawater because the salinity reaches 55,000 mg/kg (Baba *et al.*, 2008). The geothermal fluid of Tuzla field has 32,000 mg/kg Cl and 22,400 mg/kg Na content. Conductivity and total hardness of the geothermal fluid is approximately 100,000 μS/cm and 9000 mg/kg, respectively. Even though the hardness is high, the precipitation of $CaCO_3$ has been prevented in the wellbores by inhibitors (Inanli and Atilla, 2011).

Metal silicate scaling is the main problem in the wellhead equipment, brine pipelines, and heat exchangers in the Tuzla geothermal power plant (Fig. 11.23). Inanli and Atilla (2011) studied metal silicate formation at Tuzla geothermal brine lines. According to the study, composition of scale mostly consists of silis, iron, and magnesium. The metal silicate scaling occurs from the geothermal wells to the power plant (separator walls, capillary tubing, brine pumps, valve seats, thermowells, orifice tubings, preheater and vaporizer tubes, etc.). Scaling is not observed on the reinjection line and equipment. After many inhibitor tests to prevent metal silica scaling, 8–10 mg/kg iron dispersant was selected and it has been injecting into the separated brine on the ground (Fig. 11.24).

The performance of the inhibitor is mainly controlled by using coupons, water analysis, and heat exchangers parameters. Stainless steel rods, which have small holes, are used as coupons (Fig. 11.24). If there is a precipitation problem in the geothermal fluid, initially the small holes

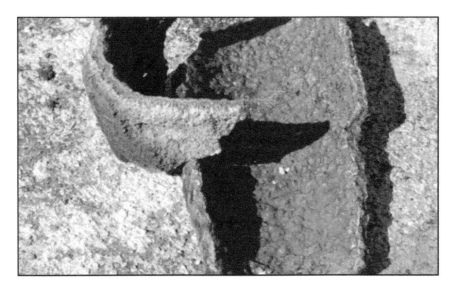

Figure 11.23.   Metal silicate formation at separator brine outlet (Inanli and Atilla, 2011).

Figure 11.24.   Iron dispersant dosage point (Inanli and Atilla, 2011).

of the coupons are blocked. In this way, precipitation is physically monitored on the wellheads, brine lines, and before the plant inlet. The pH, electrical conductivity, alkalinity, iron, silica, Ca hardness, and total hardness are measured during the chemical analysis (Inanli and Atilla, 2011). Pressure and mass flow rate of the brine is also monitored at the inlet of the vaporizer and the preheater to evaluate performance of the inhibitor. When scaling occurs in the tubes of the heat

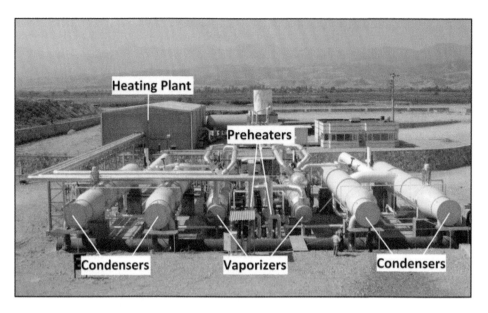

Figure 11.25.   Back view of Bereket geothermal power plant.

exchangers, the pressure inlet of the brine increases and the mass flow rate of the brine decreases. Therefore, scaling decreases the plant performance.

Although iron dispersant decreased the rate of scaling, it did not totally prevent scaling in the plant. Besides the chemical analysis and monitoring of coupons and plant parameters, tubing of the heat exchangers are mechanically cleaned almost every month by using high-pressure water jetting method at 1250 bar (125 MPa).

11.5.5   *Other geothermal power plants*

11.5.5.1   *Bereket geothermal power plant*

The Bereket geothermal power plant was installed in Denizli-Sarayköy field, near to Kizildere geothermal power plant, in 2007. The plant is a binary cycle and it has used waste geothermal fluid of Kizildere geothermal power plant at 145°C. The plant is designed as ITLU by Ormat, and the working principle is the same as Dora-1, except the condensers. The plant has a cooling tower and two horizontal cylindrical shell & tube condensers which are used for each level.

A general view of the Bereket geothermal power plant is shown in Figures 11.25 and 11.26.

After being used for power generation, the geothermal fluid was sent to a heating plant at 70–80°C for district heating of the town of Sarayköy during winters. Sarayköy is 10 km away from the plant, where 5000 houses have been heated by waste geothermal fluid of the Bereket geothermal power plant. Thereafter, cooled geothermal fluid was sent to the Menderes river with drainage, which caused environmental problems due to the heat load and the chemicals in the fluid. The plant had been in operation between 2007 and 2010. Currently, it is in partial operation because the Kizildere geothermal power plant has fully reinjected waste geothermal fluid since 2009. However, only waste geothermal fluid of Kizildere geothermal power plant is sent for district heating in Sarayköy during winters. Therefore, the plant only generates electricity during winter seasons.

11.5.5.2   *Dora-2 geothermal power plant*

The Dora-2 geothermal power plant with 11.2 MW$_e$ capacity was installed in March 2010 in the Aydın-Salavatlı field by MEGE. This plant has an ITLU binary system and manufactured by

Figure 11.26.   Front view of Bereket geothermal power plant.

Figure 11.27.   General view of Dora-2 geothermal power plant.

Ormat (Fig. 11.27). Two production wells (AS-3, 4) and two reinjection wells (ASR-4, 5) are used to operate Dora-2. The production wells produce 853.74 tons/h geothermal fluid with 18.74 tons/h steam, which has 52% NCG content. The plant generates 11.2 MW$_e$ gross, 9.8 MW$_e$ net power at 18.5°C ambient air temperature. The *n*-pentane is used as a working fluid. The working principle of the plant is the same as Dora-1 geothermal power plant. Due to the heat transfer between the geothermal fluid and the *n*-pentane, the temperature of the fluid decreases from 169°C to 71.2°C. 100% of cooled geothermal fluid, except NCG content, is reinjected into the ground at 71.2°C. NCGs are sent to another facility (HABAŞ) for liquid CO_2 production. The facility has 45,000 tons/year liquid CO_2 production capacity.

11.5.5.3   *Irem geothermal power plant*
Irem geothermal power plant was installed in Germencik-Hidirbeyli geothermal field in November 2011 by MAREN. The plant generates 20 MW$_e$ gross, 15.5 MW$_e$ net power at 17.5°C ambient air

Figure 11.28.   General view of Irem geothermal power plant (Kipaş Holding, 2012).

temperature. The Irem geothermal power plant works with three production wells with a total flow rate of 1200 tons/h (7 tons/h steam and NCGs) at 160°C. All brine and condensate is reinjected by three wells at 75°C. The plant is manufactured as ITLU binary system by Ormat (Maren, 2012, *pers. commun.*). A general view of the plant is shown in Figure 11.28.

11.6   CONCLUSION

The present study focuses on the rapid development and present status of geothermal power generation in Turkey. Geothermal exploration activities in Turkey started in 1960s and several high-temperature geothermal fields have been discovered until today. Kizildere geothermal power plant was the first geothermal power plant of Turkey, installed in 1984 with a gross capacity 17.4 MW$_e$, and was also the only one until 2006. From 2006 to 2013, the total installed power plant capacity reached to166.35 MW$_e$ (by June 2013) with a 90% increase. The reason of the boom was the enaction of the laws and regulations on power generation from renewables. The increasing trend is expected to continue and to reach, approximately, 500 MW$_e$ by the year 2016.

Geothermal studies in Turkey mainly concentrated on the Aegean region; therefore, all installed power plants are located in Aegean region. A brief historical development of geothermal fields, reservoir characteristics, power generation systems, and performance of plants are explained for each power plant in this study.

This chapter reveals that there is now dynamic progress in the development of geothermal resources for power generation, at different locations. Furthermore, it could be developed by discovering new geothermal fields and installation of new geothermal power plants.

ACKNOWLEDGMENTS

The authors would like to thank the Management of Menderes Geothermal Inc., Gurmat Electric Generation Co. Inc., Zorlu Geothermal Energy Electricity Generation Inc., and Maren Inc. in providing plant data for this chapter. In addition, the contributions of Ms. Füsun Tut, Mr. Ömer Bergamalı, Mr. Kadir M. Sancak, Mr. Levent Aka, Mr. Özgür Erol, and Mr. Bayram Erkan are appreciated.

REFERENCES

Aneke, M., Agnew, B. & Underwood, C.: Performance analysis of the Chena binary geothermal power plant. *Appl. Therm. Eng.* 31 (2011), pp. 1825–1832.
Baba, A., Ozcan, H. & Deniz, O.: Environmental impact by spill of geothermal fluids at the geothermal field of Tuzla, Canakkale-Turkey. *Proceedings World Geothermal Congress 2005,* 24–29 April 2005, Antalya, Turkey, 2005.

Baba, A., Ozan, D., Ozcan, H., Erees, S.F. & Cetiner, S.Z.: Geochemical and radionuclide profile of Tuzla geothermal field, Turkey. *Environ. Monit. Assess.* 125 (2008), pp. 361–374.

Bronicki, L.: Innovative geothermal power plants, fifteen years experience. *Proceedings World Geothermal Conference*, Florence, Italy, 1995.

Coskun, C., Oktay, Z. & Dincer, I.: Performance evaluations of a geothermal power plant. *Appl. Therm. Eng.* (2011a), pp. 4074–4082.

Coskun, C., Oktay, Z. & Dincer, I.: Modified exergoeconomic modeling of a geothermal power plant. *Energy* 36 (2011b), pp. 6358–6366.

Diner, C.: ENDA Energy Holding Tuzla Geothermal Energy Co. Presentation. *Geothermal potential of Turkey Meeting-German Near and Middle East Association (NUMOV)*, 23–26 February 2010, Offenburg, Germany, 2010.

DiPippo, R.: Geothermal power plants: principles, applications, case studies and environmental impact. *Elsevier Advanced Technology*, Oxford, UK, 2005.

DiPippo, R.: *Geothermal power plants: principles, applications and case studies.* Elsevier, Oxford, UK, 2012.

EMRA, Republic of Turkey Energy Market Regulatory Authority, http://lisans.epdk.org.tr/epvys-web/faces/pages/lisans/elektrikUretim/elektrikUretimOzetSorgula.xhtml (accessed June 2013).

Franco, A.: Power production from a moderate temperature geothermal resource with regenerative organic Rankine cycles. *Energy Sustain. Develop.* 15 (2011), pp. 411–419.

Gokcen, G., Ozturk, H.K. & Hepbasli, A.: Overview of Kizildere geothermal power plant in Turkey. *Energy Convers. Manage.* 45 (2004a), pp. 83–98.

Gokcen, G., Ozturk, H.K. & Hepbasli, A.: Geothermal fields suitable for power generation. *Energy Sources* 26 (2004b), pp. 441–451.

Inanli, M. & Atilla, V.: Metal silicate formation at Tuzla geothermal brine lines. *Proceedings International Workshop on Mineral Scaling 2011*, 25–27 May 2011, Manila, Philippines, 2011.

Kaplan, U. & Serpen, U.: Developing geothermal power plants for geothermal fields in western Turkey. *Proceedings World Geothermal Congress 2010*, 25–29 April 2010, Bali, Indonesia, 2010.

Karadaş, M.: Modeling of a binary cycle geothermal power plant for geothermal resources of Turkey. *IZTECH Energy Engineering Workshop Study*, June 2012 (unpublished), 2012.

Kipaş Holding, Maraş Elek. Ürt. San. Ve Tic. A.Ş.: Irem geothermal power plant. http://www.kipas.com.tr/index.php?option=com_content&view=article&id=132&Itemid=209&lang=tr, 01.10.2012 (accessed October 2012).

MENR, The Ministry of Energy and Natural Resources: Energy information about Ministry. http://www.enerji.gov.tr/index.php?dil=en&sf=webpages&b=elektrik_EN&bn=219&hn=&nm=40717&id=40732 (accessed August 2012).

Mutlu, H. & Gulec, N.: Hydrogeochemical outline of thermal waters and geothermometry applications in Anatolia (Turkey). *J. Volcanol. Geoth. Res.* 85 (1998), pp. 495–515.

Ormat: Salavatlı geothermal project final acceptance performance tests. 28 February, 2005.

Sarikurt, H.: Report on Kizildere geothermal power plant. Turkish Electricity Generation and Transmission Corporation (TEAS), 1983 (unpublished).

Serpen, U. & Satman, A.: Reassessment of the Kizildere geothermal reservoir. *Proceedings World Geothermal Congress 2000*, 28 May–10 June 2000, Kyushu-Tohoku, Japan, 2000.

Serpen, U. & Aksoy, N.: Turkey's Dora I geothermal plant. *Geoth. Bull.* 37:6 (2008), pp. 21–23.

Serpen, U. & Turkmen, N.: Reassessment of Kizildere geothermal power plant after 20 years of exploitation. *Proceedings World Geothermal Congress 2005*, 24–29 April 2005, Antalya, Turkey, 2005.

Shokouhmand, H. & Atashkadi, P.: Performance improvement of a single, flashing, binary, combined cycle for geothermal power plants. *Energy* 22:7 (1997), pp. 637–643.

Simsek, S., Yildirim, N. & Gülgör, A.: Development and environmental effects of the Kızıldere geothermal power project, Turkey. *Geothermics* 34 (2005), pp. 239–256.

Swandaru, R.B.: *Modeling and optimization of possible bottoming units for general single-flash geothermal power plants.* MSc Thesis, Department of Mechanical and Industrial Engineering University of Iceland, Reykjavík, Iceland, 2009.

Tarcan, G.: Mineral saturation and scaling tendencies of waters discharged from wells (>150°C) in geothermal areas of Turkey. *J. Volcanol. Geoth. Res.* 142 (2005), pp. 263–283.

TGNA_1, Turkish Grand National Assembly, Law Number: 5084, Official Newspaper: 25365, 06.02. 2004. http://www.sanayi.gov.tr/Files/Mevzuat/yatirimlarin_ve_istihdaml-13042010211153.pdf (accessed December 2012).

TGNA_2, Turkish Grand National Assembly, Law Number: 6094, Official Newspaper: 27809, 08.01.2011. http://www.enver.org.tr/modules/mastop_publish/files/files_4ed5f2332448e.pdf (accessed December 2012).

Toksoy, M., Serpen, U. & Aksoy, N.: Performance monitoring in the geothermal power plants. *Proceedings Geothermal Energy Seminar-TESKON 2007*, İzmir, Turkey, 2007.

Wallace, K., Dunford, T., Ralph, M. & Harvey, W.: Aegean steam: the Germencik dual flash plant. *GeoFund – IGA Geothermal Workshop "Turkey 2009"*, 16–19 February 2009, Istanbul, Turkey, 2009.

Yildirim, N. & Olmez, E.: Hydrochemical relationship between new drilled wells and production wells in the Kizildere field. *Proceedings Raw Material Sources in Western Anatolia Symposium Baksem 1999*, İzmir, Turkey, 1999.

CHAPTER 12

Scaling problem of the geothermal system in Turkey

Irmak Doğan, Mustafa M. Demir & Alper Baba

12.1 INTRODUCTION

In recent years, the energy needs of countries such as Turkey has gradually increased in parallel with economic and population growth. Almost 90% of energy production in the world relies on fossil fuels such as petroleum, coal, natural gas, and nuclear energy (Birol, 2011; Chandrasekharam and Bundschuh, 2008). However, the increasing awareness of hazardous effects of fossil fuels on the environment entails usage/production of clean and renewable energy source such as geothermal. The geothermal energy is one of the most reliable, clean, and safe way to produce high-yield energy (Etemoglu *et al.*, 2007).

Geothermal energy is the thermal energy generated and stored in the earth and the geothermal source provides the most reliable, sustainable, cost-effective, and environment friendly method among the other energy sources. It is widely used for many applications such as power generation, district heating system, chemical production, snow melting, fish industry, and thermal tourism. However, geothermal brine can be extremely difficult to handle in geothermal operations. This mixture of water, elements, and gases contains enormous amounts of energy for power production. The high temperature solution of elements and compounds, however, causes operational limitations in geothermal power plants. These limitations are due to the severe scaling and corrosion of geothermal brine and steam (Figs. 12.1 and 12.2). Therefore, power plants face curtailment and even complete shutdown. Geothermal brine causes a variety of operational problems. The main and dominant one is scaling. As a rough approximation, it can be defined as the occurrence of an insoluble ionic solid, called deposit, when certain cations and anions are combined. Apart from the accumulation of deposits in the geothermal systems, scaling triggers additional operational problems including equipment damage and failure, equipment repair and replacement, brine leaks and spills, well and line plugging, reduced steam/brine flow, power production losses, and reduction of heat transfer effectiveness. Different types of scaling are found in various geothermal areas and sometimes, even within the various wells of the same site. Calcium carbonate is the most common form of scale deposition attributable to ground water. However, the major types of scaling products in geothermal wells in Turkey are typically calcium sulfate (Akyol, 2009; Atamanenko *et al.*, 2002), calcium carbonate (Chong *et al.*, 2001; Gal *et al.*, 1996), barium sulfate (Barouda *et al.*, 2007; Jones *et al.*, 2003), calcium oxalate (Akin *et al.*, 2008), strontium sulfate (He *et al.*, 1995; Yeboah *et al.*, 1994), and colloidal iron oxides (Dubin *et al.*, 1984) and silica (Mavredaki *et al.*, 2005) scales, all of which can cause consequential losses of equipment efficiency. Calcium compounds frequently encountered are calcium carbonate and calcium silicate. Metal silicate and metal sulfide scales are often observed in higher temperature resources. Typical metals associated with silicate and sulfide scales include zinc, iron, lead, magnesium, antimony, and cadmium. Silica forms an amorphous silica scale that is not associated with other cations.

Geothermal fluids are very close to saturation with respect to calcite. It becomes more soluble when the temperature of the solution decreases and when dissolved carbon dioxide (CO_2) is present in the solution. Degassing of such water causes a sharp increase in pH of the

Figure 12.1.   Some representative views from geothermal fields in Turkey.

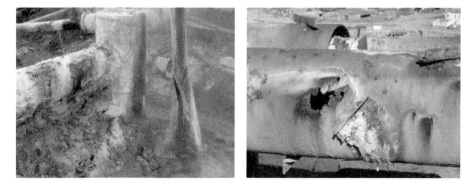

Figure 12.2.   Some representative corrosion and scaling examples from geothermal fields in Turkey.

medium. The activity of the carbonate ion (CO_3^{2-}) increases significantly, leading to oversaturation (Arnórsson, 1989):

$$CaCO_3\ (s) \rightleftharpoons Ca^{2+}\ (aq) + CO_3^{2-}\ (aq)$$

Cooling, which occurs during depressurization boiling, counteracts the effect of CO_2 degassing with respect to the state of calcite saturation (Arnórsson, 1989). The release of CO_2 and deposition of calcite are conveniently expressed as:

$$Ca^{2+}\ (aq) + 2HCO_3^-\ (aq) \rightleftharpoons CaCO_3\ (s) + CO_2\ (g) + H_2O(l) \tag{12.1}$$

Carbonate-based deposits have frequently been observed in geothermal sites all over the world.

Siliceous scales are commonly observed in Indonesia, Awibengkok, Silangkitang, Philippines, and Mak-Ban (Bulalo) geothermal fields (Benevidez *et al.*, 1988; Gunderson *et al.*, 1995; Murray *et al.*, 1995) and metal silicate scales are observed in Djibouti, Japan, Iceland, California and Dixie Valley, Greece, Coso Hot Springs, Salton Sea, and Nevada geothermal fields (Gallup and Reiff, 1991; Yokoyama *et al.*, 1993).

12.2 GEOTHERMAL ENERGY IN TURKEY

Turkey, placed in one of the most seismically active regions in the world, has considerably high level of geothermal energy potential due to its geological and tectonic setting. Although, Turkey is among the top 10 countries in the world in the abundance of geothermal energy potential, only 3% of its potential is used to produce power. In this respect, to provide sustainable development and to reduce the foreign-energy dependency of the country, the usage of alternative energy sources should be generalized and stimulated (Baba *et al.*, 2006). Turkey is located within the Mediterranean seismic zone, whose complex deformation results from the continental collision between the African and Eurasian plates (Bozkurt, 2001). The border of these plates constitutes seismic belts marked by younger volcanics and active tectonism while the latter allows the circulation of water, as well as geothermal heat. The distribution of hot springs in Turkey roughly parallels the distribution of the fault systems, young volcanism, and hydrothermally altered areas (Simsek *et al.*, 2002). The first geothermal researches and investigations were started in Turkey by the General Directorate of Mineral Research and Exploration (MTA) in the 1960s. There are more than 1500 thermal and mineral water spring groups in the country (Baba, 2012; MTA, 1980; Şimşek, 2009; Simsek *et al.*, 2002) (Fig. 12.3). Most of the these geothermal fluid originates from Menderes massif metamorphic, which discharges from the rims of east–west-trending faults that form Büyük Menderes (BMG), Küçük Menderes (KMG), Gediz (GG), and Simav grabens (SG) in the western part of Turkey. The bottom temperatures of geothermal wells reach up to 287°C in Alaşehir in Gediz graben (GG) and 242°C in Denizli-Kızıldere site of Büyük Menderes graben (BMG). Also there are numerous hot springs that align along E to W direction of Küçük Menderes graben (KMG), GG, and BMG with discharge temperatures between 25 and 100°C (Baba, 2012; Karakus and Simşek, 2012).

Geothermal resources are used in various applications such as electricity generation, greenhouse, district heating, industrial processes, thermal tourism, and balneology in Turkey. Most of the geothermal sites have low- to medium-enthalpy fields, which are suitable for direct-use applications such as district heating, greenhouse heating, thermal facilities, and balneology (Baba, 2012; Baba and Armannsson, 2006; Şimşek, 2009). The installed power capacity is 2705 MW_t for direct-use and 243.35 MW_e for power production in Turkey (Baba, 2012; 2013; Şimşek, 2009).

Many geothermal power plants such as Kızıldere (77.4 MW_e), Dora-1 (7.36 MW_e), Dora-2 (11.2 MW_e), Gürmat (47.4 MW_e), Bereket (7.5 MW_e), Maren (18.5 MW_e), and Tuzla (7.5 MW_e) have been producing energy in the western Anatolia since 1968 (Fig. 12.4). Geothermal energy has been used extensively for thermal application in western Anatolia since Romans (Fig. 12.5).

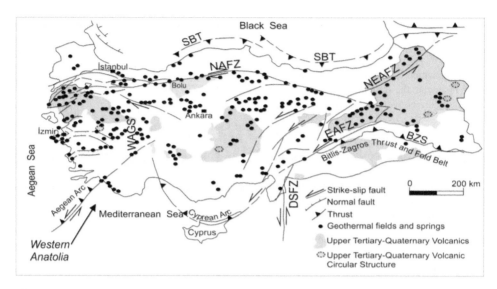

Figure 12.3.   Tectonic map of the eastern Mediterranean region showing structures developed during the Miocene to Holocene time and distribution of geothermal areas around Turkey (compiled from Simsek *et al.*, 2002; Yigitbas *et al.*, 2004). (SBT, Southern Black Sea Thrust; NAFZ, North Anatolian Fault Zone; NEAFZ, Northeast Anatolian Fault Zone; EAFZ, Eastern Anatolian Fault Zone; WAGS, Western Anatolian Graben System; DSF, Dead Sea Fault Zone; BZS, Bitlis-Zagros Suture) (Baba and Armannsson, 2006).

Figure 12.4.   Geothermal power plant in western Turkey.

Figure 12.5.   Application of geothermal system for district heating in western Turkey.

Figure 12.6. Greenhouse application in western Turkey.

Most of the development is achieved in geothermal direct usage in Turkey. 201,000 residences have been using geothermal heating (2084 MW$_t$) including district heating, thermal facilities and 2,300,000 m^2 geothermal greenhouse heating (Fig. 12.6) (Şimşek, 2009). Some cities are heated by geothermal energy, such as Afyon, (Diyadin) Ağrı, (Kizilcahamam) Ankara, (Gonen) Balikesir, (Balcova) Izmir, Kirsehir, (Simav) Kutahya, (Kozakli) Nevsehir, (Salihli) Manisa, and (Saraykoy) Denizli. Medium- and high-enthalpy geothermal fields of Turkey contain 1.0–2.5% of CO_2 by weight of fluid. CO_2 production from geothermal resources reached 120,000 kilotons per year (Serpen *et al.*, 2009a). All these geothermal systems have scaling problem.

12.3 SCALING IN GEOTHERMAL SYSTEM OF TURKEY

Scaling problem is seen at different geothermal sites/areas in Turkey. Calcite, aragonite, dolomite, partly amorphous silica, strontianite, and barite are the mineral scales most often seen (Aktan, 2002; Baba and Armansson, 2006; Baba and Sözbilir, 2012; Baba *et al.*, 2009; Buyuksagis and Erol, 2013; Gökgöz, 1998; Lindal and Kristmannsdottir, 1989; Serpen *et al.*, 2009b; Simsek, 1985; Tarcan *et al.*, 2012). Among them, siliceous scaling is arguably one of the most difficult ones to manage in geothermal systems. Compared to commonly observed carbonate-based deposits in Turkey, siliceous scale is harder, less soluble, and tenacious. The source of silicate deposits is, not surprisingly, the reservoir that is usually rich in terms of silicic acid (Zhang, 2011).

As brine flows through the well to the surface, the temperature and pressure of the brine decreases. The pressure decreases as CO_2 is released as a vapor phase. Removal of CO_2 makes the medium slightly basic and causes to increase pH, therefore the silica solubility decreases correspondingly. At flash vessel, steam flashes and the temperature of the brine further decreases. The brine phase becomes more concentrated and the silica, already unstable, becomes even more unstable. Under these conditions, silica either precipitates as amorphous silica or it reacts with available cations (e.g., Fe, Mg, Ca, and Zn) and forms silicate deposits. Scale forms *via* condensation polymerization. Silicic acid undergoes cascade condensation reactions as the brine approaches to well head. This sol–gel reaction is catalyzed in basic medium that is the case in geothermal systems. In consequence of the formation of critical mass, silicate scale precipitates. The reaction is catalyzed by hydroxide ion so the reaction rate increases at pH 6.0–8.0 (Demadis *et al.*, 2005).

Polymerization of silica requires a balance of un-ionized and ionized silicate species. At higher pH, this balance is shifted toward the ionization of silicic acid, and the polymerization of silica tends to slow down while the potential for the formation of calcium-magnesium silicates increases (Gill, 1993).

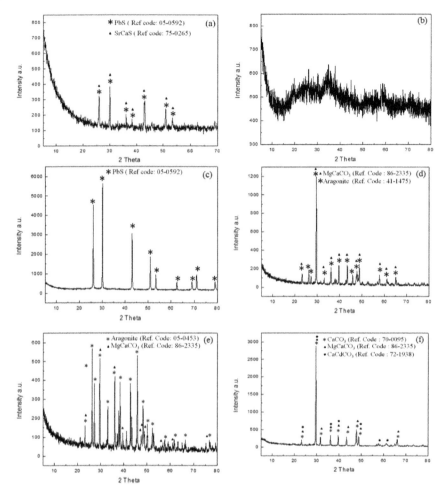

Figure 12.7.   XRD profiles of representative scales derived from (a) Tuzla (Canakkale) well, (b) Tuzla vaporizer, (c) Tuzla capillary pipeline, (d) Afyon well, (e) Afyon well, and (f) Kızıldere (Denizli) in BMG.

The results of crystallographic and morphological investigation of representative deposits obtained from three different geothermal regions of Western Turkey are presented. The samples from Afyon, Kızıldere (Denizli), and Tuzla (Çanakkale) were analyzed employing X-ray diffraction (XRD, Fig. 12.7) and scanning electron microscopy (SEM, Fig. 12.8), respectively. The elemental composition of the scales was determined by XRF (Table 12.1). Table 12.1 shows that samples from Kızıldere (Denizli) and Afyon mostly contain calcite ($CaCO_3$). In addition,

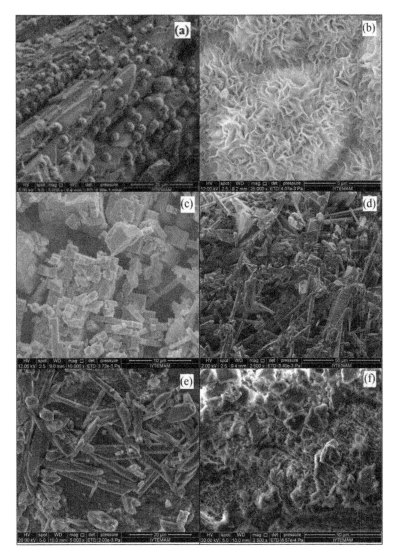

Figure 12.8.   SEM images of scales derived from (a) Tuzla well, (b) Tuzla vaporizer, (c) Tuzla capillary pipeline, (d) Afyon well I, (e) Afyon well II, and (f) Kızıldere (Denizli) in BMG.

the samples from Afyon contain magnesium silicate scaling. On the other hand, the samples from Tuzla (Çanakkale) contain silicate and calcite. X-ray diffractograms show sharp reflections indicating large crystallites in the structure of deposit. Based on the detailed investigation of the reflections, it can be claimed that none of the deposits is composed of pure compound; rather they are a mixture of different inorganic minerals. Table 12.2 presents the types of scales seen at different geothermal fields. While $CaCO_3$-type scales are seen in Afyon and Kızıldere (Denizli), siliceous scales are obtained in Tuzla (Çanakkale) geothermal field where reservoir rocks of geothermal system are volcanic. Kizildere geothermal field (Denizli) that is located on BMG system consists of 1–1.5% dissolved CO_2 (Simsek, 1985). When the partial pressure of CO_2 drops, calcite-scaling forms in this region where reservoir rock is marble. Calcite scale causes a reduction in well-bore radius, thereby decreasing the production of the well bore in these regions. XRD analysis of these graben shows that calcite scaling is dominate in this region (see Fig. 12.7f).

Table 12.1.   Elemental composition of scaling products (XRF results) from (a) Tuzla (Canakkale) well, (b) Tuzla vaporizer, (c) Tuzla capillary pipeline, (d) Afyon well I, (e) Afyon well II, and (f) Kızıldere (Denizli) in BMG.

Element	Concentration (%)					
	a	b	c	d	e	f
Ca	6.63	1.67	1.42	66.10	62.00	53.32
Fe	11.92	16.54	5.98	1.62	3.12	0.12
Mg	1.35	4.03	1.58	1.40	0.02	5.63
Al	0.58	0.57	0.66	0.07	0.15	0.68
Na	0.66	1.04	1.97	0.69	0.85	1.08
Mn	0.68	1.40	0.12	0.06	0.02	0.00
Si	7.00	17.77	7.44	0.00	0.55	4.22
Pb	5.10	0.00	49.42	0.00	0.00	0.00
S	0.23	0.07	3.80	0.26	0.13	0.11
Sr	0.17	0.04	0.02	0.80	1.59	2.78

Table 12.2.   Types of scales seen in geothermal fields of Turkey.

Geothermal field	Types of scales (major/minor)
Tuzla (Çanakkale)	$PbS/SrCaCO_3$ (well) Fe–Mg silicate/$CaCO_3$ (vaporizer) Fe–Mg–Ca silicate (capillary pipeline)
Afyon	Calcite, aragonite/$MgCaCO_3$
Kızıldere (Denizli)	Calcite, aragonite/Mg–Al silicate, $CaCdCO_3$, $MgCaCO_3$

In addition, calcite and aragonite scaling can be seen in central part of Turkey such as Afyon region, where geothermal fluid is extensively used for thermal and district heating system (see Figs. 12.7d,e). On the other hand, silicate-type scaling (Figs. 12.7a–c) is seen in geothermal system in northwestern Turkey (around Tuzla region), where geothermal fluid is composed of hypersaline brine.

SEM images of the scales support the results obtained from XRD. The morphology of PbS (panel a), Fe–Mg layered hydroxides (panel b), and $CaCO_3$ (panel c–e) of Figure 12.8 is revealed by micrographs. SEM-EDX analyses indicate that the samples from Tuzla well include PbS and SrCaS (Figs. 12.7a and 12.8a), Tuzla vaporizer mostly composes of Fe–Mg silicate and includes $CaCO_3$ and PbS (Figs. 12.7b and 12.8b), Tuzla capillary pipeline mostly composed of PbS and includes Fe–Mg–Ca silicate and $CaCO_3$ (Figs. 12.7c and 12.8c). Indeed, the samples from Afyon well are composed of aragonite and $MgCaCO_3$ (Figs. 12.7d,e and 12.8d,e). On the other hand, the sample from BMG (around Denizli) composes of Mg–Al silicate, $CaCO_3$, $MgCaCO_3$, and $CaCdCO_3$ (Figs. 12.7f and 12.8f).

12.4   CONCLUSION

Different types of scales have been found in various geothermal areas of western Turkey. The major species of scale in geothermal brine typically include calcium, silica, and sulfide compounds. Kızılıdere (Denizli), mainly the formation of carbonate scaling can be observed. Calcite and aragonite are the main scaling around Afyon region. Saponite/hectorite like amorphous structure as main component, layered double hydroxide, and iron silicide is seen in Tuzla (Çanakkale) geothermal system. Carbonate-rich scaling can be readily reduced by commercially available inhibitors. Since this type of scaling has been commonly observed in Turkey as well as all around the world, the knowledge on this scaling seems to have matured, and efficient inhibitors are currently available. However, the formation of siliceous scaling is not as well understood as that of carbonate. There is room to acquire more knowledge about the formation mechanism of siliceous scaling and accordingly the methods for its inhibition and/or suppression.

ACKNOWLEDGMENTS

The authors acknowledge The Centers of Materials Research at Izmir Institute of Technology for assistance in all measurements.

REFERENCES

Akin, B., Öner, M., Bayram, Y. & Demadis, K.D.: Effects of carboxylate-modified "green" inulin biopolymers on the crystal growth of calcium oxalate. *Crystal Growth Design* 8:6 (2008), pp. 1997–2005.
Akyol, E., Öner, M. & Demadis, K.D.: Systematic structural determinants of the effects of tetraphosphonates on gypsum crystallization. *Crystal Growth Design* 9:12 (2009), pp. 5145–5154.
Amjad, Z.: *The science and technology of industrial water treatment.* CRC Press, Boca Raton, FL, 2010.
Arkan, S., Akin, S. & Parlaktuna, M.: Effect of calcite scaling on pressure transient analysis of geothermal wells. *Proceedings Twenty-Seventh Workshop on Geothermal Reservoir Engineering*, 28–30 January 2002, Stanford University, Stanford, CA, 2002, Sgp-Tr-171.
Arnórsson, S.: Deposition of calcium carbonate minerals from geothermal waters–theoretical considerations. *Geothermics* 18:1–2 (1989), pp. 33–39.
Atamanenko, I. & Kryvoruchko, A.: Study of the $CaSO_4$ deposits in the presence of scale inhibitors. *Desalination* 147:1–3 (2002), pp. 257–262.
Baba, A.: Present energy status and geothermal utilization in Turkey. *IAH 2012 Congress*, Niagara, Canada, 2012, p. 401.
Baba, A.: Environmental impact of the utilization of geothermal areas in Turkey. *Geopower TURKEY Conference*, Istanbul, Turkey, 2013, pp. 1–61.
Baba, A. & Ármannsson, H.: Environmental impact of the utilization of a geothermal area in Turkey. *Energy Sources* 1 (2006), pp. 267–278.
Baba, A. & Sözbilir, H.: Source of arsenic based on geological and hydrogeochemical properties of geothermal systems in western Turkey. *Chem. Geol.* 334 (2012), pp. 346–377.
Baba, A., Yuce, G., Deniz, O. & Ugurluoglu, Y.D.: Hydrochemical and isotopic composition of Tuzla Geothermal (Canakkale-Turkey) field and its environmental impacts. *Environ. Forensic* 10 (2009), pp. 144–161.
Barouda, E., Demadis, K.D., Freeman, S.R., Jones, F. & Ogden, M.I.: Barium sulfate crystallization in the presence of variable chain length aminomethylenetraphosphonates and cations (Na^+ or Zn^{2+}). *Crystal Growth Design* 7 (2007), pp. 321–327.
Benevidez, P.J., Mosby, M.D., Leong, J.K. & Navarro, V.C.: Development and performance of the Bulalo geothermal field. *Proceedings 10th New Zealand Geothermal Workshop*, Auckland, New Zealand, 1988, pp. 55–60.
Birol, F.: World Energy Outlook, International Energy Agency, IEA, Paris, France, 2011.
Bozkurt, E.: Neotectonics of Turkey – A synthesis. *Geodin. Acta* 14 (2001), pp. 3–30.
Buyuksagis, A. & Erol, S.: The examination of Afyonkarahisar's geothermal system corrosion. *J. Mater. Eng. Perform.* 22 (2013), pp. 563–573.
Chandrasekharam, D. & Bundschuh, J.: *Low-enthalpy geothermal resources for power generation.* Taylor & Francis Group, London, UK, 2008.
Chong, T.H. & Sheikholeslami, R.: Thermodynamics and kinetics for mixed calcium carbonate and calcium sulfate precipitation. *Chem. Eng. Sci.* 56 (2001), pp. 5391–5400.
Demadis, K.D. & Mavredaki, E.: Green additives to enhance silica dissolution during water treatment. *Environ. Chem. Lett.* 3 (2005), pp. 127–131.

Dubin, L. & Fulks, K.E.: The role of water chemistry on iron dispersant performance. *CORROSION 84*, NACE, New Orleans, LA, 1984, Paper no. 118.

Etemoglu, A.B. & Can, M.: Classification of geothermal resources in Turkey by exergy analysis. *Renew. Sustain. Energy Rev.* 11:7 (2007), pp. 1596–1606.

Gal, J.Y., Bollinger, J.C., Tolosa, H. & Gache, N.: Calcium carbonate solubility: a reappraisal of scale formation and inhibition. *Talanta* 43 (1996), pp. 1497–1509.

Gallup, D.L. & Reiff, W.M.: Characterization of geothermal scale deposits by Fe-57 Mrssbauer spectroscopy and complementary X-ray diffraction and infrared studies. *Geothermics* 20 (1991), pp. 207–224.

Gill, J.S.: Inhibition of silica—silicate deposit in industrial waters. *Colloid Surf. A: Physicochem. Eng. Asp.* 74 (1993), pp. 101–106.

Gunderson, R.P., Dobson, P.F., Sharp, W.D., Pudjianto, R. & Hasibuan, A.: Geology and thermal features of the Sarulla contract area, north Sumatra, Indonesia. *Proceedings World Geothermal Congress 1995*, Florence, volume 2, 1995, pp. 687–692.

He, S., Oddo, J.E. & Tomson, M.B.: The nucleation kinetics of strontium sulfate in NaCl solutions up to 6 m and 90°C with or without inhibitors. *J. Colloid Interface Sci.* 174 (1995), pp. 327–335.

Jones, F., Stanley, A., Oliveira, A., Rohl, A.L., Reyhani, M.M., Parkinson, G.M. & Ogden, M.I.: The role of phosphonate speciation on the inhibition of barium sulfate precipitation. *J. Crystal Growth* 249 (2003), pp. 584–593.

Karakuş H. & Şimşek, Ş.: Spatial variations of carbon and Helium isotope rations in geothermal fluids of Buyuk Menderes Graben. *5th Geochemistry Symposium*, 23–25 May 2012, Pamukkale University, Denizli, Turkey, 2012, pp. 82–83.

Mavredaki, E., Neofotistou, E. & Demadis, K.D.: Inhibition and dissolution as dual mitigation approaches for colloidal silica fouling and deposition in process water systems: functional synergies. *Ind. Eng. Chem. Res.* 44 (2005), pp. 7019–7026.

MTA: Hot and mineral water inventory. General Directorate of Mineral Research and Exploration (MTA), MTA Rap., Ankara, Turkey, 1980 (in Turkish).

MTA: Geothermal inventory of Turkey. General Directorate of Mineral Research and Exploration (MTA) publication. Ankara, Turkey, 2005 (in Turkish).

Murray, L.E., Rohrs, D.T., Rossknecht, T.G., Aryawijaya, R. & Pudyastuti, K.: Resource evaluation and development strategy, Awibengkok field. Proc. *Proceedings World Geothermal Congress 1995*, Florence, volume 3, 1995, pp. 1525–1530.

Serpen, U., Aksoy, N. & Öngür, T.: Geothermal industry's 2009 present status in Turkey. *Proceedings of TMMOB 2nd Geothermal Congress of Turkey*, Ankara, Turkey, 2009a, pp. 55–62.

Serpen, U., Aksoy, N., Ongur, T., Yucel, M. & Kayan, I.: Geoscientific investigations on north of Balçova geothermal system in Turkey. *Int. J. Geol.* 4:3 (2009b), pp. 87–96.

Simsek, S.: Geothermal energy development possibilities in Turkey. *NUMOW Conference on "Geothermal Energy in Turkey"*, 1 October 2009, Potsdam, Germany, 2009, pp. 1–6.

Simsek, S., Yildirim, N., Simsek, Z.N. & Karakus, H.: Changes in geothermal resources at earthquake regions and their importance. *Proceedings of Middle Anatolian Geothermal Energy and Environmental Symposium*, Ankara, 2002, pp. 1–13.

Yeboah, Y.D., Saeed, M.R. & Lee, A.K.K.: Kinetics of strontium sulfate precipitation from aqueous electrolyte solutions. *J. Crystal Growth* 135 (1994), pp. 323–330.

Yigitbas, E., Elmas, A., Sefunc, A. & Ozer, N.: Major neotectonic features of eastern Marmara region, Turkey: development of the Adapazari-Karasu corridor and its tectonic significance. *Geol. J.* 39:2 (2004), pp. 179–198.

Yokoyama, T., Sato, Y., Maeda, Y., Tarutani, T. & Itoi, R.: Siliceous deposits formed from geothermal water. I. The major constituents and the existing states of iron and aluminum. *J. Geochem.* 27 (1993), pp. 375–384.

Zhang, B.R., Chen, Y.N. & Li, F.T.: Inhibitory effects of poly(adipic acid/amine-terminated polyether D230/diethylenetriamine) on colloidal silica formation. *Colloid Surf. A Physicochem. Eng. Asp.* 385 (2011), pp. 11–19.

CHAPTER 13

Exergetic and exergoeconomic aspects of ground-source (geothermal) heat pumps in Turkey

Arif Hepbasli & Ebru Hancioglu Kuzgunkaya

13.1 INTRODUCTION

The demand for energy has recently increased rapidly as a result of the world's population increase and industrial growth. On the contrary, fossil fuel supply is declining due to depletion of the resources. Moreover, utilization of fossil fuel energy resources has resulted in adverse effects on the global environment. In this regard, there is an urgent need to implement the use of sustainable and environmentally clean energy sources. In this context, renewable energy sources, including geothermal sources, can play a critical role in meeting the energy demands of societies (Ganjehsarabi *et al.*, 2012).

Among the various renewable energy systems, geothermal heat pump or ground-coupled heat pump (GCHP) systems have been spotlighted as an efficient building energy system because of its great potential for reducing energy utilization in building air-conditioning and hence decreasing CO_2 emissions. Ground-heat exchangers (GHE), earth heat exchangers (EHE), or GCHPs are heat exchangers used for the exploitation of the ground thermal capacity and the difference in temperature between ambient air and ground. A GCHP is usually an array of buried pipes installed either horizontally or vertically into the ground. They use the ground as a heat source when operating in the heating mode and as a heat sink when operating in the cooling mode, with a fluid, usually air, water, or a water-antifreeze mixture, to transfer the heat from or to the ground. Generally, GCHP systems are much more energy efficient than conventional air-source heat pump systems. A higher coefficient of performance (COP) can be achieved by a GCHP because the terrestrial source and sink temperature is relatively constant compared to air temperature. The GCHP system mainly comprises a heat pump unit and a system for exchanging heat with the ground. The heat exchanging system can be configured as either a closed (GCHP) or open loop (groundwater heat pump, GWHP) and the loop itself can be either vertical or horizontal (Chunga and Choib, 2012; Florides *et al.*, 2013).

Dincer (2002) reported the linkages between energy and exergy, exergy and the environment, energy and sustainable development, and energy policy making and exergy in detail. He provided the following key points to highlight the importance of the exergy and its essential utilization in numerous ways: (i) it is a primary tool in best addressing the impact of energy-resource utilization on the environment, (ii) it is an effective method using the conservation of mass and conservation of energy principles together with the second law of thermodynamics for the design and analysis of energy systems, (iii) it is a suitable technique for furthering the goal of more efficient energy-resource use, for it enables the locations, types, and true magnitudes of wastes and losses to be determined, (iv) it is an efficient technique revealing whether or not and by how much it is possible to design more efficient energy systems by reducing the inefficiencies in existing systems, and (v) it is a key component in obtaining a sustainable development.

An exergy analysis (second-law of thermodynamics) has proven to be a powerful tool in the simulation thermodynamic analyses of energy systems. In other words, it has been widely used in the design, simulation, and performance evaluation of energy systems. Exergy analysis method is employed to detect and to evaluate quantitatively the causes of the thermodynamic imperfection of the process under consideration. It can, therefore, indicate the possibilities of thermodynamic improvement of the process under consideration, but only an economic analysis can decide the expediency of a possible improvement (Rosen and Dincer, 2003a; Szargut *et al.*, 1998).

The concepts of exergy, available energy, and availability are essentially similar. The concepts of exergy destruction, exergy consumption, irreversibility, and lost work are also essentially similar. Exergy is also a measure of the maximum useful work that can be done by a system interacting with an environment which is at a constant pressure P_0 and a temperature T_0. The simplest case to consider is that of a reservoir with heat source of infinite capacity and invariable temperature T_0. It has been considered that maximum efficiency of heat withdrawal from a reservoir that can be converted into work is the Carnot efficiency (Rosen and Dincer, 2003b; Rosen *et al.*, 2005).

In this chapter, the main relations used to analyze ground-source heat pump (GSHP) systems both exergetically and exergoeconomically are given. The studies previously conducted in Turkey are also presented in tabulated forms for comparison purposes.

13.2   ENERGETIC, EXERGETIC, AND EXERGOECONOMIC RELATIONS

For a general steady-state, steady-flow process, the four balance equations (mass, energy, entropy, and exergy) are applied to find the work and heat interactions, the rate of exergy decrease, the rate of irreversibility, and the energy and exergy efficiencies (Balkan *et al.*, 2005; Cornelissen, 1997; Dincer *et al.*, 2004; Wall, 2003).

13.2.1   *Mass, energy, entropy, and exergy balances*

The mass balance equation can be expressed in the rate form as:

$$\sum \dot{m}_{in} = \sum \dot{m}_{out} \tag{13.1}$$

where $\dot{m}$ is the mass flow rate and the subscripts "in" and "out" stand for inlet and outlet, respectively.

The general energy balance can be expressed below as the total energy input rates equal to total energy output rates:

$$\sum \dot{E}_{in} = \sum \dot{E}_{out} \tag{13.2a}$$

with all energy rate terms as follows:

$$\dot{Q} + \sum \dot{m}_{in} h_{in} = \dot{W} + \sum \dot{m}_{out} h_{out} \tag{13.2b}$$

In the absence of electricity, magnetism, surface tension, and nuclear reaction, the total exergy rate of a system $\dot{E}x$ can be divided into four components, namely (i) physical exergy rate, $\dot{E}x^{PH}$, (ii) kinetic exergy rate, $\dot{E}x^{KN}$, (iii) potential exergy rate, $\dot{E}x^{PT}$, and (iv) chemical exergy rate, $\dot{E}x^{CH}$ (Dincer et al., 2004):

$$\dot{E}x = \dot{E}x^{PH} + \dot{E}x^{KN} + \dot{E}x^{PT} + \dot{E}x^{CH} \tag{13.3}$$

Although exergy is extensive property, it is often convenient to work with it on a unit of mass or molar basis. The total specific exergy on a mass basis may be written as follows:

$$ex = ex^{PH} + ex^{KN} + ex^{PT} + ex^{CH} \tag{13.4}$$

The general exergy balance can be written in the rate form as follows:

$$\sum \dot{E}x_{in} - \sum \dot{E}x_{out} = \sum \dot{E}\dot{x}_{dest} \tag{13.5a}$$

or

$$\dot{E}\dot{x}_{heat} - \dot{E}\dot{x}_{work} + \dot{E}\dot{x}_{mass,in} - \dot{E}\dot{x}_{mass,out} = \dot{E}\dot{x}_{dest} \tag{13.5b}$$

with

$$Ex_{heat} = \Sigma \left(1 - \frac{T_0}{T_k}\right) \dot{Q}_k \qquad (13.6a)$$

$$Ex_{work} = \dot{W} \qquad (13.6b)$$

$$Ex_{mass,in} = \Sigma \dot{m}_{out} \psi_{out} \qquad (13.6c)$$

$$Ex_{mass,out} = \Sigma \dot{m}_{out} \psi_{out} \qquad (13.6d)$$

where $\dot{Q}_k$ is the heat transfer rate through the boundary at temperature T_k at location k and $\dot{W}$ is the work rate.

The flow (specific) exergy is calculated as follows:

$$\psi = (h - h_0) - T_0(s - s_0) \qquad (13.7)$$

where h is enthalpy, s is entropy, and the subscript zero indicates properties at the restricted dead state of P_0 and T_0.

The rate form of the entropy balance can be expressed as:

$$\dot{S}_{in} - \dot{S}_{out} + \dot{S}_{gen} = 0 \qquad (13.8)$$

where the rates of entropy transfer by heat transferred at a rate of $\dot{Q}_k$ and mass flowing at a rate of $\dot{m}$ are $\dot{S}_{heat} = \frac{\dot{Q}_k}{T_k}$ and $\dot{S}_{mass} = \dot{m}s$, respectively.

Taking the positive direction of heat transfer to be to the system, the rate form of the general entropy relation given in Equation (13.8) can be rearranged to give

$$\dot{S}_{gen} = \dot{m}_{out}s_{out} - \dot{m}_{in}s_{in} - \frac{\dot{Q}_k}{T_k} \qquad (13.9)$$

It is usually more convenient to find $\dot{S}_{gen}$ first and then to evaluate the exergy destroyed or the irreversibility rate $\dot{I}$ directly from the following equation, which is called Gouy-Stodola relation (Szargut, 2005):

$$\dot{I} = \dot{Ex}_{dest} = T_0 \dot{S}_{gen} \qquad (13.10)$$

The specific exergy (flow exergy) of an incompressible substance (i.e., water) is given by (Szargut, 2005):

$$\psi_w = C \left(T - T_{0-} - T_0 \ln \frac{T}{T_0}\right) \qquad (13.11)$$

The total flow exergy of air is calculated from (Wepfer, 1979):

$$\psi_{a,t} = (C_{p,a} + \omega C_{p,v})T_0[(T/T_0) - 1 - \ln(T/T_0)] + (1 + 1.6078\omega)R_a T_0 \ln(P/P_0)$$
$$+ R_a T_0\{(1 + 1.6078\omega) \ln[(1 + 1.6078\omega_0)/(1 + 1.6078\omega)] + 1.6078\omega \ln(\omega/\omega_0)\} \qquad (13.12)$$

where the specific humidity ratio is:

$$\omega = \dot{m}_v/\dot{m}_a \qquad (13.13)$$

Assuming air to be a perfect gas, the specific physical exergy of air is calculated by the following relation (Kotas, 1995):

$$\psi_{a,per} = C_{p,a} \left(T - T_0 - T_0 \ln \frac{T}{T_0}\right) + R_a T_0 \ln \frac{P}{P_0} \qquad (13.14)$$

13.2.2   *Energy and exergy efficiencies*

Numerous ways of formulating exergetic (or exergy or second-law) efficiency (effectiveness or rational efficiency) for various energy systems are given in detail elsewhere (Cornelissen, 1997). It is very useful to define efficiencies based on exergy (sometimes called *second law efficiencies*). While there is no standard set of definitions in the literature, two different approaches are generally used—one is called "*brute-force (or universal)*" while the other is called "functional" (DiPippo, 2004):

1. A "*brute-force*" *exergy efficiency* for any system is defined as the ratio of the sum of all output exergy terms to the sum of all input exergy terms.
2. A "*functional*" *exergy efficiency* for any system is defined as the ratio of the exergy associated with the desired energy output to the exergy associated with the energy expended to achieve the desired output.

Here, in a similar way, exergy efficiency is defined as the ratio of total exergy output to total exergy input, i.e.:

$$\varepsilon = \frac{\dot{E}x_{\text{output}}}{\dot{E}x_{\text{input}}} = 1 - \frac{\dot{E}x_{\text{dest}}}{\dot{E}x_{\text{input}}} \tag{13.15}$$

where "output or out" stands for "net output" or "product" or "desired value" or "benefit," and "input or in" stands for "given" or "used" or "fuel."

It is clear that the brute-force definition can be applied in a straightforward manner, irrespective of the nature of the component, once all exergy flows have been determined. The functional definition, however, requires judgment and a clear understanding of the purpose of the system under consideration before the working equation for the efficiency can be formulated (DiPippo, 2004).

13.2.3   *Exergetic improvement potential*

Van Gool (1997) has also proposed that maximum improvement in the exergy efficiency for a process or system is obviously achieved when the exergy loss rate or irreversibility rate ($\dot{E}x_{\text{in}} - \dot{E}x_{\text{out}}$) is minimized. Consequently, he suggested that it is useful to employ the concept of an exergetic "*improvement potential*" when analyzing different processes or sectors of the economy. This improvement potential in the rate form, denoted $I\dot{P}$, is given by:

$$I\dot{P} = (1 - \varepsilon)(\dot{E}x_{\text{in}} - \dot{E}x_{\text{out}}) \tag{13.16}$$

13.2.4   *Some thermodynamic parameters*

Thermodynamics analysis of GCHP systems may also be performed using the following parameters (Xiang and Cali, 2004):

- Fuel depletion ratio:

$$\delta_i = \frac{\dot{I}_i}{\dot{F}_T} \tag{13.17}$$

- Relative irreversibility:

$$\chi_i = \frac{\dot{I}_i}{\dot{I}_T} \tag{13.18}$$

- Productivity lack:

$$\xi_i = \frac{\dot{I}_i}{\dot{P}_T} \tag{13.19}$$

- Exergetic factor:

$$f_i = \frac{\dot{F}_i}{\dot{F}_T} \tag{13.20}$$

13.2.5 *Exergoeconomic analysis relations*

Exergoeconomics is a combination of exergy and economics. This analysis is one of the best suitable methods for design, simulation, analysis, and performance improvement studies of energy conversion systems. There are various types of exergoeconomic analysis methods. Here, exergy, cost, energy, and mass (EXCEM) method was presented. In this regard, thermodynamic losses are considered to be either one of two types. These are described in Equations (13.2a) and (13.5a) as differential forms of the thermodynamic balances.

Energy losses can be identified directly from the energy balances in Equations (13.2a) and (13.2b). For convenience, the energy loss rate for a system is denoted in the present analysis as $\dot{L}_{en}$ (loss rate based on energy). As there is only one loss term, the "waste energy output" in Equation (13.2b) (Rosen and Dincer, 2003b) is given by:

$$\dot{L}_{en} = \text{Waste energy output rate} \tag{13.21}$$

Exergy losses can be identified from the exergy balance in Equation (13.5a). There are two types of exergy losses, namely the "waste exergy output," which represents the loss associated with exergy that is emitted from the system, and the "exergy consumption," which represents the internal exergy loss due to process irreversibilities. These two exergy losses sum to the total exergy loss. Hence, the loss rate based on exergy $\dot{L}_{ex}$ is defined as (Rosen and Dincer, 2003b):

$$\dot{L}_{ex} = \text{Exergy consumption rate} + \text{waste exergy output rate} \tag{13.22}$$

The capital cost is defined here using the cost balances in Equation (13.23) and is denoted by K. Capital cost is simply that part of the cost generation attributable to the cost of equipment:

$$K = \text{Capital cost of equipment} \tag{13.23}$$

For a thermal system operating normally in a continuous steady-state, steady-flow process mode, the accumulation terms in Equations (13.1)–(13.5a) are zero. Hence all losses are associated with the already discussed terms $\dot{L}_{en}$ and $\dot{L}_{ex}$. The energy and exergy loss rates can be obtained through the following equations (Rosen and Dincer, 2003b):

$$\dot{L}_{en} = \sum_{\text{inputs}} \dot{E} - \sum_{\text{products}} \dot{E} \tag{13.24}$$

and

$$\dot{L}_{ex} = \sum_{\text{inputs}} \dot{E}x - \sum_{\text{products}} \dot{E}x \tag{13.25}$$

where the summations are over all input streams and all product output streams.

Another parameter, $\dot{R}$, is used as the ratio of thermodynamic loss rate $\dot{L}$ to capital cost K as follows (Tsatsaronis and Park, 2002):

$$\dot{R} = \frac{\dot{L}}{K} \tag{13.26}$$

The value of $\dot{R}$ generally depends on whether it is based on energy loss rate in which case it is denoted ($\dot{R}_{en}$), or exergy loss rate ($\dot{R}_{ex}$), as follows:

$$\dot{R}_{en} = \frac{\dot{L}_{en}}{K} \tag{13.27}$$

and

$$\dot{R}_{ex} = \frac{\dot{L}_{ex}}{K} \tag{13.28}$$

Values of the parameter $\dot{R}$ based on energy loss rate and on total internal and external exergy loss rates are considered.

13.3   EXERGETICALLY AND EXEGOECONOMICALLY ANALYZED GSHPS

13.3.1   *Exergetically analyzed GSHP systems*

Heat pumps are preferred and widely used in many applications due to their high utilization efficiencies compared to conventional heating and cooling systems. There are two common types of heat pumps: air-source heat pumps and ground-source (geothermal) heat pumps (GSHPs). The main components of these systems are a compressor, a condenser, an expansion valve, and an evaporator. In addition, circulating pumps and fans are used. GSHPs are exergetically studied in more detail and relevant exergy relations are presented in a tabulated form, while studies conducted by various investigators are compared. The relations given in these sections may be used in modeling and exergetic evaluation for heat pump systems.

GSHPs have been used for years in the developed countries due to their higher energy utilization efficiencies compared with conventional heating and cooling systems. There are two main types of GSHP system: (i) ground-coupled (vertical or horizontal) closed loop and (ii) water source open loop. GCHPs are known by a variety of names. These include GSHPs, earth-coupled heat pumps, earth energy heat pumping systems, earth energy systems, ground-source systems, geothermal heat pumps, closed-loop heat pumps, solar energy heat pumps, geoexchange systems, geosource heat pumps, and a few other variations (Hepbasli, 2004).

Figure 13.1 illustrates a schematic diagram of a GSHP system in the heating mode as an illustrative example (Hepbasli, 2005). The system was designed and installed in the Solar Energy

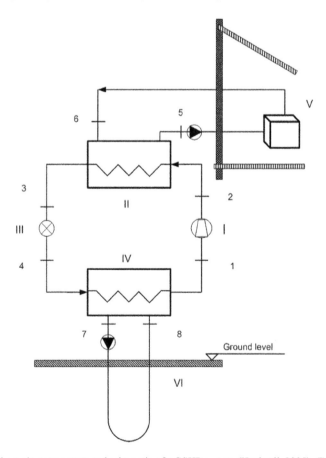

Figure 13.1.   The main components and schematic of a GSHP system (Hepbasli, 2005). (I) compressor; (II) condenser; (III) capillary tube; (IV) evaporator; (V) fan-coil unit; and (VI) GHE.

Table 13.1. Mass, energy, exergy, and entropy balance relations of a GSHP system (Hepbasli and Akdemir, 2004).

No	Element	Mass analysis using Equation (13.1) (the conversion of mass principle)	The irreversibility or the exergy destroyed using Equation (13.5a) (the exergy balance)	The irreversibility or the exergy destroyed using entropy balance and Equation (13.10) (the decrease of exergy principle)
I	Compressor	$\dot{m}_1 = \dot{m}_2 = \dot{m}_r$ (13.29)	$\dot{E}x_1 + \dot{W}_{comp,e} = \dot{E}x_{2a} + \dot{I}_1$ (13.30) $\dot{I}_1 = \dot{m}_r (\psi_1 - \psi_{2a}) + \dot{W}_{comp}$ (13.31)	$\dot{I}_1 = T_0 \dot{m}_r (s_1 - s_{2a})$ (13.32)
II	Condenser	$\dot{m}_1 = \dot{m}_2 = \dot{m}_r$ (13.33) $\dot{m}_5 = \dot{m}_6 = \dot{m}_w$ (13.34)	$\dot{E}x_{2a} + \dot{E}x_6 = \dot{E}x_3 + \dot{E}x_5 + \dot{I}_{II}$ (13.35) $\dot{I}_{II} = \dot{m}_r (\psi_{2a} - \psi_3) + \dot{m}_w (\psi_6 - \psi_5)$ (13.36)	$\dot{I}_{II} = T_0 [\dot{m}_r (s_3 - s_{2a}) + \dot{m}_w (s_5 - s_6)]$ (13.37)
III	Throttling valve (capillary tube)	$\dot{m}_3 = \dot{m}_4 = \dot{m}_r$ (13.38)	$\dot{E}x_3 = \dot{E}x_4 + \dot{I}_{III}$ (13.39) $\dot{I}_{III} = \dot{m}_r (\psi_3 - \psi_4)$ (13.40)	$\dot{I}_{III} = T_0 \dot{m}_r (s_4 - s_3)$ (13.41)
IV	Evaporator	$\dot{m}_4 = \dot{m}_1 = \dot{m}_r$ (13.42)	$\dot{E}x_4 + \dot{E}x_8 = \dot{E}x_1 + \dot{E}x_7 + \dot{I}_{IV}$ (13.43) $\dot{I}_{IV} = \dot{m}_r (\psi_4 - \psi_1) + \dot{m}_{wa} (\psi_8 - \psi_7)$ (13.44)	$\dot{I}_{IV} = T_0 [\dot{m}_r (s_1 - s_4) + \dot{m}_{wa} (s_7 - s_8)]$ (13.45)
V	Fan-coil units	$\dot{m}_{air,in} = \dot{m}_{air,out} = \dot{m}_{air}$ (13.46)	$\dot{E}x_5 = \dot{E}x_6 + \dot{Q}_{fc} \left(\frac{T_{in,air}-T_0}{T_{in,air}}\right) + \dot{I}_V$ (13.47) $\dot{I}_V = \dot{m}_w (\psi_5 - \psi_6) - \dot{Q}_{fc} \left(1 - \frac{T_0}{T_{in,air}}\right)$ (13.48)	$\dot{I}_V = T_0 \left[\dot{m}_w (s_6 - s_5) + \frac{\dot{Q}_{fc}}{T_{in,air}}\right]$ (13.49)
VI	Ground-heat exchanger	$\dot{m}_7 = \dot{m}_8 = \dot{m}_{wa}$ (13.50)	$\dot{E}x_7 + \dot{Q}_{gh} \left(\frac{T_{soil}-T_0}{T_{soil}}\right) = \dot{E}x_8 + \dot{I}_{VI}$ (13.51) $\dot{I}_{VI} = \dot{m}_{wa} (\psi_7 - \psi_8) + \dot{Q}_{gh} \left(1 - \frac{T_0}{T_{soil}}\right)$ (13.52)	$\dot{I}_{VI} = T_0 \left[\dot{m}_{wa} (s_8 - s_7) - \frac{\dot{Q}_{gh}}{T_{soil}}\right]$ (13.53)

Institute of Ege University, Izmir, Turkey. The system was commissioned in May 2000, while GSHP systems have been applied to the Turkish residential buildings since 1997.

The performance tests have also been performed since then. This system has been recently integrated with a solar collector and a greenhouse (Ozgener and Hepbasli, 2005a). It mainly consists of three separate circuits: (i) the ground coupling circuit (brine circuit or water-antifreeze solution circuit), (ii) the refrigerant circuit (or a reversible vapor compression cycle), and (iii) the fan-coil circuit (water circuit).

Equations (13.29)–(13.53) used in modeling and analyzing GSHP systems are summarized in Table 13.1 (Hepbasli and Akdemir, 2004). These relations have been obtained for each of the GSHP components illustrated in Figure 13.1 using mass, energy, entropy, and exergy balance equations as well as the exergy destructions obtained using exergy balance equations given in Section 13.2.1. For further details, some relevant studies published should be taken into account (Akpinar and Hepbasli, 2007; Hepbasli, 2005; Hepbasli and Akdemir, 2004; Ozgener and Hepbasli, 2005a).

Using Equation (13.15), exergy efficiency relations for the GSHP unit and the whole system are written as follows:

$$\varepsilon_{o,R,HP} = \frac{\dot{Ex}_{heat}}{\dot{W}_{a,in}} = \frac{\dot{Ex}_{cond,in} - \dot{Ex}_{cond,out}}{\dot{W}_{a,in}} \tag{13.54}$$

$$\varepsilon_{o,R,sys} = \frac{\dot{Ex}_{heat}}{\dot{W}_{a,in} + \dot{W}_{pumps} + \dot{W}_{fan-coil}} \tag{13.55}$$

$$\varepsilon_{HP} = \frac{COP_{a,h}}{COP_{C,h}} \tag{13.56}$$

$$\varepsilon_{HP} = \frac{\dot{P}_{T,HP}}{\dot{F}_{T,HP}} \tag{13.57}$$

$$\varepsilon_{sys} = \frac{\dot{P}_{T,sys}}{\dot{F}_{T,sys}} \tag{13.58}$$

Esen et al. (2007) constructed a horizontal GCHP system for space heating, which is illustrated in Figure 13.2. The depths of the horizontal GHE1 (or HGHE1) and HGHE2 are 1 and 2 m,

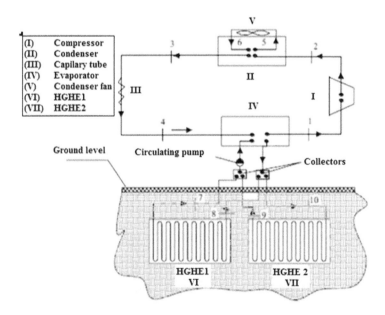

Figure 13.2.　Schematic diagram of the experimental apparatus (Esen *et al.*, 2007).

respectively. The experimental setup consists of three main components, namely (i) horizontal GHEs, (ii) heat pump unit equipment, and (iii) auxiliary equipment. The energy efficiency of GCHP systems were obtained to be 2.5 and 2.8, respectively, while the corresponding exergetic efficiencies of the overall system were found to be 53.1% and 56.3% for HGHE1 and HGHE2, respectively.

13.3.2 *Greenhouses*

Good plant-growth conditions can be achieved by using greenhouses. A greenhouse can be managed to protect the plants by creating a favorable environment, allowing intensive use of soil and helping sanitary plant control. A variety of heating systems are being used in present-day greenhouses to meet the heating and cooling requirements. As an alternative, there is great potential in employing a heat pump system for greenhouse air-conditioning based on its ability to perform the multifunction of heating, cooling, and dehumidification (Ozgener and Hepbasli, 2005a).

Although various studies were undertaken to evaluate the performance of solar-assisted heat pump systems, to the best of author's knowledge, two studies (Ozgener and Hepbasli, 2005a; 2007a) have appeared in the open literature on the experimental performance testing of a solar-assisted vertical GSHP system for greenhouse heating. The first one (Ozgener and Hepbasli, 2005a) was related to an exergetic assessment at a fixed dead state temperature, while the second one (Ozgener and Hepbasli, 2007a) included a parametric study investigating the effect of varying dead state temperature on the exergy efficiencies.

A schematic diagram of the constructed experimental system is illustrated in Figure 13.3 (Ozgener and Hepbasli, 2005a; 2007a). This system mainly consists of three separate circuits as follows: (i) the ground coupling circuit with solar collector (brine circuit or water-antifreeze solution circuit), (ii) the refrigerant circuit (or a reversible vapor compression cycle), and (iii) the fan-coil circuit for greenhouse heating (water circuit).

The exergetic results of these two studies are illustrated in Table 13.2. The highest irreversibility on a system basis occurred in the greenhouse fan-coil unit, followed by the compressor, condenser,

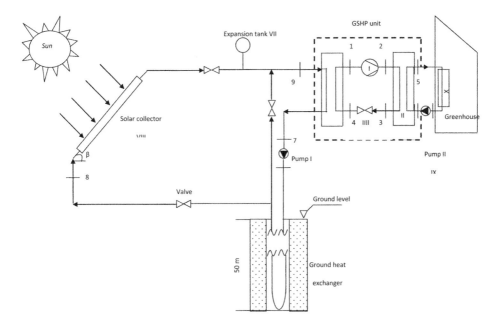

Figure 13.3. A schematic diagram of a solar assisted vertical GSHP system for a greenhouse heating (Ozgener and Hepbasli, 2005a; 2007a).

Table 13.2. Various ground source (geothermal) heat pumps analyzed using exergy analysis method (Akpinar and Hepbasli, 2007; Esen, *et al.*, 2007; Hepbasli, 2005; Hepbasli and Akdemir, 2004; Kuzgunkaya and Hepbasli, 2007; Ozgener and Hepbasli, 2005a; 2007a; 2007b).

Location of the system built/country	Type of heat pump/operating mode	Capacity in the mode operated (kW)	Dead state temperature (°C)	Date of data used	Exergy destructions (rate of improvement potential) (% of the whole heat pump system)[a]								Eq. no for exergy eff.	Exergy efficiency (%)		Reference
					Ground source heat pump system (whole system)[a]											
					Ground source heat pump unit				Other components							
					I Com.	II Con.	III Thr. valve	IV Evap.	V Fan-coil/ Con. WT	VI GHE/ PHEx	VII Cir. pumps	VIII Sol. col.		Heat pump	Overall	
Izmir/ Turkey	Vertical GSHP/heating (for space heating)	3.41	20	February 2001	16.54	27.91	21.62	7.83	21.14				(54) (56)	2.94	3.84	Hepbasli and Akdemir (2004)
Izmir/ Turkey	Vertical GSHP/heating (for space heating)	4.27	2.2	January 7, 2004	51.83 (57.12)	8.00 (3.03)	8.36 (2.43)	9.42 (14.30)	17.67 (17.21)				(54) (55) (57) (58)	9.07 66.8	8.4 66.6	Hepbasli (2005)
Izmir/ Turkey	Solar assisted vertical GSHP/heating (for GH heating)	3.98	10.93	December 17, 2003 to March 17, 2004	28.33	13.85	11.34	8.19	30.22	2.52	1.83 and 3.09	0.63	(57) (58)	71.8	67.7	Ozgener and Hepbasli (2005a)
Izmir/ Turkey	Solar-assisted vertical GSHP/heating (for GH heating)	4.15	−0.69	January 7, 2004	23.06	10.45	6.53	9.79	43.11	2.61	1.24 and 1.89	1.30	(57) (58)	76.2	75.6	Ozgener and Hepbasli (2007a)

Location	System														Reference
Izmir/Turkey	Solar assisted vertical GSHP/Heating (for GH heating)	3.97	1.00	17 Dec. 2003 to 17 March 2004	3.13	17.7	13.68	6.43	49.08	8.61	0.8	0.57	(57) (58)	91.8 86.13	Ozgener and Hepbasli (2007b)
Erzurum/Turkey	GSHP with water souce/Heating (for space heating)[b]	7.02	25		45.92	15.83	6.27	11.69	14.64	5.66			(57) (58)	4.64 1.44	Akpinar and Hepbasli (2007)
Izmir/Turkey	Ground-source heat pump dryer	3.3	25		1.05	0.42	0.26	0.16	0.03	0.04	0.08		(59) (60)	87.48 81.35 / 15.48 9.11	Kuzgunkaya and Hepbasli (2007)
Elazğ/Turkey	Ground-coupled heat pump system	4.279	1	25—26 January 2004	71.8 84	63.3 69	52.3 85.6	12.5 14.5	24.2 27.8	82.71 81.76			(63) (64)	62.95 53.1 / 65.45 56.3	Esen et al. (2007)

[a] I-Com: Compressor, II-Con: Condenser, III-Thr. valve: Throtle (expansion) valve, IV-Evap: Evaporator, V: Fan-coil,/Con. WT: Condenser watertank, VI-GHE: Ground-heat exchanger/PHEx: Plate heat exchanger, VII-Cir. Pumps: Circulating pumps, VIII-Sol. col.: Solar collector; GH: greenhouse.

[b] This system was designed and constructed by Kara (1999) on the base of a PhD thesis in the Mechanical Engineering Department, Ataturk University, Erzurum in Turkey in order to evaluate geothermal resources with low temperatures. The performance of the system was evaluated by Kara and Yuksel (2000) using the energy analysis method only, while this system was evaluated by Akpinar and Hepbasli (2007) using exergy analysis method.

expansion valve and evaporator, subregions I and V for the GSHP unit, and the whole system, respectively. Besides this, the remaining system components have a relatively low influence on the overall efficiency of the whole system. Experiments have also shown that monovalent central heating operation (independent of any other heating system) could not meet the overall heat loss of greenhouse if ambient temperature was very low. The bivalent operation (combined with other heating system) could be suggested as the best solution for Mediterranean and Aegean regions of Turkey, if peak load heating could be easily controlled (Ozgener and Hepbasli, 2005a).

13.3.3 Drying

Large quantities of food products are dried to improve shelf life, reduce packing costs, lower ship-ping weights, enhance appearance, encapsulate original flavor, and maintain nutritional value. In evaluating the performance of food systems, energy analysis method has been widely used, while studies on exergy analysis, especially on exergetic assessment of drying process, are relatively few in numbers (Akpinar, 2004; Akpinar et al., 2005a; 2006; Dincer and Sahin, 2004; Midilli and Kucuk, 2003). In these previous studies, the drying process was thermodynamically modeled by Dincer and Sahin (2004), while drying of different products such as pistachio (Midilli and Kucuk, 2003), red pepper slices (Akpinar, 2004), potato (Akpinar et al., 2005a), pumpkin (Akpinar et al., 2006), wheat kernel (Syahrul et al., 2002), and apple slices (Akpinar et al., 2005b) was evaluated in terms of energetic and exergetic aspects using various drying devices, such as a fluidized bed dryer, a solar drying cabinet, and cyclone type dryers (Kuzgunkaya and Hepbasli, 2007). Based on the previous studies, the first study on exergetic evaluation of a drying process using a GSHP drying system appears to be done by Kuzgunkaya and Hepbasli (2007), who used laurel leaves as a product being dried, and made a comparison of exergetic efficiency values obtained from various studies on food drying systems.

The main objective of this contribution was to perform exergy analyses of the drying process in terms of drying of laurel leaves under different operating conditions for the assessment of the drying performance.

Figure 13.4 illustrates a schematic diagram of this system, which consists of mainly three separate circuits, namely (Kuzgunkaya and Hepbasli, 2007): (I) the ground coupling circuit (brine circuit or water-antifreeze solution circuit), (II) the refrigerant circuit (or a reversible vapor compression cycle), and (III) the drying cabinet circuit (air circuit). The main components of the heat pump system are an evaporator, a condenser, a compressor, and an expansion valve. To avoid freezing the water under the working condition and during the winter, a 10% ethyl glycol mixture by weight was prepared. The refrigerant circuit was built on the closed loop copper tubing. The working fluid was R-22.

There are mainly two ways of formulating exergetic efficiency for drying systems (Akpinar et al., 2005a; Kuzgunkaya and Hepbasli, 2007; Midilli and Kucuk, 2003). The first one can be defined as the ratio of the product exergy to exergy inflow as follows (Kuzgunkaya and Hepbasli, 2007; Midilli and Kucuk, 2003):

$$\eta_{ex} = \frac{\text{Exergy inflow-exergy loss}}{\text{exergy inflow}} = 1 - \frac{\dot{E}x_{loss}}{\dot{E}x_{in}} \tag{13.59}$$

The second one may be defined on the product/fuel basis, as given in Equation (13.60). The product is the rate of exergy evaporation ($\dot{E}x_{evap}$) and the fuel is the rate of exergy drying air entering the dryer chamber ($\dot{E}x_{da}$). In this regard, exergy efficiency may be written as follows (Syahrul et al., 2003):

$$\eta_{ex} = \frac{\dot{E}x_{evap}}{\dot{E}x_{da}} \tag{13.60}$$

with

$$\dot{E}x_{evap} = \left[\left(1 - \frac{T_0}{T_{m2}} \right) \dot{Q}_{evap} \right] \tag{13.61}$$

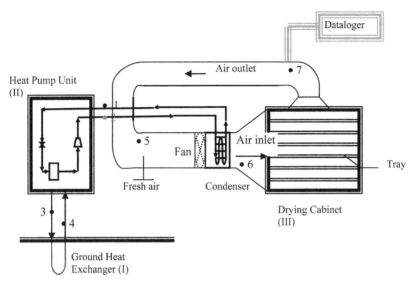

Figure 13.4.   Schematic diagram of the heat pump tray drying system (Kuzgunkaya and Hepbasli, 2007). 1: Refrigerant inlet to the expansion valve, 2: Refrigerant outlet from the compressor, 3: Water inlet to the GHE, 4: Water outlet from ground-heat exchanger, 5: Air inlet to the condenser, 6: Air outlet from the condenser, and 7, Air outlet from the drying chamber.

and

$$\dot{Q}_{evap} = \dot{m}_w h_{fg} \qquad (13.62)$$

where $\dot{Q}_{evap}$ is the heat transfer rate due to phase change, T_{m2} is the exit temperature of the material, $\dot{m}_w$ is the mass flow rate of water, and h_{fg} is the vaporization latent heat.

The laurel leaves were sufficiently dried at the temperatures ranging from 40°C to 50°C with relative humidity varying from 16% to 19% and a drying air velocity of 0.5 m/s during the drying period of 9 h. The exergy efficiency values were obtained to range from 81.35% to 87.48% using Equation (13.58) based on the inflow, outflow, and loss of exergy, and 9.11% to 15.48% using Equation (13.59) based on the product/fuel basis between the same drying air temperatures with a drying air mass flow rate of 0.12 kg/s (Kuzgunkaya and Hepbasli, 2007).

Exergy efficiency values in drying various products are given in Table 13.2 (Kuzgunkaya and Hepbasli, 2007). It is obvious from this table that the exergy efficiency values calculated using Equation (13.60) is quite higher than those using Equation (13.59). It may be concluded that it is more meaningful, objective, and useful to assess the performance of the drying process relative to the performance of similar drying processes on the product (or benefit)/fuel basis (Kuzgunkaya and Hepbasli, 2007).

Table 13.2 illustrates various ground source (geothermal) heat pumps analyzed using exergy analysis method (Akpinar and Hepbasli, 2007; Esen, *et al.*, 2007; Hepbasli, 2005; Hepbasli and Akdemir, 2004; Kuzgunkaya and Hepbasli, 2007; Ozgener and Hepbasli, 2005a; 2007a; 2007b). As can be seen from this table, exergy efficiency values change depending on the equation used. In other words, there is not only one exergy relation to be used. In this regard, when making a comparison between exergy efficiency values of various types of GSHP systems, the relations utilized should be taken into consideration.

13.3.4   *Exergoeconomically analyzed GSHP systems*

Ozgener and Hepbasli (2005b) examined the relations between thermodynamic losses and capital costs for devices and suggested possible generalizations in the relation between thermodynamic losses and capital costs. A systematic correlation appears to exist between exergy loss rate (total or

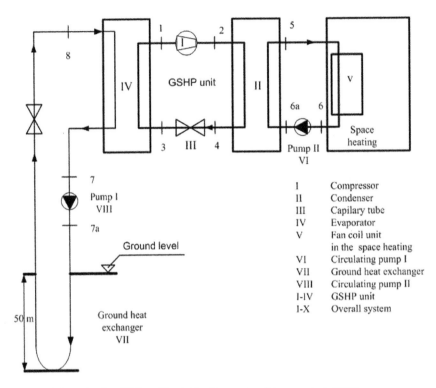

Figure 13.5.   Schematic of a GSHP residential heating system (Ozgener *et al.*, 2007).

internal) and capital cost for solar-assisted ground-source heat pump greenhouse heating system (SAGSHPGHS). Furthermore, a correlation appears to exist between the mean thermodynamic loss rate-to-capital-cost ratios for all of the devices in a SAGSHPGHS and the ratios for the overall system, when the ratio is based on total or internal exergy losses, but not when it is based on energy losses. This correlation may imply that devices in successful SAGSHPGHSs are configured so as to achieve an overall optimal design, by appropriately balancing the thermodynamic (exergy-based) and economic characteristics of the overall system and its devices (Fig. 13.5).

The work discussed in this chapter can likely be extended to marginal costs. Here, the marginal cost would be the cost increase resulting from saving one unit of energy or exergy (i.e., from reducing the energy or exergy loss by one unit). The results would be expected to indicate that marginal costs based on exergy for many devices have similar values, while marginal costs based on energy vary widely.

Figure 13.5 illustrates a schematic diagram of the experimental setup. To avoid freezing the water under the working condition and during the winter, a 10% ethyl-glycol mixture by weight was used (Ozgener *et al.*, 2007). The refrigerant circuit was built on the closed-loop copper tubing (1–4 lines). The working fluid is R-22. The GSHP residential heating system is an air/refrigerant vapor compression heat pump and consists mainly of a compressor with a capacity of 1.4 kW, a 6.66 kW condenser, a 8.2 kW evaporator, an expansion device equipped with a series of capillary tubes with a length of 1.5 m and an inside diameter of 0.0015 m. Besides this, the system has three separate circuits, namely (i) the ground-coupling circuit (brine circuit or water-antifreeze solution circuit), (ii) the refrigerant circuit (or a reversible vapor compression cycle), and (iii) the fan-coil circuit for resident heating (water circuit). In the analysis, thermodynamic quantities were obtained using actual data from the experimental setup. The COP and exergy efficiency of the overall system were determined to be 2.38 and 67.7%, respectively. The loss-to-capital-cost ratio based on exergy, Rex, for the overall GSHP was about 0.30.

13.4   CONCLUDING REMARKS

Exergetic and exergoeconomic directions of GSHPs designed, constructed, and tested in Turkey were considered in this chapter. The relations used to evaluate their performances were given while their exergetic values were listed in the tabulated forms. The main conclusions drawn may be listed as follows.

Although exergy analysis has been used for years in assessing the performance of various energy-related systems, the first exergetic evaluation of a GSHP system in Turkey was performed in 2004 based on the experimental values (Hepbasli and Akdemir, 2004).

There are various ways to describe exergy efficiency in the literature. In this regard, the use of the efficiency definition on the benefit/fuel basis is more convenient than that on the output/input basis.

Exergy analysis is a useful tool for determining the locations, types, and true magnitudes of energy losses, and therefore help in the design of more efficient energy systems. It is also a way to a sustainable development and reveals whether or not (and by how much) it is possible to improve GSHP systems by reducing inefficiencies.

As a conclusion, the authors expect that the analyses and assessments reported here will provide the investigators, government administration, and engineers working in the area of GSHP systems as well as sustainable energy technologies with knowledge about how these systems may be evaluated from the exergetic and exergoeconomic points of view.

REFERENCES

Akpinar, E.K.: Energy and exergy analyses of drying of red pepper slices in a convective type dryer. *Int. Commun. Heat Mass Transfer* 31:8 (2004), pp. 1165–76.
Akpinar, E.K. & Hepbasli, A.: A comparative study on exergetic assessment of two ground-source (geothermal) heat pump systems for residential applications. *Build. Environ.* 42 (2007), pp. 2004–2013.
Akpinar, E.K., Midilli, A. & Bicer, Y.: Energy and exergy of potato drying process via cyclone type dryer. *Energy Convers. Manage.* 46:15–16 (2005a), pp. 2530–2452.
Akpinar, E.K., Midilli, A. & Bicer, Y.: Thermodynamic analysis of the apple drying process. *Proc. IMechE* Part E: *J. Process Mech Eng.* 219 (2005b), pp. 1–14.
Akpinar, E.K., Midilli, A. & Bicer, Y.: The first and second law analyses of thermodynamic of pumpkin drying process. *J. Food Eng.* 72:4 (2006), pp. 320–331.
Balkan, F., Colak, N. & Hepbasli, A.: Performance evaluation of a triple effect evaporator with forward feed using exergy analysis. *Int. J. Energy Res.* 29 (2005), pp. 455–470.
Chunga, J.T. & Choib, J.M.: Design and performance study of the ground-coupled heat pump system with an operating parameter. *Renew. Energy* 42 (2012), pp. 118–124.
Cornelissen, R.L.: *Thermodynamics and sustainable development: the use of exergy analysis and the reduction of irreversibility*. PhD Thesis, University of Twente, Twente, The Netherlands, 1997.
Dincer, I.: The role of exergy in energy policy making. *Energy Policy* 30 (2002), pp. 137–149.
Dincer, I. & Sahin, A.Z.: A new model for thermodynamic analysis of a drying process. *Int. Commun. Heat Mass Transfer* 47 (2004), pp. 645–652.
Dincer, I., Hussain, M.M. & Al-Zaharnah, I.: Energy and exergy use in public and private sector of Saudi Arabia. *Energy Policy* 32:141 (2004), pp. 1615–1624.
DiPippo, R.: Second Law assessment of binary plants generating power from low-temperature geothermal fluids. *Geothermics* 33 (2004), pp. 565–586.
Esen, H., Inalli, M., Esen, M. & Pihtili, K.: Energy and exergy analysis of a ground-coupled heat pump system with two horizontal ground heat exchangers. *Build. Environ.* 42 (2007), pp. 3606–3615.
Ganjehsarabi, H., Gungor, A. & Dincer, I.: Exergetic performance analysis of Dora II geothermal power plant in Turkey. *Energy* 46 (2012), pp. 101–108.
Florides, G., Pouloupatis, P.D., Kalogirou, S., Messaritis, V., Panayides, I., Zomeni, Z., Partasides, G., Lizides, A., Sophocleous, E. & Koutsoumpas, K.: Geothermal properties of the ground in Cyprus. and their effect on the efficiency of ground coupled heat pumps. *Renew. Energy* 49 (2013), pp. 85–89.
Hepbasli, A.: Ground-source heat pumps. In: J. Cutler & C.J. Cleveland (eds): *The encyclopedia of energy*. Academic Press/Elsevier Inc., Vol. 3, 2004, pp. 97–106.

Hepbasli, A.: Thermodynamic analysis of a ground-source heat pump system for district heating. *Int. J. Energy Res.* 7 (2005), pp. 671–687.

Hepbasli, A. & Akdemir, O.: Energy and exergy analysis of a ground source (geothermal) heat pump system. *Energy Convers. Manage.* 45 (2004), pp. 737–753.

Kara, Y.A.: *Utilization of low temperature geothermal resources for space heating by using GHPs.* PhD Thesis, Ataturk University, Erzurum, Turkey, 1999 (in Turkish).

Kara, Y.A. & Yuksel, B.: Evaluation of low temperature geothermal energy through the use of heat pump. *Energy Convers. Manage.* 42 (2000), pp. 773–781.

Kotas, T.J.: *The exergy method of thermal power plants.* Krieger Publishing Company, Malabar, FL, 1995.

Kuzgunkaya, E.H. & Hepbasli, A.: Exergetic evaluation of drying of laurel leaves in a vertical ground-source heat pump drying cabinet. *Int. J. Energy Res.* 31:3 (2007), pp. 245–258.

Midilli, A. & Kucuk H.: Energy and exergy analysis of solar drying process of pistachio. *Energy* 28 (2003), pp. 539–556.

Ozgener, O. & Hepbasli, A.: Experimental performance analysis of a solar assisted ground-source heat pump greenhouse heating system. *Energy Build.* 37 (2005a), pp. 101–110.

Ozgener, O. & Hepbasli, A.: Exergoeconomic analysis of a solar assisted ground-source heat pump greenhouse heating system. *Appl. Therm. Eng.* 25 (2005b), pp. 1459–1471.

Ozgener, O. & Hepbasli, A.: A parametrical study on the energetic and exergetic assessment of a solar assisted vertical ground-source heat pump system used for heating a greenhouse. *Build. Environ.* 42:1 (2007a), pp. 11–24.

Ozgener, O. & Hepbasli, A.: Modeling and performance evaluation of ground source (geothermal) heat pump systems. *Energy Build.* 39:1 (2007b), pp. 66–75.

Ozgener, O., Hepbasli, A. & Ozgener, L.: A parametric study on the exergoeconomic assessment of a vertical ground-coupled (geothermal) heat pump system. *Build. Environ.* 42 (2007), pp. 1503–1509.

Rosen, M.A. & Dincer, I.: Exergy methods for assessing and comparing thermal storage systems. *Int. J. Energy Res.* 27:4 (2003a), pp. 415–430.

Rosen, M.A. & Dincer, I.: Exergoeconomic analysis of power plants operating on various fuels. *Appl. Therm. Eng.* 23 (2003b), pp. 643–658.

Rosen, M.A., Le, M.N. & Dincer, I.: Efficiency analysis of a cogeneration and district energy system. *Appl. Therm. Eng.* 25:1 (2005), pp. 147–159.

Syahrul, S., Hamdullahpur, F. & Dincer, I.: Exergy analysis of fluidized bed drying of moist particles. *Exergy* 2 (2002), pp. 87–98.

Syahrul, S., Dincer, I. & Hamdullahpur, F.: Thermodynamic modeling of fluidized bed drying of moist particles. *Int. J. Therm. Sci.* 42 (2003), pp. 691–701.

Szargut, J.: *Exergy method: technical and ecological applications.* WIT Press, Southampton, Boston, MA, 2005.

Szargut, J., Morris, D.R. & Stewart, F.R.: *Exergy analysis of thermal, chemical, and metallurgical processes.* Edwards Brothers Inc., USA, 1998.

Tsatsaronis, G. & Park, M.: On avoidable and unavoidable exergy destructions and investment costs in thermal systems. *Energy Convers. Manage.* 43 (2002), pp. 1259–1270.

Van Gool, W.: Energy policy: fairly tales and factualities. In: O.D.D. Soares, A. Martins da Cruz, G. Costa Pereira, I.M.R.T. Soares & A.J.P.S. Reis (eds); *Innovation and technology-strategies and policies.* Kluwer, Dordrecht, The Netherlands, 1997, pp. 93–105.

Wall, G.: Exergy tools. *Proceedings of the Institution of Mechanical Engineers, Part A: Journal of Power and Energy* 217 (2003), pp. 125–136.

Wepfer, W.J., Gaggioli, R.A. & Obert, E.F.: Proper evaluation of available energy for HVAC. *ASHRAE Transactions* 85:1 (1979), pp. 214–230.

Xiang, J.Y., Cali, M. & Santarelli, M.: Calculation for physical and chemical exergy of flows in systems elaborating mixed-phase flows and a case study in an IRSOFC plant. *Int. J. Energy Res.* 28 (2004), pp. 101–115.

CHAPTER 14

Application of geophysical methods in Gulbahce geothermal site, Urla-Izmir, western Anatolia

Oya Pamukçu, Tolga Gönenç, Petek Sındırgı & Alper Baba

14.1 INTRODUCTION

The western Anatolian region is considered to be one of the most tectonically active, rapidly deforming, and extending areas in the world (Bozkurt 2001; Dewey and Sengor 1979; Jackson and McKenzie 1984; Şengör *et al.* 1985; Seyitoğlu and Scott 1992) (Fig. 14.1). The region is rich with geothermal potential. Systematic geothermal exploration of the region began in 1960s. Medium- and high-temperature fields in and around Izmir city (Fig. 14.1) have been identified. There are a number of district heating systems, greenhouses, and spa complexes commercially utilizing geothermal energy in the region.

Different geophysical methods have been used to determine the locations of geothermal systems in Turkey. The geophysical studies in a geothermal field are generally undertaken:

- to determine the structure of the field,
- to determine the hydrogeological conditions and thickness of the overburden, and
- to give information about the location and distribution of the geothermal system.

This study is the first geophysical exploration carried out in the region (see the map in Fig. 14.1). Geophysical studies have been performed to assess geothermal resources in the Gulbahce Bay. Gravity, magnetic, self-potential (SP), and resistivity methods were used. The gravity method is controlled by subsurface density contrast. The magnetic method depends on the susceptibility contrast and is highly affected by high temperatures. SP and resistivity methods are related with the electrical conductivity of the medium.

Cretaceous limestone, which has karstic properties and cavities, outcrops in the southeast of study area (Fig. 14.2). Fractured karst limestone and dolomite deeply buried in the Mesozoic system contain the targeted geothermal reservoirs of this region. Hot springs emerge along the fault zones. These springs have been controlled by the fault system. Furthermore, these springs have been affected by the sea intrusion. The surface temperature of these springs range from 30 to 35°C. The main karstic reservoir is cut by tectonic zones. Volcanic units, outcropped in the region, are the heat source of this resource, which is tectonically controlled. Determination of geophysical methods is related to geological features. All parameters such as karstic features, hot spring along the fault zone, and volcanic rocks affect the control of intensity values in microgravity, resistivity values of electrical method, SP anomalies, and magnetic susceptibility effects within the magnetic method. Consequently, gravity, magnetic, self-potential (SP), and resistivity methods were applied around Çiflik plain (Gulbahce) (Fig. 14.1).

14.2 GEOLOGY AND TECTONIC PROPERTIES OF STUDY AREA

The Neogene stratigraphy of the Karaburun peninsula is represented by a volcano-sedimentary succession, including several sedimentary and volcanic units. These units rest on a basement

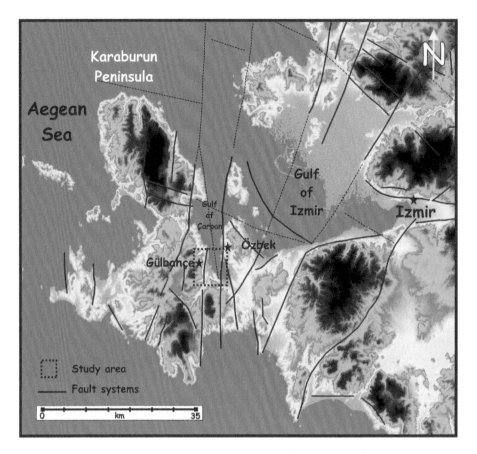

Figure 14.1.   The general tectonic features and the study area within the rectangular region.

comprising non-metamorphic and intensely sheared Paleozoic to Mesozoic rocks of the Karaburun belt (Baba and Sözbilir, 2012; Çakmakoğlu and Bilgin, 2006; Erdoğan, 1990; Tatar-Erkül *et al.*, 2008) (see Figs. 14.2 and 14.3). The Neogene volcano-sedimentary units laterally and vertically pass into the limestone composed of mainly white-colored freshwater limestones (Baba and Sözbilir, 2012). In general, five different types of geologic units have been distinguished in the campus area and its surroundings (see Fig.14.3).

Guvercinlik formation, which consists of laminated stromatolitic dolomite, limestone with megalodons, and red sandstone lenses, is located at the bottom of a stratigraphic sequence (see Fig. 14.3). The Nohutalan formation, composed of dolomitic and grey limestone formation with smooth and well bedding, is overlain on this formation. Nohutalan formation can be seen in the north of Balıklıova, around Barbaros village and İçmeler (Urla). Nohutalan formation was formed in shallow marine platform conditions (Erdoğan, 1990). Neogene-aged Kavaklı formation consists of lacustrine limestones and argillaceous limestones and has tectonic boundary with Nohutalan formation. Kavaklı formation can be seen on high topography of the study area. These rocks are densely jointed. This formation has overlain angular unconformity by Gulbahce volcanic, which is composed of tuff, riyodasite, agglomerate, and latite andesite. These volcanic units were observed around the campus area and are available at relatively high elevations outstretched in direction N–S. All these geological units are overlain by fanglomerate and alluvium in the study area.

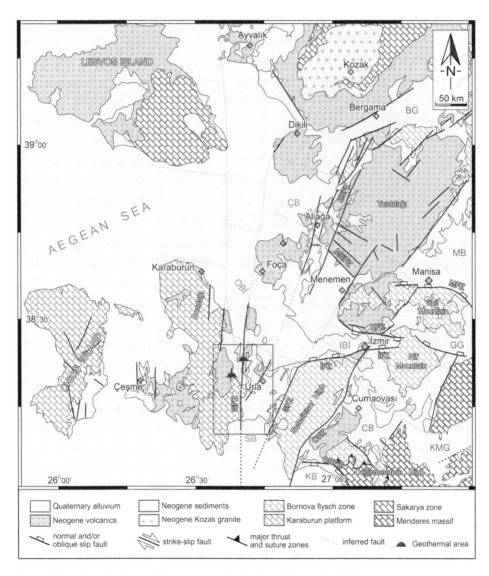

Figure 14.2.   Geological map of Karaburun peninsula (from Uzel *et al.*, 2012).

Karaburun peninsula has a very complex and rapidly changing tectonic structure due to relative motions. This area is characterized by N–S and NE–SW trending active-strike slip faults. Most of the earthquake epicenters are concentrated along the southern part of the Gülbahce-Karaburun. Figure 14.2 shows the direction of fault system in study area. In addition, most of the geothermal springs are located on these tectonic zones (Fig. 14.4). There are some geothermal springs venting on the onshore areas of the Karaburun peninsula. Gülbahce is placed in the western part of the Izmir city where there is a high potential geothermal resource expectation. Gulbahce water types are of Na–Cl type. One of the bore well was done by General Directorate of Mineral Research and Exploration (MTA) in Gulbahce geothermal site and reached 33°C at 300 m in karstic limestone (Fig. 14.4).

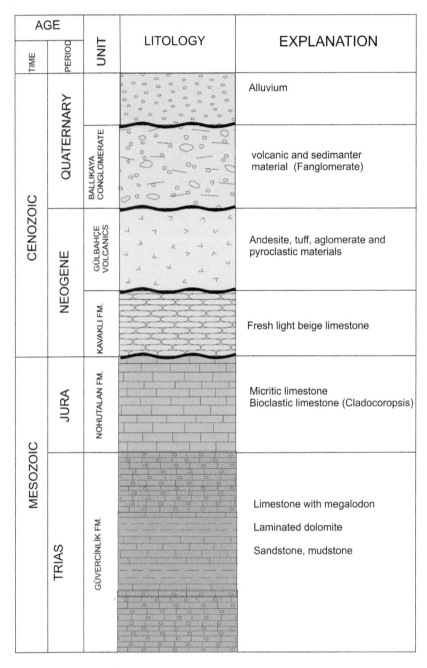

AGE		UNIT	LITOLOGY	EXPLANATION
TIME	PERIOD			
CENOZOIC	QUATERNARY	BALLIKAYA CONGLOMERATE		Alluvium
				volcanic and sedimanter material (Fanglomerate)
	NEOGENE	GÜLBAHÇE VOLCANICS		Andesite, tuff, aglomerate and pyroclastic materials
		KAVAKLI FM.		Fresh light beige limestone
MESOZOIC	JURA	NOHUTALAN FM.		Micritic limestone Bioclastic limestone (Cladocoropsis)
	TRIAS	GÜVERCİNLİK FM.		Limestone with megalodon Laminated dolomite Sandstone, mudstone

Figure 14.3.   Cross section of study area.

14.3   GEOPHYSICAL STUDIES

14.3.1   *Gravity and magnetic*

The study area is approximately 1000 m E–W and 1500 m N–S trending in Çiftlik plain (Gulbahce) located in the boundary of Izmir Institute of Technology (IYTE) (Fig. 14.5).

Figure 14.4.   Hot water spring emerge along the fault zones.

During the data collection phase, gravity measurements were performed by Scintrex CG-5 gravity device. Scintrex Envi-Mag proton magnetometer was used for magnetic measurements. Field sampling was conducted within 50 m. In addition, the distance between each profile was selected as 200 m according to the geological structure. A total of 105-point data were taken from the study area. After gravity work, latitude, free air, and terrain corrections were applied on measured data (Fig. 14.6). The calculated data were used with the first degree trend to determine regional anomaly map (Fig. 14.7) and then the local Bouguer gravity anomaly map was formed (Fig. 14.8b,c). In the next stage of the study, the overall magnetic anomaly map was obtained utilizing the measured values (Fig. 14.9).

14.3.2   *Self-potential*

Self potential profile measurements were performed in the study area by using the gradient technique. The profile lengths are A—615 m, B—255 m, C—660 m, D—220 m, E—200 m, and F—585 m. The sampling interval of profiles A, B, C, and F are 15 m, profiles D and E are 10 m. The extension of profiles A, D, C, and F are NW–SE, D and E are NE–SW (Fig. 14.10).

The SP measurements performed along the six profiles as shown in Figure 14.10 are transformed to a contour map in Figure 14.11. Profiles A, B, D, and E are contoured separately. There is not any profile between Profile C and F because of the rough terrain conditions; so these profiles could not be contoured. The approximate data values at the head, middle, and end of the profile C and F were taken a place in the Figure 14.11. It is observed that the SP anomaly increases to NW and decreased down toward to S and SE and reached the negative values. Profiles A, D, and F, which are suitable for interpretation, structure parameters, were determined by using the sphere model (Equation (14.1)). Profile D was interpreted into two parts. The general equation of the

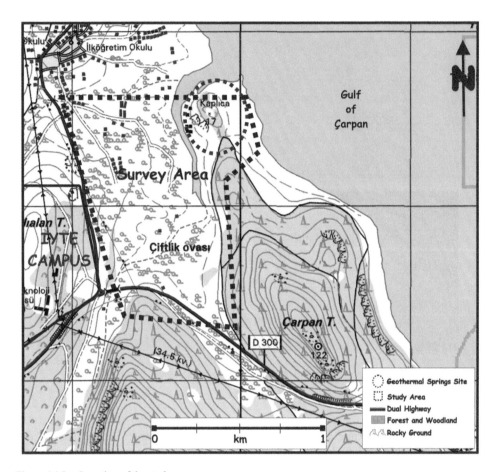

Figure 14.5.   Location of the study area.

SP sphere model (electrical potential occurs at a P(x) point) is defined by (Bhattacharyya and Roy, 1981):

$$V(x) = K\frac{(x - x_o)\cos\alpha - h\sin\alpha}{((x - x_o)^2 + h^2)^{3/2}} \qquad (14.1)$$

where K is the electrical dipole moment ($K = I\rho/2\pi$), h is the depth to the center of sphere, x_0 is the distance from the origin and is the angle between polarization and vertical axes (Fig. 14.12). Figure 14.13a–c shows data interpreted by the inversion technique. Determined parameters are also displayed in Table 14.1.

14.3.3   Vertical electrical sounding method

Vertical electrical sounding (VES) method has been applied by using VES1, VES2, VES3, and VES4 points (Fig. 14.14), choosing $AB/2 = 500$ m. The graphics, observed field VES values, are shown respectively in Figure 14.15. Pseudo and resistivity cross-sections of VES1 and VES2, VES3 and VES4 points evaluations are also presented in Figure 14.16.

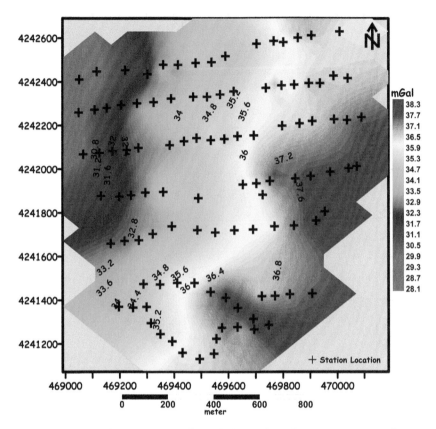

Figure 14.6. Bouguer gravity anomaly map of the calculated values of the study area (in this figure, profiles 1, 2, 3 and 4 – gravity stations between 4242600N-4241600N latitudes – are related with VES profiles and SP measurement points).

14.4   RESULT AND CONCLUSION

Gulbahce geothermal system is located on N–S and NE–SW trending active-strike slip faults. Surface temperature of the springs in the area range between 30°C and 36°C. Gulbahce water types, which have been affected by the seawater, are of Na–Cl type.

Bouguer gravity anomaly map shows that the anomaly values range from 29.9 to 37.7 mgal in the study area (Fig. 14.6). The anomaly values decrease from southeast to the northwest. According to the result of the regional gravity anomaly map (Fig. 14.6) and Bouguer gravity map, which is generated by the calculated values (Fig. 14.7) the anomaly values decrease from southeast to the northwest.

Local (residual) Bouguer gravity anomaly map (Fig. 14.8b) indicates that higher values have been observed approximately N–S in the mid-way directionality in the study area whereas low anomaly values of Bouguer gravity anomaly can be seen approximately northwest–southeast to the northwest (Fig. 14.8b,c). In addition, the total magnetic anomaly values obtained from bath surroundings are reduced in the northwest of study area (Fig. 14.9).

Possible structure, which approximately ranges from 15 to 163 m (Table 14.1), can also be seen in the study of SP (Fig. 14.10). In Figure 14.11, Profile C, the north east of the site, has high negative SP anomaly (ca. −50 mV) that has been affected by geothermal fluid and seawater.

More negative SP anomalies have been observed on Profile F (ca. −24 mV) and the northwest of study area in Profile A in Figures 14.10 and 14.11. Table 14.1 shows SP anomaly values. These

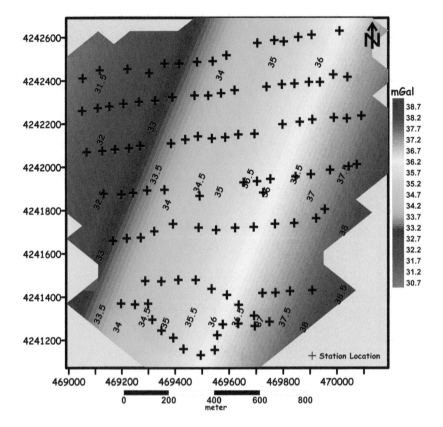

Figure 14.7.   Bouguer gravity anomaly map of the regional values of the study area.

values that are presented in Figure 14.11 show that the depth increases from the southeast to the northwest. These values also support high negative SP anomalies that may have been affected by the deep structure.

According to the VES3 resistivity (Fig. 14.14), points of VES4 anomalies are similar (Fig. 14.15) and in a certain depth of the amplitude decreases. The decreases of VES point may be the result of the Neogene limestones and volcanic rocks (Taşkıran, 2004) that contain geothermal fluid. The distance between points VES1 and VES2 is 195 m, and between VES3 and VES4 is 170 m. VES graphics are shown in Figure 14.15. Pseudo and resistivity cross-sections are also shown in Figure 14.16.

VES1 and VES2 points are very close to the seashore and a geothermal spring. Therefore, resistivity values are very small especially at VES1 point. All resistivity points include alluvium (clay, sand, and gravel) unit in the first meters. According to the VES method, it can be said that alluvium and limestone units are present under VES1 and VES2 points and probably these units. VES3 and VES4 points consist of alluvium and gravel units and gravel units may include probably hot and salt water.

All geophysical data (gravity in Fig. 14.8, magnetic in Fig. 14.9, SP in Fig 14.11, and vertical electrical sounding in Fig. 14.16 applications) point out that those anomaly amplitudes are low especially northwest of study area. These results indicate that possible geothermal system is expanding and deepening in the western block of Gulbahce fault (Fig. 14.17).

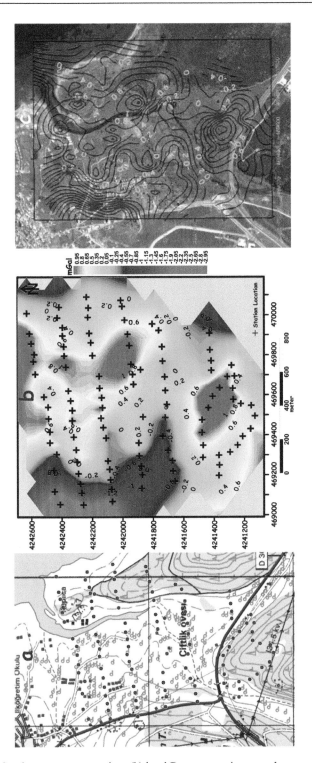

Figure 14.8.   (a) Gravity measurement points, (b) local Bouguer gravity anomaly map, and (c) the overall view of anomalies in field.

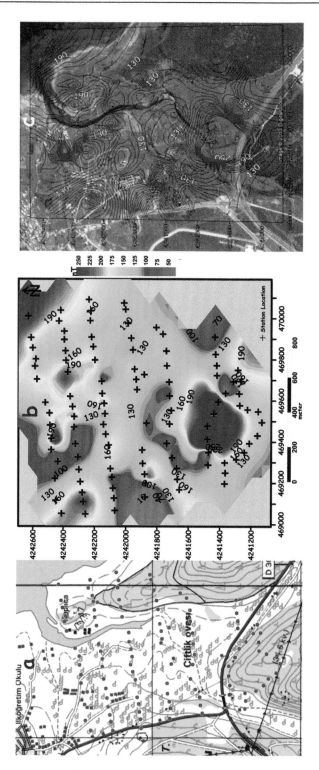

Figure 14.9.   (a) Magnetic measurement points, (b) total magnetic anomaly map, and (c) the overall view of anomalies in field.

Figure 14.10. Self-potential (SP) measurement profiles (A, B, C, D, E, and F).

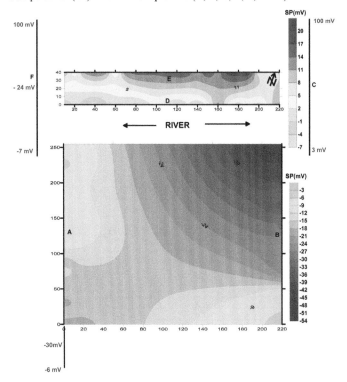

Figure 14.11. SP anomaly map and profiles.

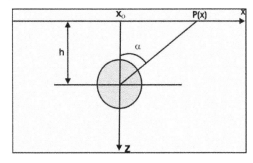

Figure 14.12. Sphere model in self-potential method.

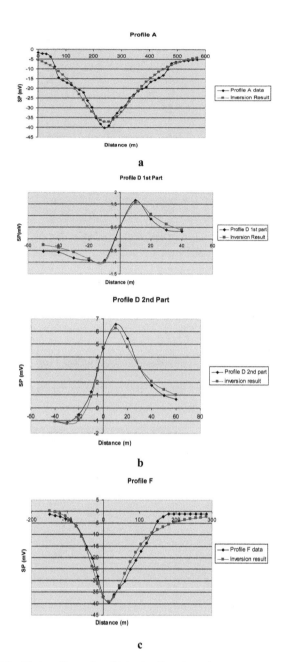

Figure 14.13. (a) SP Profile A and inversion, (b) SP Profile D inversion was evaluated by dividing the curve into two pieces, and (c) SP Profile F and inversion.

Table 14.1. Determined parameters from SP anomalies.

Parameter/Profile name	Profile A	Profile D		Profile F
Polarization angle (α)(°)	98	350	328.5	120
Depth (h) (m)	162.6	15	21.9	79
Distance from origin (m)	240	70	150	150
K (Electric dipole moment)	984977	748.8	4395	266147

Figure 14.14. Location of VES points.

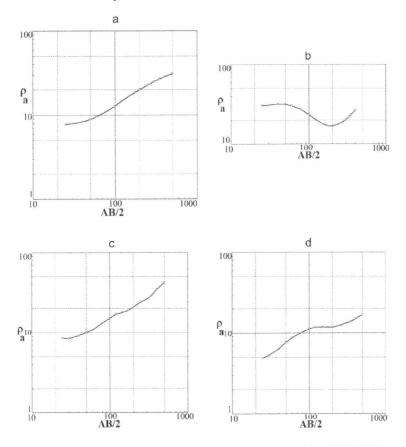

Figure 14.15. The observed field VES values: (a) VES1, (b) VES2, (c) VES3, and (d) VES4 curves.

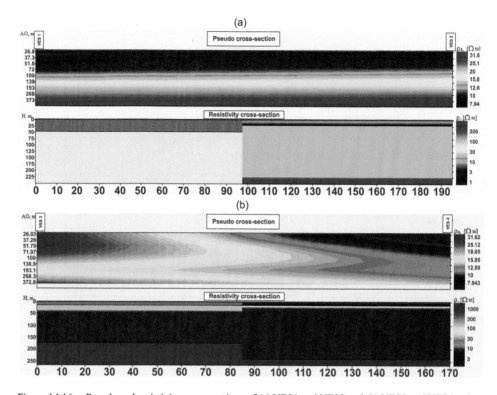

Figure 14.16.   Pseudo and resistivity cross-sections of (a) VES1 and VES2 and (b) VES3 and VES4 points.

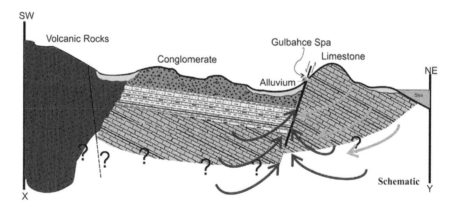

Figure 14.17.   Schematic cross-section of geothermal site.

ACKNOWLEDGMENTS

This study has been achieved under the scope of No: 108Y285 The Scientific and Technological Research Council of Turkey (TUBİTAK) project. We thank our undergraduate students for their field performance and academic staff of Izmir Institute of Technology for the technical support.

REFERENCES

Baba, A. & Sözbilir, H.: Source of arsenic based on geological and hydrogeochemical properties of geothermal systems in western Turkey. *Chem. Geol.* 334 (2012), pp. 346–377.

Bhattacharyya, B.B. & Roy, N.: A note on the use of a nomogram for self-potential anomalies. *Geophys. Prospect.* 29 (1981), pp. 102–107.

Bozkurt, E.: Neotectonics of Turkey – A synthesis. *Geodin. Acta* 14 (2001), pp. 3–30.

Çakmakoğlu, A. & Bilgin, Z.R.: Karaburun Yarımadası' nın Neojen Öncesi Stratigrafisi. *General Directorate of Mineral Research and Exploration (MTA) Journal* 132, 2006, pp. 33–62.

Dewey, J.F. & Şengör, A.M.C.: Aegean and surrounding regions: complex multi-plate and continuum tectonics in a convergent zone. *Geol. Soc. America Bull.* Part 1, 90 (1979), pp. 84–92.

Erdoğan, B.: İzmir-Ankara Zonu İle Karaburun Kuşağının Tektonik İlişkisi. *General Directorate of Mineral Research and Exploration (MTA) Journal Magazine* 110, 1990, pp. 1–15.

Jackson, J. & Mc Kenzie, D.P.: Active tectonics of the Alpine-Himalayan belt between western Turkey and Pakistan. *Geophys. J. Royal Astr. Soc.* C 77 (1984), pp. 185–264.

Şengör, A.M.C., Görür, N. & Şaroğlu, F.: Strike-slip faulting and related basin formation in zones of tectonic escape: Turkey as a case study. In: K. Bıddle & N. Chrıstıe-Blıck (eds): Strike-slip deformation, basin formation and sedimentation. *Society of Economic Paleontologists and Mineralogists, Special Publications*, 37, 1985, pp. 227–264.

Seyitoğlu, G. & Scott, B.C.: Late Cenozoic volcanic evolution of the northeastern Aegean region. *J. Volcan. Geotherm. Res.* 54 (1992), pp. 157–176.

Taşkıran, A.: İzmir Urla Gülbahçe sıcaksu sondajı (GU-1) kuyu bitirme raporu. *General Directorate of Mineral Research and Exploration (MTA) Journal*, MTA 201, 2004, p. 432.

Tatar-Erkül, S., Sözbilir, H., Erkül, F., Helvacı, C., Ersoy, E.Y. & Sümer, Ö.: Geochemistry of I-type granitoids in the Karaburun Peninsula, west Turkey: evidence for Triassic continental arc magmatism following closure of the Palaeotethys. *Island Arc* 17 (2008), pp. 394–418.

Uzel, B., Sözbilir, H. & Özkaymak, Ç.: Neotectonic evolution of an actively growing superimposed basin in western Anatolia: The Inner Bay of İzmir, Turkey. *Turkish J. Earth Sci.* 21 (2012), pp. 439–471.

CHAPTER 15

Palaeoenvironmental and palynological study of the geothermal area in the Gülbahçe Bay (Aegean Sea, western Turkey)

Mine Sezgül Kayseri-Özer, Bade Pekçetinöz & Erdeniz Özel

15.1 INTRODUCTION

The collision of the Arabian and African plates dominates the tectonic framework of the eastern Mediterranean (Jackson and McKenzie, 1984, 1988). In Turkey, this collision had induced the westward escape of the Anatolian block, which is accommodated by the right-lateral North Anatolian Fault on the north and the left-lateral East Anatolian Fault on the south. Back-arc spreading behind the Hellenic and Cyprean arcs combined with the westward escape of the wedge-shaped Anatolian block results in a complex regional extension and transtension in western Turkey, where abundant geothermal activity is focused (Aydin et al., 2005; Şengör et al., 1984; Westaway, 2003) (Fig. 15.1a). Hence the majority of the geothermal activity is observed in the tectonically active Aegean and Marmara regions of western Turkey which are located along or near major E–W striking normal faults and including major graben-bounding faults (Çağlar et al., 2005; Drahor and Berge, 2006; Faulds et al., 2006, 2009; Karamanderesi and Helvacı, 2003; Pfister et al., 1998; Şimşek, 1997; Şimşek and Demir, 1991; Şimşek and Güleç, 1994; Tezcan, 1995; Şimşek et al., 2000a; 2000b) (Fig. 15.1b). It is therefore critical to evaluate the type and position of the faults that are the most favorable for geothermal activity in such regions (Dewey and Şengör, 1979; McKenzie, 1972; Şengör, 1976).

İzmir is part of the graben systems in western Anatolia, and the geothermal potential of the İzmir region and its surroundings is high. There are hot water springs, especially on the Karaburun Peninsula and lands surrounding the İzmir Gulf (provide a map with all the structural features and the locations of the springs). Various geothermal studies suggest that the deposition basin of the thermal area is expanded towards the sea. For this reason, İzmir Gulf and Gülbahçe Bay are important regions for the thermal potential research. The Gülbahçe Bay is a geothermal area in the Aegean Sea (Agostini et al., 2007; Canbolat, 1986; Condrad et al., 1997; Eşder, 1990; Eşder and Şimşek, 1975; 1977; Eşder et al., 1995; Filiz, 1982; Filiz and Tarcan, 1993; Filiz et al., 1997; Innocenti et al., 2005; Tarcan et al., 1999; Yilmaz, 2001; Yılmazer, 2001; 1984). It is located on the north of the İzmir Gulf and it is N–W trending inner gulf (17 km in length and 11 km in width of north, 6.5 km in width of south) (Fig. 15.1c). The deepest part of the bay is approximately 35 m. There are a numerous studies, relating to the geophysics, geology, and palynology, of the Quaternary sediments of Black Sea, Marmara Sea, and Aegean Sea (Aksu et al., 2002; Mudie et al., 2004). However, firstly in this study, palynological evidence of the Quaternary sediments in the Gülbahçe Bay where geothermal activity has been recorded based on the seismic data and bathymetric study is defined (Fig. 15.1c). In addition the vegetational interpretations based on the palynomorphs and nonpollen palynomorphs are summarized.

15.2 GEOLOGICAL SETTING AND HIGH-RESOLUTION SHALLOW SEISMIC STUDY (3.5 KHZ)

There are eight geological units around the Gülbahçe Bay (Fig. 15.2). These units are İçmeler limestone (Jurassic-Early Cretaceous), Demircili Melange (Late Cretaceous), Yağcılar unit,

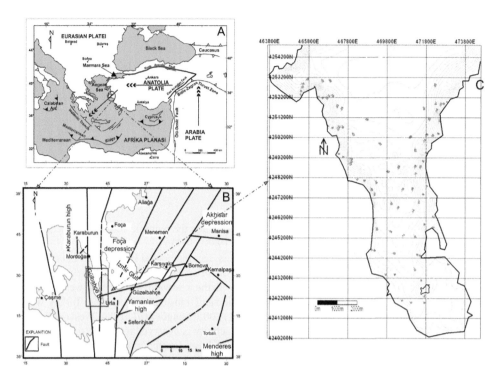

Figure 15.1.   (a) The active tectonic map of Turkey (Okay *et al.*, 1996), (b) structural-stratigraphic segment of the Middle Eastern Aegean Depression (Kaya, 1979), and (c) the bathymetric map of Gülbahçe Bay (Pekçetinöz, 2010).

Kızıltepe and Gülbahçe volcanics, Ortatepe basalts (Neogen), and Ballıkaya conglomerates and alluvium (Quaternary) (Tarcan, 2001).

In Gülbahce Bay, two stratigraphic units were observed as a result of seismic studies (Fig. 15.3). These units were shown as units A and B on seismic sections and interpreted as units A and B due to their acoustic features. Unit A indicates thin layering that is parallel to each other. This unit is divided into subunits A1 and A2 on the basis of acoustic features. Unit A1 (Late Pleistocene-Holocene), the youngest unit, is a poorly regular reflective package with continuous reflection surface. Unit A1 thickens approximately to 4–5 m in the middle of the bay and thins approximately to 0.5–1 m thick towards the coast of the bay. Unit A2 is separated from unit B by nonuniform, wavy reflectors. Units A and B are separated from each other with a high reflector. According to land geology studies, unit A is an alluvial soft sediment package exhibiting thin and smooth layering and is permeable. Unit B, interpreted as acoustic basement, to the bottom has a dispersive reflection effect when compared to its upper reaches and a rough surface. With respect to geological studies in the vicinity, it is considered that unit B, which is characteristic of acoustic basement and has a corrugated appearance, consists of karst limestone (Figs. 15.4 and 15.5). Its corrugated appearance was probably formed in the last glacier period, and the depths of Quaternary soft sediments increased in zones close to shore.

Units A and B are evidently observed in the seismic sections in bay entrance (G5). However, it is followed that unit B has a more regular structure in the outer part of the bay than in the inner part. Accordingly, it is determined that the bay entrance is quite different with regard to the inner and the central part of bay. Because of this difference, the entrance of the bay (G5) is characterized as nonhydrothermal activity and is indicated as "reference zone".

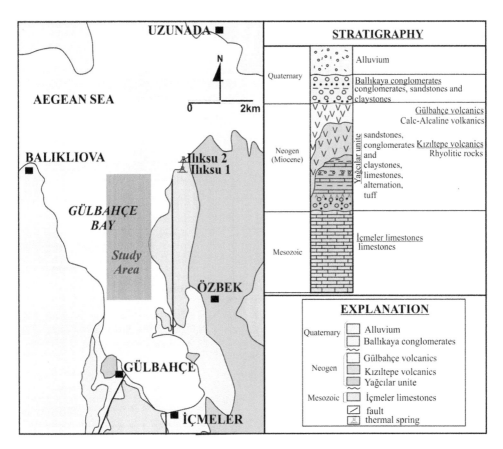

Figure 15.2.   Simplified geological map and stratigraphy of Gülbahçe Bay and the surroundings (Tarcan, 2001).

The most striking feature in seismic sections and bathymetric map is highs situated on the sea bottom and they constitute sudden clustering on bottom morphology (Figs. 15.1c, 15.4, and 15.5) (Pekçetinöz, 2010). These highs are referred to as "Morphological Highs." They prevent the emission of acoustic signal and also occasionally provide deeper penetration of the signal. This situation and the ability to observe of sea bottom reflector under clusters indicate that these highs are formed on sea bottom. It was understood from the sediment samples that these clustering on sea bottom originated from the corals (*Cladocora caespitosa*) that have accumulated. It was detected that the dimensions of these highs representing the conical propagation are between 0.4 and 7.5 m in height and 1.26 and 101 m in width. The structure of morphological highs shows a spreading in an area of approximately 16 km^2.

15.3   IMPORTANT PLANTS AND NONPOLLEN PALYNOMORPHS OF QUATERNARY IN GÜLBAHÇE BAY

15.3.1   *Pollen*

Generally, Fagaceae-*Quercus* are abundant in the palynofloral diagrams of the eastern Mediterranean (van Zeist and Bottema, 1988; 1991). There are two types of Fagaceae-*Quercus*: evergreen (i.e., *Quercus ilex, Q. coccifera* subsp. *Q. calliprinos*) and deciduous (i.e., *Quercus cerris*-type,

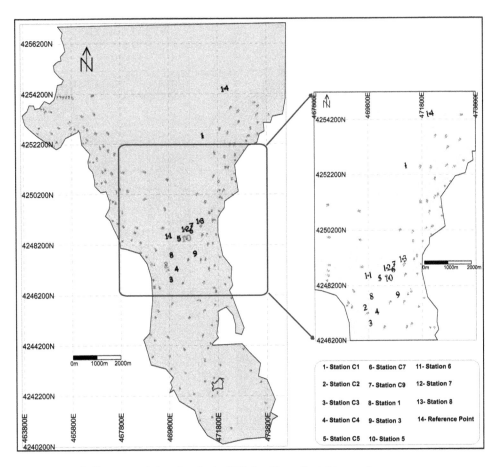

Figure 15.3.   Sediment sample locations in the Gülbahçe Bay collected by gravity corer.

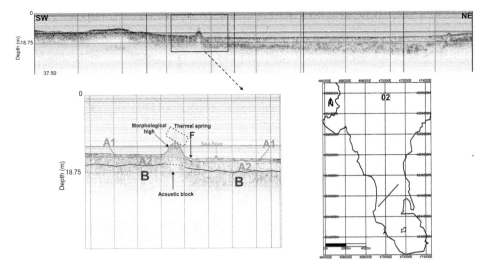

Figure 15.4.   Seismic line 02 and stratigraphic units of the sea floor in the Gülbahçe Bay.

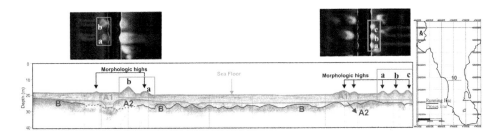

Figure 15.5.   Seismic line 10 in the Gülbahçe Bay.

Q. pubescens, *Q. robur*). Evergreen *Quercus* can tolerate drought stress but it requires significant rainfall in the winter. In addition, *Quercus* type is affected by low winter temperature. On the other hand, deciduous *Quercus* needs arid, semi-arid, and relatively high soil moisture in summer, but this plant can endure the cold, dry winters.

Chenopodiaceae is a typical marker of the steppe vegetation and plant species of this family and shows a widespread distribution in the very dry condition and halophytic marshes. In addition, without drought, the lack of moisture can improve Chenopodiaceae in halophytic marsh around the lake.

Abundance of the *Ephedra* and Chenopodiaceae shows treeless desert–steppe vegetation. Overabundant *Ephedra* can indicate the beginning of the climatic deterioration. Poaceae (Gramineae) produces abundant pollen, easily spread, and it needs spring and summer moisture (300 mm and above). In addition, abundance of this pollen makes up herbaceous steppe.

Gymnosperms (conifers) include *Pinus* spp., *Abies*, *Cedrus libani*, and Cupressaceae (*Juniperus* spp.). In the Mediterranean region, *Pinus*, which can withstand the barrenness of the summer (e.g., *Pinus pinea*, *P. brutia*, *P. halepensis*), is observed in cooler and mountainous areas (*P. nigra*, *P. sylvestris*) and *Abies* that grow in the mountains are common. *Picea* is an element of the European temperate forest and it is the most obvious indicator of a very long distance deployment by air and water.

15.3.2   *Nonpollens*

Laboratory experiments show that *Lingulodinium* populations respond more to nutrients and low turbulence than to salinity, they are extremely euryhaline (Lewis and Hallett, 1997). Cyst populations of the autotroph *Lingulodinium machaerophorum* respond first to a temperature increase for breaking winter dormancy. *Spiniferites* species are common in the Mediterranean-Aegean seas based on the environmental changes (*S. mirabilis, S. hyperacanthus, S. bulloideus, S. delicatus*, and *S. bentorii*) (Mudie *et al.*, 2002).

Pseudoschizaea rubina (colonial algae) may be a zygospore of the zygnematacean alga (Grenfell, 1995), and it is grouped with the acritarchs (Mudie *et al.*, 2011). This sphaeromorphic acritarch was first described as *Sporites circulus* in the Pliocene brown coals, and then as *Concentricystes rubinus* in marine sediments of Israel. *Pseudoschizaea rubinus* is distinguished from the similar species *Pseudoschizaea circula* by an irregular, maze-like polar complex up to one quarter of the vesicle diameter (Christopher, 1976). *Concentricystes* sp. is usually considered a freshwater alga (Christopher, 1976). However, both *P. circula* and *P. rubina* are irregularly present in the Late Holocene marine sediments of the Black Sea (Mudie *et al.*, 2011).

The other important nonpollen palynomorphs are fungi and various types of planktonic or benthic zooplankton. Several spore types (e.g., *Tilletia*, *Ustilago*) are produced by parasites of specific native plants and domestic plants (Mudie *et al.*, 2011). *Glomus*-type fungal spores are extremely resistant to fire and biological degradation (Bryant and Holloway, 1983; Leroy *et al.*,

2009), and they may survive very long-distance transport by river water. The organic-walled palynomorph *Halodinium* was first recorded and described as an acritarch of unknown affinity occurring in subarctic marine sediments of the Bering Sea, and it is widely distributed in the Arctic (Matthiessen *et al.*, 2000). *Halodinium* is scarcely recorded in the Black Sea (Mudie *et al.*, 2011).

Microforaminiferal linings are the organic inner linings of juvenile foraminifers, and they actually represent the chitinous inner test (Limaye *et al.*, 2007). The size of these fossils ranges from 10 to 100 µm, and may be even more. They are mostly elongated, unilocular, uniseriate, biseriate, and triseriate and coiling-type. Spiral forms characterize hyposaline lagoons and estuaries, and may also be found intertidal and shallow water in the coastal area. According to Mudie *et al.* (2011), microforaminiferal linings might be expected to be good markers of sustained marine connection with the Mediterranean, and this palynomorph is also recorded in the hypersaline water.

The presence of *Isoetes* spores indicates summer dryness and a wet winter half-year, features well known for the typical Mediterranean (Bottema and Sarpaki, 2003).

15.4   CORALS IN GÜLBAHÇE BAY

Corals are significant species in marine life, and these are used as bioindicators to assess the impact of climate change on marine communities (Okamoto *et al.*, 1998; Özalp and Alpaslan, 2011 and Wilkinson, 2000). The Scleractinian *Cladocora caespitosa* is the only colonial and zooxanthellate coral in the Mediterranean (Peirano *et al.*, 1999; Veron and Staff, 2000; Zibrowius, 1980). The species is the major carbonate producers among the Mediterranean organisms (Montagna *et al.*, 2007; Peirano *et al.*, 2001). It lives on rocky and sandy bottoms and is rarely found below 30 m of depth (Kružić *et al.*, 2008).

Sediment samples were collected from the unit A, which is the younger unit in Gülbahçe Bay. Coral colonies have been observed on the morphological highs (Fig. 15.6) and *Cladocora caespitosa* has been defined in these colonies. This species has been previously recorded in the northwestern Aegean Sea at Gökçeada island (Öztürk, 2004) (Zibrowius, 1979, for the Greek coast), in Edremit Bay (Çınar, 2003 and Zibrowius, 1979, for the Greek coast), and in the Marmara Sea in the strait of Çanakkale (Özalp and Alpaslan, 2011). Most of these studies provide a more expanded geographical distribution of the species. Fossil records indicated that *Cladocora caespitosa* was more abundant and reef forming in warmer phases of the Quaternary (Bernasconi *et al.*, 1997; Peirano, 2007). Today, however, the coral is less abundant since it suffered mortality events during the heat waves over the last two decades (Metalpa *et al.*, 2005). Geothermal springs in the Gülbahçe Bay enable the seawater to reach temperatures similar to the warmer phases of the Quaternary, and it is under these warm conditions that *Cladocora caespitosa* flourish.

15.5   PALYNOLOGY

Of the 41 sediment samples collected from the sea floor in the Gülbahçe Bay, 40 samples are suitable for palynological analysis (Table 15.1). Forty-three palynomorphs, constituting 37 taxa of spores and pollen, 4 taxa of the dinoflagellate species, and 2 taxa of siliceous algae, were defined.

15.5.1   *Reference zone*

The reference zone is selected from the entrance of the Gülbahçe Bay in which the geothermal spring is not observed (Fig. 15.7). Palynomorph content of the reference point is poor (Fig. 15.8), and this content is characterized by the less abundantly Polypodiaceae, Osmundaceae, Apiaceae, Chenopodiaceae, Taxodiaceae, Fagaceae-*Quercus* evergreen type, *Castanea,* Nyssaceae-*Nyssa*,

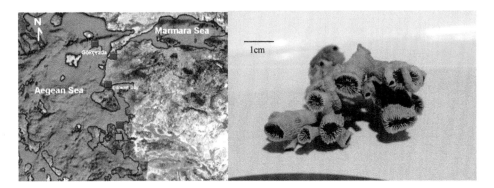

Figure 15.6.   The corals (*Cladocora caespitosa*) observed in Gülbahçe sediment samples and red squares are records from the northern Aegean Sea (Özalp and Alpaslan, 2011); the green square shows the new records from the Gülbahçe Bay.

Table 15.1.   Core samples and sample numbers.

Sample location	Water depth (m)	Core length (m)	Sample number
1	18.00	1.30	bottom, middle, top = 3
3	20.00	1.45	bottom, middle, top = 3
6	19.50	0.96	bottom, middle, top = 3
5	18.80	1.50	bottom, middle = 2
7	18.50	1.38	bottom, middle, top = 3
C8	20.00	1.20	bottom, middle, top = 3
C1	27.00	1.55	bottom, middle, top = 3
C2	15.00	1.45	bottom, middle, top = 3
C3	13.50	2.00	bottom, middle, top = 3
C4	16.80	1.45	bottom, middle, top = 3
C5	19.20	1.45	bottom, middle, top = 3
C7	19.50	1.05	bottom, middle, top = 3
C9	21.00	1.90	bottom, middle, top = 3
References location	29.50	1.85	bottom, middle, top = 3

Oleaceae, Asteraceae, Cichoriaceae, and more abundantly Pinaceae-*Pinus* species (Fig. 15.9). Nonpollen palynomorphs are rare in this palynospectra, and these are represented by the siliceous algae (*Coscinodiscus nodulifer* and *Auliscus punctatus*), Dinoflagellate spp., microforaminiferal lignin, *Glomus*, and fungal spores (Fig. 15.9).

15.5.2   *Defining palynomorphs of thermal spring locations from Gülbahçe Bay*

The other core locations (13 point; Location 1, 3, 5, 6, 7, C8, C1–5, C7, and C9) take place in the middle part of the Gülbahçe Bay, and samples are collected from the "A1" sediments. Generally, in all Gymnosperm pollen cores (Pinaceae-*Pinus haploxylon*) and Fagaceae-*Quercus*-evergreen type are abundant (Figs. 15.10–15.12). Other pollen and spores (*Abies, Pinus dyploxylon* type, *Podocarpus, Sciadopits, Cedrus,* Taxodiaceae, Myricaceae, Cyrillaceae, Fagaceae-*Quercus*-deciduous type, Nyssaceae-*Nyssa*, Oleaceae, Ulmaceae-*Ulmus/Zelkova, Salix, Ostrya, Castanea,* Betulaceae-*Alnus*, Ashopdelaceae, Sapotaceae, Poypodiaceae, Davaliaceae, Schizaceae, *Riccia* sp., *Botryhium* sp., Selaginellaceae, and Sterculiaceae) are defined in low percentage (Figs. 15.10–15.12). Besides, herb species (Poaceae, Chenopodiaceae, Apiaceae, Ephedraceae, Geraniaceae, Compositae-Tubuliflorae/Liguliflorae) are less abundantly accompanied with these

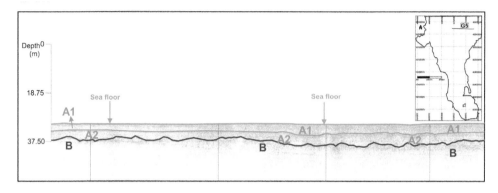

Figure 15.7.   Seismic line G5 in the Gülbahçe Bay and references point.

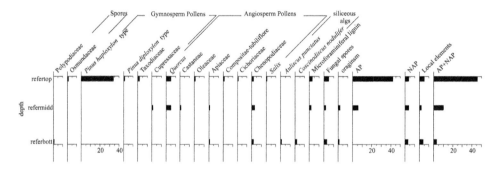

Figure 15.8.   Palynomorphs distribution of references line.

angiosperm pollens. Especially, nonpollen palynomoprhs are various in all samples, and these are represented by microforaminiferal lignin, *Tilletia*, *Halodium*, Pseudoschizceae, *Isoetes*, *Lingulodonium machaerophorum*, *Cymatiosphaera globulosa*, *Spiniferites ramosus*, *Spiniferites* spp., and zooclasts.

15.6   PALAEOENVIRONMENT

Defined pollen and spores are grouped according to the palaeovegetational characteristics.

- Swamp Palaeovegetation: Polypodiaceae, Taxodiaceae, Cyrillaceae, Davaliaceae, Schizaceae, Osmundaceae, *Riccia* sp., *Botryhium* sp., Selaginellaceae, Sterculiaceae, Nyssaceae-*Nyssa*, and Myricaceae.
- Mixed Mesophytic Forest Palaeovegetation (conifers and evergreen plants): Pinaceae-*Pinus haploxylon* and *diploxylon* types, *Abies, Cathaya, Podocarpus, Sciadopits*, *Cedrus,* Fagaceae-*Quercus*-evergreen and deciduous types, Fabaceae, Ulmaceae-*Ulmus/Zelkova*, Sapotaceae, Betulaceae-*Alnus,* Tiliaceae-*Tilia,* Oleaceae, Betulaceae-*Carpinus*, *Ostrya*, *Salix*, Cyrillaceae, and *Castaneae*.
- Open Palaeovegetation and Shrubs: Chenopodiaceae, Ephedraceae, Apiaceae, Compositae-Tubuliflorae/Liguliflorae types, Asophadelaceae, Poaceae, and Geraniaceae.
- Marine Palynomorphs: *Lingulodinium machaerophorum, Cymatiosphaera globulosa, Spiniferites ramosus, Spiniferites* spp., *Tilletia, Halodium*, microforaminiferal lignin, Aquatic fungal spore (Ingoldian type), *Isoetes*, and Pseudoschizaea.

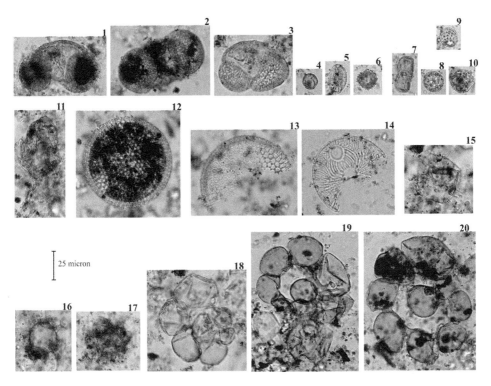

Figure 15.9. Palynomorphs figures of references line samples (1–3, *Pinus*; 4, Oleaceae; 5, Fagaceae-*Quercus*; 6, Asteraceae-Tubuliflorae; 7, Apiaceae; 9, 10, Chenopodiaceae; 11, *Glomus*; 12–14, Siliceous algae; 15–17, Dinoflagellate spp.; and 18–20, microforaminiferal lignin).

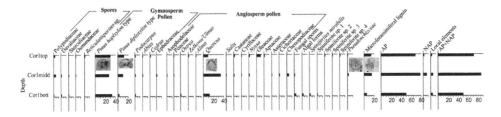

Figure 15.10. Palynomorphs distribution of location 1 "core1."

The terrestrial and marine paleoenvironmental conditions (Late Holocene period) are determined by the palynological results obtained from the sediment samples in Gülbahçe Bay.

15.6.1 *Terrestrial condition in the Gülbahçe Bay*

High palaeotopographic and lowland areas surrounding the Gülbahçe Bay are covered with *Pinus haploxylon* and *diploxylon* types, *Abies, Cathaya, Podocarpus, Sciadopits, Cedrus, Abies,* Fagaceae-*Quercus*-evergreen and deciduous types, Ulmaceae-*Ulmus/Zelkova*, Sapotaceae, Betulaceae-*Alnus*, Tiliaceae-*Tilia*, Oleaceae, Betulaceae-*Carpinus*, *Ostrya*, *Salix*, Cyrillaceae, and *Castaneae*. Polypodiaceae, Taxodiaceae, Cyrillaceae, Davaliaceae, Schizaceae, Osmundaceae, *Riccia* sp., *Botryhium* sp., Selaginellaceae, Sterculiaceae, Nyssaceae-*Nyssa*, and Myricaceae are grown in the narrow areas of freshwater marsh and between these areas. There

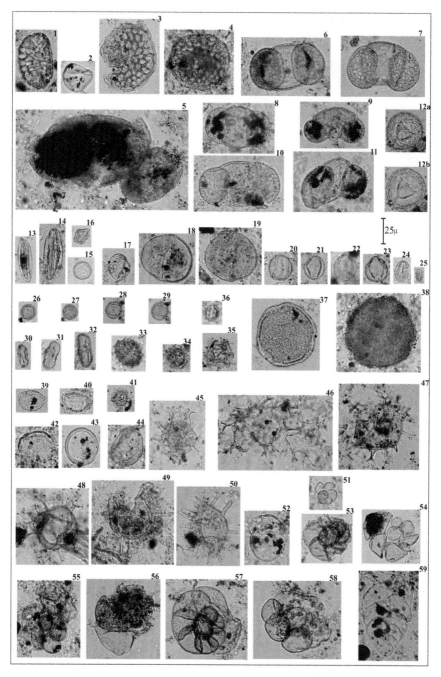

Figure 15.11.   Palynomorphs figures of location 1, 3, 5, 6, 7, C8, C1-5, C7, and C9 samples (1, Davaliaceae; 2, Sterculiaceae; 3, 4, *Riccia* sp.; 5, *Abies*; 6–11, *Pinus*; 12a,b, *Sciadopits*; 13, 14, Ephedraceae; 15, Poaceae; 16, *Alnus*; 17, *Ulmus/Zelkova*; 18, 19, Asophadelaceae; 20–24, *Quercus*; 25, *Castaneae*; 26–29, Oleaceae; 30–32, Apiaceae; 33, Asteraceae-Tubuliflorae type; 34, 35, Asteraceae-Liguliflorae type; 36, Chenopodiaceae; 37, 38, Geraniaceae; 39, 40, *Isoetes*; 41, *Tiletia*; 42–44, Pseudoschizaceae; 45, *Spiniferites mirabilis*; 46–48, *Spiniferite* spp.; 49, 50, *Lingulodinium macherophorum*; and 52–59, microforaminiferal lignin).

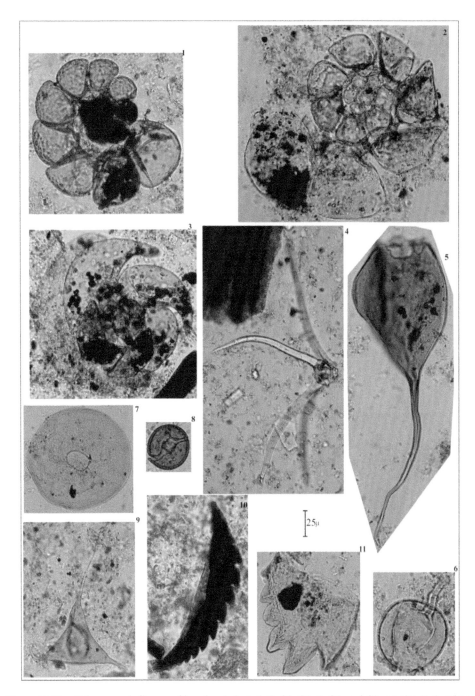

Figure 15.12.   Palynomorph figures of location 1, 3, 5, 6, 7, C8, C1–5, C7, and C9 samples (1–3, microforaminiferal lignin; 4, Aquatic fungal spore (Ingoldian type); 5, 6, *Glomus*; 7, *Halodium*;
8, *Sigmapollis*; 9, Acritarc; and 10, 11, Zooclast).

are defined constricted open vegetation areas that are characterized by the Chenopodiaceae, Ephedraceae, Apiaceae, Compositae-Tubuliflorae/Liguliflorae types, Asophadelaceae, Poaceae, and Geraniaceae.

15.6.2   *Marine condition in the Gülbahçe Bay*

Planktonic and benthic foraminifera are abundant in the Marmara Sea and Aegean Sea, but are absent in the anoxic Black Sea basins (Meriç *et al.*, 2000). Besides, warm water conditions have affected the proliferation of foraminifera (Meriç *et al.*, 2000). These findings are coherent with the distributions of microforaminiferal lignin, the organic inner linings of juvenile foraminifers (Aksu *et al.*, 2002; Limaye *et al.*, 2007; Mudie *et al.*, 2002), for the low oxygen bottom sediments of the Aegean and Marmara seas, and the high benthic productivity of the Marmara gateway (Mudie *et al.*, 2002). Microforaminiferal lignin, which is nonpollen palynomorph, is observed in abundance in the Gülbahçe Bay sediments, and this greater abundance could be related to the locally changing the seawater temperature and salinity (Na, Cl, Ca, K, Mg, SO_4; Tarcan, 2001) based on the geothermal activity in the Gülbahçe Bay (Aksu *et al.*, 2002; Meriç *et al.*, 2000; Mudie *et al.*, 2002). Other nonpollen palynomorphs (*Tilletia, Halodium*, Pseudoschizceae, *Isoetes, Lingulodonium machaerophorum, Cymatiosphaera globulosa, Spiniferites ramosus*, and *Spiniferites* spp.) recorded abundantly in Aegean Sea were found to be less abundant in this geothermal area.

Cladocora caespitosa is defined in the Gülbahçe Bay, and this coral species is widespread on the morphological highs. This species is often observed in the warm phases of the Late Holocene in the Mediterranean region. At the present time, the presence of this species in the geothermal Gülbahçe area could be related to the warm seawater conditions.

ACKNOWLEDGMENTS

The authors thank Dr. Mustafa Eftellioğlu, Dr. Cem Günay, Dr. Muhammed Duman, Dr. Funda Akgün, Dr. Doğan Yaşar, and Dr. Cemal Tunoğlu for their valuable comments. We also thank ship captains and crew of the "Piri Reis" and the "Dokuz Eylül-1" for the aiding the seismic studies. Authors would like to express their sincere thanks to an anonymous referee whose careful comments and useful criticisms have greatly improved the earlier version of the chapter.

REFERENCES

Agostini, S., Doglioni, C., Innocenti, F., Manetti, P., Tonarini, S. & Savaşçin, M.: The transition from subduction-related to intraplate Neogene magmatism in the western Anatolia and Aegean area. In: L. Beccaluva, G. Bianchini & M. Wilson, M. (eds): Cenozoic volcanism in the Mediterranean area. *Geological Society of America Special Paper* 418, 2007, pp. 1–15.

Aksu, A.E., Yaltırak, C. & Hiscott, R.N.: Quaternary paleoclimatic–paleoceanographic and tectonic evolution of the Marmara Sea and environs. *Mar. Geol.* 190 (2002), pp. 9–18.

Aydin, I., Karat, H.I. & Koçak, A.: Curie-point map of Turkey. *Geophys. J. Int.* 162 (2005), pp. 633–640.

Bernasconi, M.P., Corselli, C. & Carobene, L.: A bank of the scleractinian coral *Cladocora caespitosa* in the Pleistocene of the Crati valley (Calabria, Southern Italy): growth *versus* environmental conditions. *Boll. Soc. Paleontol. Ital.* 36:1–2 (1997), pp. 53–61.

Bottema, S. & Sarpaki A.: Environmental change in Crete: a 9000-year record of Holocene vegetation history and the effect of the Santorini eruption. *The Holocene* 13:5 (2003), pp. 733–749.

Bryant, V.M., Jr. & Holloway, R.G.: The role of palynology in archaeology. In: *Advances in archaeological method and theory*. Academic Press, New York, 1983, pp. 191–224.

Caglar, I., Tuncer, V., Kaypak, B. & Avsar, U.: A high conductive zone associated with a possible geothermal activity around Afyon, northern part of Tauride zone, southwest Anatolia. *Proceedings World Geothermal Congress 2005*, 24–29 April, Antalya, Turkey, 2005, pp. 1–5.

Canbolat, A.: Seferihisar jeotermal sondajları bitirme raporları (Tuzla 1, G-2A, G-3, G3A, G12A, G17A. General Directorate of Mineral Research and Exploration (MTA) report JI43, İzmir, Turkey, 1986.

Christopher, R.A.: Morphology and taxonomic status of Pseudoschizaea Thiergart and Frantz ex R. Potoniéemend. *Micropaleo.* 22 (1976), pp. 143–150.

Çinar, M.E.: Ecological features of Syllidae (Polychaeta) from shallow-water benthic environments of the Aegean Sea, eastern Mediterranean. *J. Mar. Bio. Assoc. UK* 83 (2003), pp. 737–745.

Condrad, M.A., Hipfel, B. & Satır, M.: Chemical and stable isotopic characterictics of thermal waters from the Çeşme–Seferihisar area, İzmir (W. Turkey). In Ö. Piksin, M.Y. Savaşcın, M. Ergün & G. Tarcan (eds): *Proceedings International Earth Sciences Colloquium on the Aegean Region*, Volume 2, 1997, pp. 669–679.

Dewey, J.F. & Şengör. A.M.C.: Aegean and surrounding regions: complex multi-plate and continuum tectonics in a convergent zone. *Geol. Soc. America Bull.*, Part 1, volume 90, 1979, pp. 84–92.

Drahor, M.G. & Berge, M.A.: Geophysical investigations of the Seferihisar geothermal area, western Anatolia, Turkey. *Geothermics* 35 (2006), pp. 302–320.

Eşder, T.: The crust structure and convection mechanism of geothermal fluids in Seferihisar geothermal area. In M.Y. Savaþcın & H. Eronat (eds): *Proceedings International Earth Sciences Congress on Aegean Regions*, İzmir, Turkey, Volume 1, 1990, pp. 135–147.

Eşder, T. & Şimşek, Ş.: Geology of İzmir–Seferihisar geothermal area, western Anatolia of Turkey, determination of reservoirs by means of gradient drilling. *Proceedings 2nd UN Symposium on the Development and Use of Geothermal Resources*, San Francisco, CA, 1975, pp. 349–360.

Eşder, T. & Şimşek, Ş.: İzmir-Seferihisar alanı Çubukludağ Grabeni ile dolayının jeolojisi ve jeotermal enerji olanakları. General Directorate of Mineral Research and Exploration (MTA) unpublished report 5842, Ankara, Turkey, 1977.

Eşder, T., Ölmez, E., Aydın, H. & Gür, Ş.: Doğanbey Ilıcası (Seferihisar–İzmir) jeotermal enerji kuyusunun bitirme raporu. General Directorate of Mineral Research and Exploration (MTA) report No. JT-134, Ankara, Turkey, 1995.

Faulds, J.E., Coolbaugh, M.F., Vice, G.S. & Edwards, M.L.: Characterizing structural controls of geothermal fields in the northwestern Great Basin: a progress report. *Geoth. Resour. Council Trans.* 30 (2006), pp. 69–76.

Faulds, J.E., Bouchot, V., Moeck, I. & Oğuz, K.: Structural controls on geothermal systems in western Turkey: a preliminary report. *Geoth. Resour. Council Trans.* 33 (2009), pp. 375–381.

Filiz, Ş.: *Ege Bölgesindeki önemli jeotermal alanların ^{18}O, ^2H, ^3H, ^{13}C izotoplarıyla incelenmesi.* Assoc. Prof. Thesis. E.Ü.Y.B.F., İzmir, Turkey, 1982.

Filiz, Ş. & Tarcan, G.: Seferihisar (İzmir) jeotermal alanının hidrojeolojisi. *TPJD Bült* 5:1 (1993), pp. 97–112.

Filiz, Ş., Tarcan, G. & Gemici, Ü.: Seferihisar (İzmir) jeotermal alanındaki sıcak suların Hidrojeokimyasal incelenmesi. *Su ve Çevre Sempozyumu 97*, İstanbul, 1997, pp. 117–128.

Grenfell, H.R.: Probable fossil zygnemataceae algal spore genera. *Review of Palaeo. and Paly.* 84 (1995), pp. 201–220.

Innocenti, F., Agostini, S., DiVincenzo, G., Doglioni, C., Manetti, P., Savaşcin, M.Y. & Tonarini, S.: Neogene and Quaternary volcanism in western Anatolia: magma sources and geodynamic evolution. *Mar. Geol.* 221 (2005), pp. 397–421.

Jackson, J.A. & McKenzie, D.P.: Active tectonics of the Alpine-Himalayan belt between western Turkey and Pakistan. *Geophys. J. Royal Astron. Soc.* 77 (1984), pp. 185–264.

Jackson, J.A &, McKenzie, D.P.: Rates of active deformation in the Aegean Sea and surrounding regions. *Basin Res.* 1 (1988), pp. 121–128.

Karamanderesi, I.H. & Helvaci, C.: Geology and hydrothermal alteration of the Aydın-Salavatli geothermal field, western Anatolia, Turkey. *Turkish J. Earth Sci.* 12 (2003), pp. 175–198.

Kaya, O.: Ortadoğu Ege çöküntüsünün (Neojen) stratigrafisi ve tektoniği. *Türkiye Jeoloji Kurumu Bülteni* 22 (1979), pp. 35–58.

Kružić, P., Žuljević, A. & Nikolić, V.: Spawning of the colonial coral Cladocora caespitosa (Anthozoa, Scleractinia) in the southern Adriatic Sea. *Coral Reefs* 27 (2008), pp. 337–341.

Leroy, S.A.G., Boyraz, S. & Gürbüz, A.: High-resolution palynological analysis in Lake Sapanca as a tool to detect earthquakes on the North Anatolian Fault over the last 55 years. *Quaternary Sci. Rev.* 28 (2009), pp. 2616–2632.

Lewis, J. & Hallett, R.I.: *Lingulodinium polyedrum (Gonyaulax polyedra)* a blooming dinoflagellate. In: A. Ansell, M. Barnes & R.N. Gibson, R.N. (eds): *Oceanography and marine biology: an annual review*, volume 35, 1997, pp. 96–161.

Limaye, R.B., Kumara, K.P.N., Nair, K.M. & Padmalal, D.: Non-pollen palynomorphs as potential palaeoenvironmental indicators in the Late Quaternary sediments of the west coast of India. *Current Sci.* 92 (2007), pp. 1370–1382.

Matthıessen, J., Kunz-Pırrung, M. & Mudıe, P.J.: Freshwater chlorophycean algae in recent marine sediments of the Beaufort, Laptev and Kara Seas (Arctic Ocean) as indicators of river runoff. *Int. J. Eart. Sci.* 89 (2000), pp. 470–485.

McKenzie, D.P.: Active tectonics of the Mediterranean regions. *Geophys. J. Royal Astron. Soc.* 30 (1972), pp. 109–185.

Meriç, E., Kerey, İ.E., Avşar, N., Tunoğlu, C., Taner, G., Kapan-Yeşilyurt, S., Ünsal, İ. & Rosso, A.: Geç Kuvaterner (Holosen)'de İstanbul Boğazı yolu ile Marmara Denizi-Karadeniz bağlantısı hakkında yeni bulgular. *TJB* 43:1 (2000), pp. 73–118.

Metalpa, R.R., Bianchi, C.N., Peirano, A. & Morri, C.: Tissue necrosis and mortality of the temperate coral *Cladocora caespitosa*. *Ital. J. Zool.* 72 (2005), pp. 271–276.

Montagna, P., McCulloch, M., Mazzoli, C., Silenzi, S. & Odorico, R.: Th e non-tropical coral *Cladocora caespitosa* as the new climate archive for the Mediterranean: high-resolution (weekly) trace element systematics. *Quaternery Sci. Rev.* 26 (2007), pp. 441–462.

Mudie, P.J., Rochon, A., Aksu, A.E. & Gillespie, H.: Dinoflagellate cysts, freshwater algae and fungal spores as salinity indicators in Late Quaternary cores from Marmara and Black seas. *Mar. Geol.* 190 (2002), pp. 203–231.

Mudie, P. J., Rochon A., Aksu A.E. & Gillespie H.: Late glacial, Holocene and modern dinofagellate cyst assemblages in the Aegean–Marmara Black Sea corridor: statistical analysis and re-interpretation of the early Holocene Noah's Flood hypothesis. *Rev. Palaeobot. Palyno.* 128 (2004), pp. 143–167.

Mudie, P.J., Leroy, S.A.G., Marret, F., Gerasimenko, N., Kholeif, S.E.A., Sapelko, T. & Filipova-Marinova, M.: Nonpollen palynomorphs: Indicators of salinity and environmentalchange in the Caspian-Black Sea–Mediterranean corridor. *The Geological Society of America*, volume 473, 2011, pp. 1–27.

Okamoto, M. & Yamaguchi, H.: Diving survey technique of coral cover as a bio-indicator of environmental changes. *Proceedings of the Annual Conference on the marine technology*, The Marine Technology Society. Volume 1, Baltimore, 1998, pp. 445–451.

Okay, A.I., Satır, M., Maluski, H., Siyanko, M., Metzger, R. & Akyuz, H.S.: Palaeo- and Neo-Tethyan events in northwest Turkey: Geological and geochronological constraints. In: Y. An & M. Harrison (eds.): *Tectonics of Asia,* Cambridge. Cambridge Univ., UK, 1996, pp. 420–441.

Özalp, H.B. & Aparslan, M.: The first record of *Cladocora caespitosa* (Linnaeus, 1767) (Anthozoa, Scleractinia) from the Marmara Sea. *Turk. J. Zool.* 35:5 (2011), pp. 701–705.

Öztürk, B.: Marine life of Turkey in the Aegean and Mediterranean Sea. In: *Phylum Cnidaria*, Turkish Marine Research Foundation, Turkey, 2000, p. 48.

Peirano, A., Morri, C. & Bianchi, N.C.: Skeleton growth and density pattern of the temperate, zooxanthellate scleractinian *Cladocora caespitosa* from the Ligurian Sea (NW Mediterranean). *Mar. Ecol. Prog. Ser.* 185 (1999), pp. 195–201.

Peirano, A., Morri, C., Bianchi, C.N. & Rodolfo-Metalpa, R.: Biomass, carbonate standing stock and production of the Mediterranean coral *Cladocora caespitosa* (L.). *Facies* 44 (2001), pp. 75–80.

Peirano, A.: In vivo measurements of the seasonal photosynthetic fluorescence of the Mediterranean coral *Cladocora caespitosa* (L.). *Sci. Mar.* 71 (2007), pp. 629–635.

Pekçetinöz, B.: *The Investıgatıon of geothermal areas in Izmır Bay and its surroundıngs (as an example of Gülbahçe Bay)*. PhD Thesis, Dokuz Eylül University, Izmir, Turkey, 2010.

Pfister, M., Rybach, L. & Şimşek, S.: Geothermal reconnaissance of the Marmara Sea region (NW Turkey); surface heat flow density in an area of active continental extension. *Tectonophysics* 291 (1998), pp. 77–89.

Şengör, A.M.C.: Collision of irregular continental margins: implications for foreland deformation of Alpine-type orogens. *Geology* 4 (1976), pp. 779–782.

Şengör, A.M.C., Satir, M. & Akkök, R.: Timing of tectonic events in the Menderes Massif, western Turkey: implications for tectonic evolution and evidence for Pan-African basement in Turkey. *Tectonics* 3 (1984), pp. 693–707.

Şimşek, S.: Geothermal potential in northwestern Turkey. In: C. Schindler & M. Pfister (eds): *Active tectonics of northwestern Anatolia – the MARMARA Poly-Project: a multidisciplinary approach by space-geodesy, geology, hydrogeology, geothermics and seismology.* VDF Hochschulverlag ETH, Zurich, 1997, pp. 111–124.

Şimşek, S. & Demir, A.: Reservoir and cap rock characteristics of some geothermal fields in Turkey and encountered problems based on lithology. *J. Geoth. Res. Soc. Japan* 13 (1991), pp. 191–204.

Şimşek, S. & Guleç, N.: Geothermal fields of western Anatolia, excursion guide. *International Volcanological Congress IAVECEI-94*, Middle Eastern Technical University, Ankara Special Publications No. 8, 1994, 35 p.

Şimşek, S., Dogdu, M.S., Akan, B. & Yildirum, N.: Chemical and isotopic survey of geothermal reservoirs in western Anatolia, Turkey. *Proceedings of the World Geothermal Congress 2000*, Kyushu-Tohoku, Japan, 2000a, pp. 1765–1770.

Şimşek, S., Gunay, G., Elhatip, H. & Ekmekci, M.: Environmental protection of geothermal waters and travertines at Pamukkale, Turkey. *Geothermics* 29 (2000b), pp. 557–572.

Tarcan, G.: Hydrogeology and hydrogeochemistry of the Gülbahçe Bay hydrothermal karst system, İzmir, Turkey. In: Günay, Ford, Johnson & Johnson (eds): *Proceedings of the 6th International Symposium and Field Seminar on "Present State and Future Trends of Karst Studies"*, Marmaris, Turkey, International Hydrological Programme-UNESCO, 2001, pp. 515–524.

Tarcan, G., Filiz, Ş. & Gemici, Ü.: Balçova-Seferihisar (İzmir) jeotermal alanlarında karşılaştırılmalı hidrojeokimyasal incelemeler ve jeotermometre uygulamaları. 1. Batı Anadolu Hammadde Kaynakları Sempozyumu Bildiriler Kitabı, 1999, pp. 346–358.

Tezcan, A.K.: Geothermal explorations and heat flow in Turkey In: M.L. Gupta & M. Yamano (eds): *Terrestrial heat flow and geothermal energy in Asia*. Oxford and IBH Publishing Co., Pvt. Ltd., New Delhi, India, 1995, pp. 23–42.

Van Zeist, W. & Bottema, S.: Late Quaternary vegetational and climatic history of southwest Asia. *Proc. Indian National Acad. Sci* 84:A3 (1988), pp. 461–480.

Van Zeist, W. & Bottema, S.: Late Quaternary vegetation of the Near East. *Beihefte zum Tübinger Atlas des Vorderen Orients*, Reihe A (*Naturwissenschaften*) 18. Dr. Ludwig Reichert Verlag. Wiesbaden, Germany, 1991, pp. 156.

Veron, J. & Stafford-Smith, M.: Corals of the world III. In: M. Stafford-Smith (ed): *Faviidae*. Odyssey Publishing, Australia, 2000, pp. 251–252.

Westaway, R.: Kinematics of the Middle East and eastern Mediterranean updated. *Turkish J. Earth Sci.* 12 (2003), pp. 5–46.

Wilkinson, C.R.: *Status of coral reefs of the world*. Australian Institute of Marine Science, Cape Ferguson, Queensland, Australia, 2000.

Yilmaz, Y., Genç, Ş.C., Karacik, Z. & Altunkaynak, Ş.: Two contrasting magmatic associations of NW Anatolia and their tectonic significance. *J. Geodyn.* 31 (2001), pp. 243–271

Yılmazer, S.: *Ege Bölgesi'ndeki bazı sıcak su kaynaklarının hidrojeolojisi ve jeokimyasal incelemeleri*. MSc Thesis, D.E.Ü. Fen Bilimleri Enstitüsü, İzmir, Turkey, 1984.

Yılmazer, S.: Kıyı Ege ve İzmir İlin'deki Jeotermal kaynakların değerlendirilmesi Yer altı Suları ve Çevre Sempozyumu, İzmir, Bildiriler, Turkey, 2001, pp. 371–379.

Zibrowius, H.: Campagne de la Calypso en Méditerranée nordorientale (1955, 1956, 1960, 1964). Scléractiniaires. *Ann. Inst. Oceanogr.* 55 (1979), pp. 7–28.

Zibrowius, H.: Les Scléractiniaires de la Méditerranée et de l'Atlantique nord-oriental. *Mém. Inst. Océanogr.* 11 (1980), pp. 1–284.

Subject index

Index I:
Note: Localities, stratigraphic units, tectonic and structural elements, etc., are included in Index II.

Index II:
Localities, geothermal fields and properties, stratigraphic units, tectonic and structural elements

Sustainable Energy Developments

Series Editor: Jochen Bundschuh

ISSN: 2164-0645

Publisher: CRC Press/Balkema, Taylor & Francis Group

1. Global Cooling – Strategies for Climate Protection
   Hans-Josef Fell
   2012
   ISBN: 978-0-415-62077-2 (Hbk)
   ISBN: 978-0-415-62853-2 (Pb)

2. Renewable Energy Applications for Freshwater Production
   Editors: Jochen Bundschuh & Jan Hoinkis
   2012
   ISBN: 978-0-415-62089-5 (Hbk)

3. Biomass as Energy Source: Resources, Systems and Applications
   Editor: Erik Dahlquist
   2013
   ISBN: 978-0-415-62087-1 (Hbk)

4. Technologies for Converting Biomass to Useful Energy –
   Combustion, gasification, pyrolysis, torrefaction and fermentation
   Editor: Erik Dahlquist
   2013
   ISBN: 978-0-415-62088-8 (Hbk)

5. Green ICT & Energy – From smart to wise strategies
   Editors: Jaco Appelman, Anwar Osseyran & Martijn Warnier
   2013
   ISBN: 978-0-415-62096-3

6. Sustainable Energy Policies for Europe – Towards 100% Renewable Energy
   Rainer Hinrichs-Rahlwes
   2013
   ISBN: 978-0-415-62099-4 (Hbk)

7. Geothermal Systems and Energy Resources – Turkey and Greece
   Editors: Alper Baba, Jochen Bundschuh & D. Chandrasekaram
   2014
   ISBN: 978-1-138-00109-1 (Hbk)

Printed and bound by CPI Group (UK) Ltd, Croydon, CR0 4YY
18/10/2024
01776254-0003